T0297321

THE

LIFE AND LETTERS

OF THE REVEREND

ADAM SEDGWICK

VOLUME I.

ADAM SEDGWICK, ÆT. 47.

*From a portrait painted by Thomas Phillips, R.A. 1832, in the
possession of John Henry Gurney, Jr. Esq.*

Frontispiece to Vol. I.

THE

LIFE AND LETTERS

OF

THE REVEREND

ADAM SEDGWICK,

LL.D., D.C.L., F.R.S.,

FELLOW OF TRINITY COLLEGE, CAMBRIDGE,

PREBENDARY OF NORWICH,

WOODWARDIAN PROFESSOR OF GEOLOGY, 1818—1873.

BY

JOHN WILLIS CLARK, M.A. F.S.A.

FORMERLY FELLOW OF TRINITY COLLEGE, SUPERINTENDENT OF THE UNIVERSITY
MUSEUM OF ZOOLOGY AND COMPARATIVE ANATOMY,

AND

THOMAS McKENNY HUGHES,

M.A. TRIN. COLL. F.R.S. F.S.A. F.G.S.

PROFESSORIAL FELLOW OF CLARE COLLEGE,
WOODWARDIAN PROFESSOR OF GEOLOGY.

VOLUME I.

CAMBRIDGE
AT THE UNIVERSITY PRESS
1890

CAMBRIDGE UNIVERSITY PRESS
Cambridge, New York, Melbourne, Madrid, Cape Town, Singapore,
São Paulo, Delhi, Dubai, Tokyo

Cambridge University Press
The Edinburgh Building, Cambridge CB2 8RU, UK

Published in the United States of America by Cambridge University Press, New York

www.cambridge.org
Information on this title: www.cambridge.org/9780521137706

First published 1890
This digitally printed version 2009

A catalogue record for this publication is available from the British Library

ISBN 978-0-521-13770-6 Paperback

PREFACE.

THE cause of the delay in the publication of Sedgwick's *Life and Letters* demands a short explanation. Soon after his death, in January, 1873, Professor Hughes, his successor in the Woodwardian Chair, undertook to become his biographer. No man was better qualified, by geological knowledge, and by affectionate regard for Sedgwick, to perform the task adequately; and, had he been able to command sufficient leisure, he would doubtless have produced a very admirable piece of work. He carefully studied the materials placed in his hands, and made considerable progress with different portions. But, as time went on, and the duties of his Professorship increased, he found himself, year after year, less able to cope with the difficulties of a task which demanded a long and patient research. It became gradually obvious that if the book was to be written at all, either it must be put into the hands of some one else, or Professor Hughes must obtain assistance. In 1885 Miss Sedgwick applied to me; but, as I was at that time fully occupied with my edition of Professor Willis' *Architectural History of the University and Colleges*, I was

constrained to reply that I could not even look at the
materials until that work was completed. Finally,
Miss Sedgwick agreed to wait; and it was not till
near the end of 1886 that I was able to make a start.
Professor Hughes, with great generosity and kindness,
handed over to me all his materials, together with the
portions that he had already written; but, after some
consideration, I decided to return the latter, and to
begin afresh. It was then arranged between us that
he should contribute the geological portion only,
leaving the rest of the biography to me.

I have tried to explain, in the opening paragraphs
of my first chapter, the difficulties with which Sedgwick's biographer is confronted. In addition to what
I have there written I may now notice the further
drawback arising from the deficiency of materials.
Sedgwick outlived most of his contemporaries and
intimate friends; and, since his death, many of those
younger persons with whom he delighted to corre-
spond have passed away. Hence a number of interesting letters, which he is known to have written, and
which were long carefully preserved, have either been
destroyed, or cannot now be traced. These remarks
apply specially to his earlier years; for the letters
which he wrote in profusion to his nieces, and which
so vividly illustrate the last forty years of his life,
have been placed in my hands without reserve.

In dealing with the earlier portion of his life I have
tried to draw a picture of the University of that day,
and to bring out the prominent and energetic part
taken by Sedgwick in its studies, its controversies, and

its reforms. Further, I have gone at length into the history of Dr Woodward and his foundation, with short biographical notices of the Professors who preceded Sedgwick, in order to show what he accomplished in placing the study of geology in Cambridge on a new basis, and in getting together one of the noblest geological collections in England.

No task could have been more congenial to me. Sedgwick was one of my father's oldest and most intimate friends ; and I can remember his visits to our house, and his society in our rides and walks, as far back as I can remember anything. If I have failed in my attempt to delineate a singularly genial and loveable man, my failure has not, at any rate, been due to want of interest in my subject. I have tried to set him before my readers as I knew him, and as I heard him spoken of, in his best days, by those who had known him, respected him, and loved him since they had been undergraduates together.

The biography of one who was born so far back as 1785 involves, to some extent, an archæological investigation; and has compelled me to trouble a large number of persons, too numerous to mention individually, with inquiries. I must ask them, as well as those who have allowed me to print letters by Sedgwick, to accept a collective expression of my gratitude for their assistance.

I wish, however, specially to thank the representatives of the Reverend Joseph Romilly for the kindness with which they placed his *Diary* at the disposal, first of Professor Hughes, and then of myself. From the year 1832 until his death, Mr Romilly wrote down an account, more or less minute, of each day, the names of those whom he met, and often the subject of their conversation. Unfortunately he does not say much about what was going on in the University, so that the *Diary* has a personal, rather than a public, interest. But, it is an invaluable record of what Sedgwick was about; for he never fails to record his movements, his employments, and his almost daily visits to himself.

Further, my best thanks are due to Dr Robinson, Master of S. Catharine's College, for writing a chapter on Sedgwick's life at Norwich; and to my friend Dr Jackson, Fellow of Trinity College, for revising and correcting the proof-sheets.

In conclusion, I have to thank the Syndics of the University Press for their liberality in enriching the work with numerous illustrations; and the staff of the Press for much kindness to myself personally during its progress.

JOHN WILLIS CLARK.

Scroope House, Cambridge,
26 *May*, 1890.

CONTENTS

OF THE

FIRST VOLUME.

CHAPTER I.

Introduction. Sedgwick's birth-place. Geographical position of Dent. Description of it at the beginning of this century. Ancient manners and customs. Spelling of the name Sedgwick. Origin and history of the family. Immediate ancestors of Adam Sedgwick. His brothers and sisters. His own account of his father. . . . . pp. 1—44.

CHAPTER II.

(1785—1804.)

Birth of Adam Sedgwick. Childhood and boyhood at Dent. Love of his native Dale. His sister Isabella. Goes to school at Sedbergh. His master and schoolfellows. Selection of a College at Cambridge. Reads mathematics with Mr Dawson. Account of him. . . . . pp. 45—70.

CHAPTER III.

(1804—1810.)

Begins residence at Trinity College, Cambridge (1804). The Rev. T. Jones. Christmas at Whittlesea. College Examination (1805). Summer at Dent. Fever. Life and friends at Cambridge (1806). Preparation for degree (1807). System of Acts and Opponencies. Sedgwick's first Act. University election for M.P. His Father's account of the Yorkshire election. Scholarship at Trinity College. Blindness of his Father. Senate-House Examination (1808). Revisits Dent. Reading party at Ditton. Classical work (1809). Fellowship (1810). . . . . . pp. 71—100.

CHAPTER IV.

(1810—1818.)

Work with pupils. Reading for ordination (1810). Visit to London. Society at Cambridge. Contested election for Chancellor and M.P. for University. Installation of Chancellor. Reading party at Bury St Edmunds. Christmas at Dent (1811). Winter visit to Lakes. Visit to London. Petition against Roman Catholic claims. Reading party at Lowestoft. Appointed sub-lecturer at Trinity College. Second petition against Roman Catholics (1812). Serious illness. Summer at Dent (1813). Severe frost. Visit of Marshal Blucher. Summer at Dent. Excursion in Yorkshire (1814). Projected tour on the continent. Farish's lectures. Cambridge fever. Goes home to Dent. News of Waterloo. Assistant tutorship at Trinity College (1815). Tour on continent (1816). Ordination. Summer at Dent. Hard work in Michaelmas Term (1817). Elected Woodwardian Professor of Geology (1818). . . . . pp. 101—165.

CHAPTER V.

Sketch of the life and works of Dr John Woodward. His testamentary provisions. Arrival of his cabinets. A room constructed for their reception. Sedgwick's predecessors: Conyers Middleton; Charles Mason; John Michell; Samuel Ogden; Thomas Green; John Hailstone. Orders and regulations sanctioned in 1818. . . . . pp. 166—198.

CHAPTER VI.

(1818—1822.)

Excursion to Derbyshire and Staffordshire (1818). First course of Lectures. Visit to Isle of Wight with Henslow. Foundation of Cambridge Philosophical Society. Visit to Suffolk coast. Commencement festivities. Geological tour in Devon and Cornwall. Henslow's work in the Isle of Man (1819). Geological tour in Somerset and Dorset. Acquaintance with Rev. W. D. Conybeare. Death of his mother (1820). Visit to the Isle of Wight. Geological tour in Yorkshire and Durham (1821). Controversy respecting Professorship of Mineralogy (1822—1824). . . . pp. 199—245.

CHAPTER VII.

(1822—1827.)

Geological exploration of the Lake District (1822—1824). Contested election for University (1822). Death of his sister Isabella (1823). Geological papers. Work in the Woodwardian Museum (1823—1827). Lecture to ladies. Visit to Edinburgh with Whewell (1824). Visit to Sussex with Dr Fitton (1825). Contested election for University. Visit to Paris with Whewell (1826). Elected Vice President of Geological Society. Contested election for University (1827). Social life at Cambridge. Hyde Hall. Review of Sedgwick's Geological work (1818—1827). . . . . . pp. 246—297.

CHAPTER VIII.

(1827—1831.)

The Geological Society. First acquaintance with Murchison. Tour with him in Scotland. Office of Senior Proctor (1827). Joint papers with Murchison. Summer in Cornwall. Dolcoath Mine. Visits Conybeare in South Wales (1828). Sedgwick President of Geological Society. Divinity Act. Mr Cavendish elected University representative. Summer in Germany and the Tyrol with Murchison. Joint paper on the eastern Alps (1829). Address to Geological Society. Summer in Northumberland. Contested election for President of Royal Society (1830). Addresses to Geological Society (1831).   .   .   .   .   pp. 298—371.

CHAPTER IX.

(1831—1834.)

The Reform Bill. Contested election for University. Geological papers. Tour in Wales with Charles Darwin (1831). Declines the living of East Farleigh. Mrs Somerville's visit to Cambridge. British Association at Oxford. Summer and Autumn in Wales. President of Cambridge Philosophical Society. Discourse on the Studies of the University (1832). British Association at Cambridge. The Beverley Controversy (1833). Dislocates right wrist. Petition against Tests. British Association at Edinburgh. Made Prebendary of Norwich (1834). .   .   .   .   pp. 372—437.

CHAPTER X.

(1835—1840.)

Cambridge occupations. Election at Dent. Presentation at Court. British Association at Dublin. Skeleton of Irish Elk. Visit of Agassiz to Cambridge (1835). Lectures and Society at Norwich. Geological tour in Devonshire with Murchison. Death of Mr Simeon (1836). Ill health. Paper on Geology of Devonshire. Criticism of Babbage. Death of Bishop Bathurst. Foundation of Cowgill Chapel. Geology in Devonshire. British Association at Liverpool. Inundation of the Workington Colliery (1837). Explorations at Bartlow. Devonian Paper. Queen's Coronation. British Association at Newcastle. Open-air Lecture (1838). *The Silurian System* published. Foreign tour with Murchison (1839). Ill health. Cheltenham. Paper to Geological Society (1840). .   .   .   .   .   pp. 438—528.

CHAPTER XI.

Geological work accomplished between 1828 and 1838. The Devonian System.   .   .   .   .   pp. 529—539.

CHAPTER I.

INTRODUCTION. SEDGWICK'S BIRTH-PLACE. GEOGRAPHICAL POSITION OF DENT. DESCRIPTION OF IT AT THE BEGINNING OF THIS CENTURY. ANCIENT MANNERS AND CUSTOMS. SPELLING OF THE NAME SEDGWICK. ORIGIN AND HISTORY OF THE FAMILY. IMMEDIATE ANCESTORS OF ADAM SEDGWICK. HIS BROTHERS AND SISTERS. HIS OWN ACCOUNT OF HIS FATHER.

IT is proposed, in the following pages, to give some account of the life and labours of Adam Sedgwick. To the world he is best known as Woodwardian Professor in the University of Cambridge from 1818 to 1873, and as one of the founders of modern geology. But, eminent as he was in the subject of his choice, it would be a great mistake to suppose that he was a geologist and nothing more. Geology, it is true, was rarely absent from his thoughts. So long as his health permitted, he devoted himself, with untiring energy, to the investigation of some of its least-known formations; to the establishment of fixed principles for its study; to the defence of it against bigotry and ignorance; to the instruction of students; and to the extension of the Museum connected with his chair. But, on the other hand, his mind was far too active, his nature too warmly sympathetic, to suffer him to be content with the claims of a single science, however attractive. He took a keen interest in all that was going forward, whether in his own University, or in the world at large. In his younger days he was an ardent politician on the Liberal side, and played a prominent part in the reforms, or attempts at reforms, which distinguished that period of academic history; and in later years, as a member of the Royal Commission of 1850, he was enabled to exercise a considerable influence on the legislative changes afterwards introduced. To the end of his life he watched public affairs, both at home and abroad, with unabated zeal; and his letters will show how heartily he rejoiced over a national triumph, how bitterly he mourned a national disaster. Nor did he forget that there were other fields of research besides his own. He was fond of metaphysical and moral speculations, as we shall see when we come to speak of his *Discourse on the Studies of the University of Cambridge*; and no less an authority than Dr Whewell has admitted that had not Sedgwick's life "been absorbed in struggling with many of the most difficult problems of a difficult science," he would have been his own "fellow-labourer or master" in the work which he was then publishing, *The Philosophy of the Inductive Sciences*[1]. History too—especially the history of his own country—was one of his favourite pursuits; he took a genuine interest in archaeology and architecture, though he had no leisure to study either thoroughly; and, from his earliest years to extreme old age, no visits gave him so much pleasure as those which he paid to English Cathedrals.

Again—besides the subjects here indicated—he was an omnivorous reader of general literature. Nothing came amiss to him. Travels, biography, novels, poetry, even controversial theology, were all at least looked into. Like most men born in the last century, when light reading was comparatively unknown, and the multiplicity of modern books had not yet come into being, he delighted chiefly in the masterpieces of our older literature. He revelled in Chaucer and Shakespeare; he could quote long passages

---

[1] Letter to Rev. A. Sedgwick, prefixed to *The Philosophy of the Inductive Sciences*, by the Rev. W. Whewell, ed. 1840.

from Milton, Cowley, and Dryden; and he was never tired of reading and re-reading, till he must have known them almost by heart, the novels of Walter Scott, with which, as he often said, "he had been driven half-mad as a young man." The amount of information acquired through this incessant miscellaneous reading, taken together with a rich natural humour, and a copious flow of language and illustration, lent a singular charm to his conversation. Few have ever told a story so vividly as he did, or with such a marvellous combination of the dramatic, the humorous, and the pathetic. He knew how to invest incidents which, had they been told by anybody else, would have seemed common-place, with the most thrilling interest. He placed himself in the circumstances of those whose adventures he was relating; he thought as they did, felt as they did; and whether he was speaking of an old woman buried in a snow-drift, or a dog guarding a bag of stones, his hearers were fascinated by the truth of the picture, and carried away by the rare power of the narrator. One or two of his most celebrated stories will find a place in this narrative; but it is to be feared that without his flashing eye, and the passionate earnestness of his voice, they will hardly justify their celebrity.

And herein lies the principal difficulty with which Sedgwick's biographer is confronted. He may well despair of being able to set before a reader, who never saw him, possibly never even heard of him, his remarkable personality. Engravings may give a fairly accurate idea of his features and his figure; criticism may determine the exact value of his contributions to science; his occupations from day to day may be accurately discovered; and yet the reader may still be left in ignorance of Sedgwick as a man. Some effort must be made—though the colours may be faint, and the brush unsteady—to make posterity realise his rare originality of character, which commanded the admiration of all who knew him, from the Queen to the humblest of her subjects; his absolute sincerity, which Mr Justice Maule summed up in the

forcible remark : " Sedgwick is one of those men who, if they
ceased to believe in God, would tell you so directly;" his
kindly sympathy with the pursuits, the sorrows, and the joys
of others ; his unselfishness ; his boundless liberality ; his
enthusiasm for all that was good and noble ; his hatred of
wrong-doing and oppression in whatever form they presented
themselves ; his conscientiousness in the discharge of duties
which might at first sight appear so incompatible as a
Professorial Chair at Cambridge, and a Prebendal Stall
at Norwich; and lastly, the firmness of his belief in a
personal Redeemer, which animated him through the whole
of his active life, and cheered him in the loneliness of his
declining years. Nor is this all. His humour when he
told a story has been already alluded to. But it was not
reserved for those special narratives which became insepar-
ably connected with his name; it manifested itself in all his
occupations, even the most serious. His animal spirits rarely
flagged ; he was the life and soul of every company in which
he found himself, whether it were a knot of children at a
Christmas party, or a meeting of their seniors convened for
important business. If he were present the gravest unbent
their brows—the most serious forgot their solemnity. It was
impossible to resist the infection of that boisterous laugh ;
that cheerful geniality; that persistence in looking at the
bright side of things; in a word that union of all that was
cordial and generous and friendly which gained for him the
appropriate name of " Robin Goodfellow[1]."

Fortunately he dearly loved writing letters—not short
letters after the fashion of the present day, but long compo-
sitions in which he wrote as he talked, with that combination
of playfulness and seriousness which made his conversation

[1] A paper was circulated at the Installation of the Prince Consort in 1847,
entitled: *Sporting Intelligence. University Sweepstakes, Cambridge.* Several of the
most prominent members of the University were supposed to be running horses,
the names of which were intended to hit off their obvious characteristics. We find
Dr Whewell's *Rough Diamond*; Prof. Sedgwick's *Robin Goodfellow*; Mr W. H.
Thompson's *A-don-is*; Dr Archdall's *Mrs Gamp*, etc.

so delightful. It will be remarked that the best of these letters are usually those addressed to ladies. Sedgwick compensated himself for the want of children of his own by adopting those of his intimate friends; and he probably selected the daughters instead of the sons as his correspondents, because he felt that when addressing them he could find unreserved expression for feelings which had no other outlet—and which, had his circumstances been different, would have made him the most tender of husbands, and the most considerate of fathers. It is much to be regretted that some of the most interesting of these letters are known to have been destroyed, while others cannot now be traced. Enough, however, survive to show a side of Sedgwick's character, the existence of which, without them, might have been wholly unsuspected; and, it may be added, to make us regret more bitterly the loss of the remainder.

Another of Sedgwick's most marked characteristics was his attachment to his birth-place, Dent in Yorkshire. Most men who go out into the world from a distant corner of England, and make themselves famous in new surroundings, either forget their birth-place altogether, or revisit it only at rare intervals. Sedgwick, on the contrary, did not merely feel affection for the place where he had spent the first nineteen years of his life, but he regarded it as his home, from which he might be separated for the greater part of each year, but which he was bound to revisit at the first opportunity. "For more than three-score years" he wrote in 1868, "Cambridge has been my honoured resting-place, and here God has given me a life-long task amidst a succession of intellectual friends. For Trinity College, ever since I passed under its great portal, for the first time, in the autumn of 1804, I have felt a deep and grateful sentiment of filial regard. But, spite of a strong and enduring regard for the University and the College, whenever I have revisited the hills and dales of my native country, and heard the cheerful greetings of my old friends and countrymen, I have felt a new swell of emotion, and said to myself, 'Here is the land of my birth; this was the home of my boyhood, and is still the home of my heart[1]." Nor is it a mere fancy which traces a connection between his rugged nature and the crags of that wild mountain-valley. To the end of his days he was at heart a Yorkshire Dalesman.

On this account it will be a peculiarly interesting task to dwell at some length on the natural features of Sedgwick's birth-dale, and the old-world manners of its inhabitants, as he remembered them. And here, fortunately, he can speak for himself; for, in the supplement and appendices to his *Memorial by the Trustees of Cowgill Chapel*, he amused himself by setting down, for the information of his fellow dalesmen, not only his own personal recollections, but the traditions which he had gathered from men who were old when he was himself a boy.

The dale of Dent is situated in the westernmost extremity of Yorkshire—a corner of the county which runs forward, wedge-like, into Westmoreland, by which it is bounded on the west and north. The dale descends, as does the neighbouring Garsdale, from the lofty mountain-range called "the backbone of England," towards what Sedgwick terms the "great basin or central depression," in the upper part of which stands the town of Sedbergh. Five distinct valleys are there united. Down four of them the waters descend into the central basin; down the fifth they make their final escape. The position of these valleys will be better understood from the accompanying map than from any elaborate description. The Rawthey, a stream which rises in the fells near Kirkby Stephen, is joined, at a short distance above Sedbergh, by the Clough, or Garsdale-beck, which drains Garsdale; and, at an almost equal distance below it, by the Dee, which drains the dale of Dent. The united streams, still called the Rawthey, fall into the Lune at about two miles below

---

[1] *A Memorial by the Trustees of Cowgill Chapel, with a preface and appendix, on the climate, history, and dialects of Dent.* By Adam Sedgwick, LL.D. 8vo. Cambridge, 1868. Privately printed. p. vi.

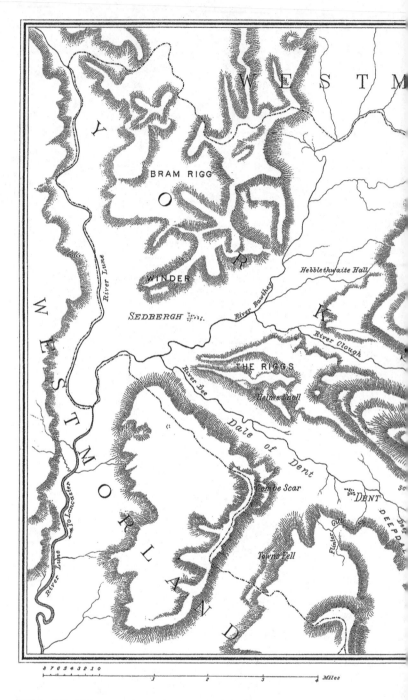

MAP OF THE DALE OF DENT AND SURROUNDING COUNTRY.

R L A N D

UGH FELL

River

Ure

DALE

River Clough

H

I   R   E

WIDDALE   FELL

HILL

KIRTHWAITE

Cowgill Ch.

Dee's Gib's

Marble

Works

WOLD

FELL

WHERNSIDE

Sedbergh. From the point of junction the name Rawthey disappears, and it is the Lune which flows past Kirkby Lonsdale into Morecambe Bay. The whole district is remarkably picturesque; but it is from the hills on the right bank of the Lune, just opposite to its junction with the Rawthey, that the finest views of the five valleys can be obtained. Those who have personally enjoyed them can best sympathise with Sedgwick's enthusiasm, when he exhorts his "younger countrymen and countrywomen" to scale these heights, and then to "warm their hearts by gazing over this cluster of noble dales, among which Providence placed the land of their fathers, and the home of their childhood[1]."

At its origin the dale of Dent is a mere gorge, leading to a mountain-pass; and, in fact, the whole upper portion, for a distance of between five and six miles (all of which is included in the hamlet of Kirthwaite) is narrow and contracted, and the boundary-hills are bare and rugged. The climate of this part of the dale is much more severe than that of the lower part; the rainfall is greater; and "it is sometimes in the winter season much obstructed by ice and snow, when the roads in the lower part of the valley are quite free[2]." It was in this hamlet that a destructive avalanche—or, as they would have said in Dent, a 'gill-brack'—took place in January, 1752, by which seven persons lost their lives[3]. As a traveller descends the dale, the ruggedness gradually disappears, the hills become less precipitous and more cultivated, and when the village of Dent, or, as the dalesmen call it, Dent's Town, is reached, they are green to their summits, and their sides are dotted with homesteads, divided, sometimes by stone walls, but more frequently by rows of trees, into plots of

---

[1] *Supplement to the Memorial of the Trustees of Cowgill Chapel*, with an appendix, etc., printed in 1868. By Rev. Adam Sedgwick, LL.D., F.R.S. Printed for private circulation only. 8vo. Cambridge, 1870. p. 48, *note*.

[2] *Memorial*, p. 4.

[3] *Ibid.* pp. 36—50, where a letter written by one of the survivors to his brother is printed. This letter will also be found in the *Philosophical Magazine* for 1866, xxxi. 80; and in the *Cambridge Chronicle*, 9 December, 1865.

pasturage. Here the dale is nearly a mile in width; and, at a short distance from the village, it is joined by a short subsidiary dale, called Deepdale. This wider part of the dale is about two miles long. Below it, through nearly the whole of the four miles which have to be traversed before Sedbergh is reached, the boundary-hills again converge, and the river-channel is deep and narrow, so that "in ancient days, when the hill-sides were covered with dense forests, Dent must have been more retired from sight, and perhaps more difficult of access, than any of the other valleys within the parish of Sedbergh[1]."

At the present day, Dent is a small picturesque village, with a single paved street; but in 1801, when the first census was taken, its population was considerably larger than that of Sedbergh[2]. Various causes have contributed to this decay, as Sedgwick's narrative will show; and, as he mournfully admitted, "there is no sign of hope that Dent may hereafter regain its lost position." His reminiscences have therefore an historical importance quite independent of his own biography.

"Dent was once a land of *statesmen*, that is, of a rural and pastoral yeomanry, each of whom lived on his own paternal glebe. The estates were small; but each of them gave a right to large tracts of mountain pasturage; and each statesman had his flock and herd. A rented farm was once a rare exception to the general rule; but now (1868) nearly the whole dale,

---

[1] *Supplement*, p. 48.

[2] Sedgwick makes the following remarks on the diminished population of Dent in the *Supplement* (1870), p. 24, "In the previous pamphlet there are several mournful notices of the gradual decay in the prosperity of Dent, and of the diminution in its population. To obtain a numerical test of this fact, during my short visit to Dent in the summer of 1868 I examined the Parish Register of Baptisms and Burials. Counting all the Baptisms from 1747 to 1766 inclusive, I found that they amounted to 985 : but counting the Baptisms from 1847 to 1866 inclusive they amounted only to 529. In like manner, taking the corresponding periods in the two centuries, I found that the Burials amounted to 671 in the last century; and in the present century to 383. In both these periods the Registers appear to have been very carefully kept : and the numbers seem to prove even a greater diminution in the population than I had stated."

from end to end, is in the occupation of farmers with very small capital, and living at a high rack-rent[1]."

The statesmen, it must be understood, were the aristocracy of the dale; they stood somewhat aloof from their fellow-dalesmen, and affected a difference in thoughts, manners, and dress. It used to be said of a lad who was leaving his father's home : "*He's a deftly farrand lad, and he'll du weel, for he's weel come, fra staëtsmen o' baith sides,*" i.e. " He is a well-mannered lad, and he will prosper, for he is well descended, from statesmen on both sides[2]." But, though the statesmen might be somewhat exclusive, " they never passed a neighbour, or even a stranger, without some homely words of kind greeting. To their Pastor, and to the Master of the grammar-school, they did not grudge any known address of courtesy; but among themselves the salutations were at once simple, frank, and kind; and they used only the Christian name to a Dalesman, no matter what his condition in life. To have used a more formal address would have been to treat him as a stranger, and unkindly to thrust him out from the Brotherhood of the Dales. And were they not right in this ? What name is so kind and loving as the dear Christian name, excepting the still dearer and more revered names of Father or Mother ? They are the names by which we speak to our brother or sister, or friend who is near our hearts.

"In former times I never returned to Dent without hearing my Christian name uttered with cheerful face and rung with merry voice by all the upgrown persons whom I encountered on the highway. But nearly all my old friends are gone; and, to my deep sorrow, I no longer hear my Christian name, but am welcomed by words that pronounce me to be a stranger, and no longer a brother living in the hearts of the Dalesmen.

"I will explain my meaning by two recent examples, which were exceptions to the above remark; but they will, I trust, prove that I am rightly interpreting the ancient

---

[1] *Memorial,* p. vii.   [2] *Memorial,* p. 65, *note.*

manners and feelings of my countrymen. There was an aged soldier in Dent, poverty-stricken and desolate, having neither wife nor daughter to cheer him. Several times I gave him a trifle by way of remembrance when I visited Dent; and for a while he had from me a small weekly allowance for tobacco. When in extreme old age he was removed to the Union Workhouse; and he then requested me to exchange the tobacco for a small daily glass of grog. In the discipline of his regiment he had learned a more smart and formal address than was usual in the Dales, but all this wore away when he tried to express his thanks to me, whenever I called on him. I was then sure to hear my Christian name sounded from his aged lips. The last time I saw him he was above ninety years of age and bedridden, yet apparently happy and in good hope; and when the master of the Union made him understand that a gentleman had called to see him, he said, 'Is it Adam?' I did not remain long with him; and as I left him he pressed my hand, and said: 'Oh, Adam, it is good of you to come and see me here!'

"The other case tells the same truth— that the Christian name was the name of loving memory—but it is told in a merrier tone. There were in my childhood two well-known, cheerful-mannered women living in Dent—a mother and daughter employed in the carrying trade[1]—old Peggy Beckett, and young Peggy Beckett. Young Peggy won my child's heart by playing with me, and helping me to leap over the tombstones in the church-yard. But she married, and disappeared from Dent; and many years, I think not less than seventy, passed away before, in extreme old age, she returned, to end her days at her son's cottage. The first time I found

---

[1] At the time of which Sedgwick is speaking all external produce came into Dent, by carriers from Hawes, Kendal, or Kirkby Lonsdale. Women were carriers as well as men, and were indeed preferred, because they could match draper's goods, and choose provisions, better. Mrs Beckett and her daughter were carriers to Kendal, 16 miles distant from Dent. They used to leave Dent on Friday, attend Kendal market on Saturday, and get back to Dent late on Saturday night.

my way to Dent, after her return, I went, along with some young nieces, to call upon her. She received our party with a bright and respectful cheerfulness; but perhaps with more formality than was usual in the Dale; and she spoke to me as a stranger. But when they told her who I was, her fine old face lighted up. She looked earnestly at me for about two seconds, and then said: 'Oh, Adam, it is lang sin' I tought ye to loup off Battersby's trough!" (Oh! Adam, it is long since I taught you to leap off Battersby's tombstone[1])! This address brought back to my memory a pleasant passage in the life of my childhood; and it proved that the young Peggy Beckett of early years, by this use of my Christian name, no longer thought me a stranger, but welcomed me again as a brother of the Dale.

"Do not these two examples prove what I contend for? That the Christian name was not used as a word of thoughtless familiarity; but as a word of confiding, brotherly, love[2]."

"Many of the old statesmen in the higher parts of Kirthwaite were numbered in the Society of Friends. Excellent men they were, and well-informed in matters of common life; lovers of religious liberty; of great practical benevolence, and of pure moral conduct; and they were among the foremost in all good measures of rural administration[3]."

"Though the population of the dale has diminished, I believe, by more than one-third since the middle of last century, yet the poor-rates are enormously increased. It was

---

[1] Sedgwick adds the following note: "Battersby was an early Master of the Chartered Grammar School, and had in his day the reputation of being a "conjurer." A large and ugly monument had been erected to his memory near the south-west angle of the new steeple; and, being partly in ruins when the builders began, some of its larger blocks of stone were placed in the new groundworks. When this desecration was discovered, an old man came in terror to my father, affirming that the steeple would never stand. My father partly allayed the old man's fears, and told him it was foolish and wrong to think that a part of God's house could not stand against the power of a dead conjurer. For a while the old man was pacified, and seemed half ashamed of himself; but shortly afterwards he returned with a blank, doubting face, and said: 'I's feard, Sir, the bells when put in the new steeple 'ill be ringing when they sud'nt.'"

[2] *Supplement*, pp. 44—47.   [3] *Memorial*, p. xii.

once a place of very active industry; well known as a great
producer of wool, which was partly carded and manufactured
on the spot for home-use; but better known for what were
then regarded as large imports of dressed wool and worsted,
and for its exports of stockings and gloves that were knit
by the inhabitants of the valley. The weekly transport of
the goods which kept this trade alive, was effected, first by
trains of pack-horses, and afterwards by small carts fitted for
mountain-work.

"Dent was then a land of rural opulence and glee. Children
were God's blessed gift to a household, and happy was the
man whose quiver was full of them. Each statesman's house
had its garden and its orchard, and other good signs of domestic comfort. But alas, with rare exceptions, these goodly
tokens have now passed out of sight; or are to be feebly
traced by some aged crab-tree, or the stump of an old plum-
tree, which marks the site of the ancient family orchard.

"The whole aspect of the village of Dent has been changed
within my memory, and some may perhaps think that it has
been changed for the better. But I regret the loss of some
old trees that covered its nakedness; and most of all the two
ancient trees that adorned the church-yard, and were cut
down by hands which had no right to touch a twig of them.
I regret the loss of the grotesque and rude, but picturesque
old galleries, which once gave a character to the streets; and
in some parts of them almost shut out the sight of the sky
from those who travelled along the pavement. For rude as
were the galleries, they once formed a highway of communication to a dense and industrious rural population which lived
on flats or single floors. And the galleries that ran before
the successive doors, were at all seasons places of free air;
and in the summer season were places of mirth and glee, and
active, happy industry. For there might be heard the buzz
of the spinning-wheel, and the hum and the songs of those
who were carrying on the labours of the day, and the merry
jests and greetings sent down to those who were passing

through the streets. Some of the galleries were gone before the days of my earliest memory, and all of them were hastening to decay. Not a trace of them is now left. I regret its old market-cross, and the stir and bustle of its

The street of Dent, looking east: from a water-colour drawing by Mr William Westall, taken about 1820.

market-days. I regret its signboards dangling across the streets; which, though sometimes marking spots of boisterous revelry, were at the same time the tokens of a rural opulence[1]."

Some traces of this peculiar feature of Dent were still in

[1] *Memorial*, pp. vii—ix.

existence so late as 1820, when the water-colour drawings
were made from which our woodcuts are taken. Both
represent the main street of Dent. The first cut shows
one of the external flights of stairs which led up to the
galleries; and from the walls on both sides of the street
project stone brackets that once supported similar structures.
In the second cut the opposite end of the gallery approached by the stairs of the former drawing occupies the
left corner of the picture; and facing it is a conspicuous row
of brackets on which imagination can place another.

In the first year of King James the First the principal
land-holders of Dent obtained a royal charter for their
grammar-school, the endowment of which had been collected
by subscription. This school "has had a very healthy
influence upon the education and manners of the valley.
The leading statesmen's sons attended it, and acquired a
smattering of classical learning; and if a statesman's younger
son, or the son of a cottager, were a lad of good promise, his
education was pushed forward into a higher course, and he
was trained for the Church. And many so trained, and
without any other collegiate education, did enter the Church,
and filled the retired curacies in the north of England.

"The necessities of the country soon led to an extension
of the course of teaching at the grammar-school. It had
large English classes, in which writing and arithmetic were
taught to young persons of both sexes; and there were also
itinerant masters, of good repute among the northern dales,
who visited certain schools in a regular cycle, and were
chiefly employed in teaching writing, arithmetic in all its
branches, and the principles of surveying.

"Nor must I omit to state that at all the knitting-schools,
where the children first learnt the art many of them were
to follow through life, the Dames always taught the art of
reading[1]."

"Trusting in the traditions of family history, we may

[1] *Memorial*, p. 54.

affirm, that after the Reformation, and down towards the concluding part of the last century, Dent was in the enjoyment of happiness and prosperity; in a humble and rustic form, it might be, but with a good base to rest upon—the intelligence and industry of its inhabitants. The statesmen were long famous for their breed of horses. The farms were

The street of Dent, looking west: from a water-colour drawing by Mr William Westall, taken about 1820.

providently managed; and the valley was well known for its exports of butter, which, from defect of ready transport, was highly salted, and packed in firkins. The art of the cooper became then of importance. Dent was supplied in abundance with the materials and the workmen; and the cooper's art flourished in it for several generations, by works both for home use and for export.

" The management and economy of the good housewives of our valley became notorious, and often was the subject of some good-humoured jest on the part of the lazy lookers on. Jests seldom bear repeating; but I will repeat one which I have heard in my boyhood. A clever lass in Dent can do four things at a time, was said of old:

> She knaws how to sing and knit,
> And she knaws how to carry the kit,
> While she drives her kye to pasture.

"Wool must have been a great staple produce of the valley, from its earliest history. The greater part of it was exported; but some of it was retained for domestic use; then worked into form by hand-cards of antique fashion (which, in my childhood, I have seen in actual use); and then spun into a very coarse and clumsy thread; and so it supplied the material for a kind of rude manufacture, that went, I think, under the elegant name of *Bump*.

" But, as art advanced, our Dalesmen gradually became familiar with the fine material prepared by the wool-comber; and, before the beginning of last century, Dent became known for its manufacture and export of yarn-stockings of the finest quality. Some of the more active and long-sighted statesmen of the Dales, taking upon themselves the part of middle-men between the manufacturers and the consumers, used occasionally to mount their horses, and ride up to London to deal personally with the merchants of Cheapside, and to keep alive the current of rural industry.

" At a further stage in the industry of our countrymen, worsted, that had been spun by machinery, came into common use; and the knit worsted-stockings were the great articles of export from the Northern Dales. Such became the importance of this export, about the middle of last century, that Government Agents were placed at Kirkby Lonsdale, Kendal, and Kirkby Stephen, during 'the seven years' war,' for the express purpose of securing for the use of the English army (then in service on the Continent), the worsted stockings knit by the hands of the Dalesmen; and in this trade Dent had an ample share.

"In the last century there was another source of industry in Dent, which I must not pass entirely over—I mean its minerals and its coal-works. All the mountains of Dent, to the east of Helm's Knot and Colm Scar, are composed of nearly horizontal beds of limestone, sandstone, and flagstone; and of dark shale, here and there showing, traces of coal. And the whole series is surmounted by a coarse gritstone, called the *Millstone Grit.* The limestone beds are arranged in six groups; of which the lowest, called the *great Scar Limestone*, is several times thicker than all the other groups put together. The top of it is seen just above the village of Flintergill; and its upper beds are finely exposed in the river-course of Kirthwaite. Its lower beds are nowhere seen in our valley; but they are grandly exposed in Chapel-le-Dale, where they rest upon the greenish slate-rocks. All the limestone groups of Dent are separated by thick masses of sandstone, flagstone, and shale; and, as the top of the *great Scar Limestone* is only seen near the river-course, the other five groups are to be looked for on the mountain-sides. The lowest of the five contains the black marble beds; and under the highest of the five, sometimes called the *upper Scar Limestone*, is the only bed of coal that has been worked in Dent for domestic use. The *upper Scar Limestone* is surmounted by a bed of shale, which is capped by the lower beds of the great group called *Millstone Grit.* This part of the *Millstone Grit* forms the flat top of the hill called Crag, and the top also of Ingleborough; and over this grit (at Great Colme, Whernside, etc.) is a shale with beds of coal that is too poor (in the hills of Dent) for domestic use, but which might, I think, be profitably employed in burning lime.

"These facts may be seen by any one who will use his senses; and indeed it seems to have been generally known, before the beginning of last century, that a profitable bed of

coal was often to be found under the *upper Scar Limestone.*
At what time the coal-beds in Dent were first opened I do
not know; but it is said that they were first considered an
object of profit in Kirthwaite. Early, I believe, in the last
century a small statesman called Buttermere found the bed
of coal under the *upper Limestone* of the Town-Fell, just
under the last rise of the Crag. The bed appeared at first
sight too thin to be worked for profit; but on examination it
proved to be free from sulphur, and well fitted for the works
of the whitesmiths in Kendal. He therefore engaged the
help of the country miners, and carried on his work for years
—conveying to Kendal, by a train of pack-horses (seventeen
miles over the mountains), the coal which he drew from a
bed not more than six or seven inches thick. And, spite of
the smallness of his produce, and the cost of its primitive
mode of transport, he went on till he had realised a fortune—
not small, according to the humble standard of his countrymen—and he ended as a public benefactor to his valley.
Joseph Buttermere's coal, as a matter of export, would now
be scouted as a mere worthless mockery. Yet I think the
tale deserves notice as a curious record of one of the primitive
modes in which our old statesmen dealt with those who, to
them, were in a kind of outer world. But I will return to the
craft more peculiar to our valley.

"It may have seemed, at first sight, almost incredible
that one of our old statesmen should have thought it worth
his while to mount his little, tough, but active horse, and to
ride up to London to make bargains with the merchants of
Cheapside for a supply of goods manufactured in his Dale.
Such however was the fact, as I have already stated; and I
well remember that, in my early boyhood, there were three
men, living at, or near, our village, who had many times
made these journeys—some before, and some more than
twenty years after, the time of the 'seven years' war.'
Changes of manners and of times had put an end to such a
primitive mode of dealing some years before I saw the light.

But I have sat upon the knee of old Leonard Sedgwick (my father's cousin) and listened to the tales of his London journeys; and how, when his horse had carried him nearly to the great city, he saw the dome of St Paul's standing up against the sky, and countless spires and steeples bristling up into the air above the houses. His homely pictures never faded from my memory. He was intelligent and honourable in his dealings; a kindhearted and mirthful man; well content to look on the brighter side of the things around him; and (a blessing on his memory!) he made all the little children near him right happy by his Christmas feasts. Such a man, and so employed, can never appear again in Dent, unless we could undo the social work of a whole century.

"And there was another man, old Thomas Waddington— a dealer in hats, cloth, drugs, and I know not what besides— who had from time to time ridden up to London to obtain a good stock of materials for the use of his countrymen. He was a statesman, and a man of high character; and a great favourite with the public, in spite of a singularly crusty and irritable manner. Upright in person, with a face glowing with the signs of good cheer—with a dark wig decorated with many curls, and with a broad-brimmed hat, looped in a way that indicated a former, and more proud, condition,—he steadily marched through his walk of life. And where is there one now to represent him? His shop was the place where all the leading statesmen met to discuss the politics of the day, and the affairs of the parish.

"And there was a third man whom I must not pass over, if I mean to give any conception of what Dent once was. I well remember Thomas Archer, the prince of rural tailors, with his wig of many curls. In my very early boyhood he was what old Chaucer would have called a 'solempne man'; and, whatever he said or did, seemed to take its tone from a feeling of inherent dignity. Ludicrous as the fact may seem, he had been in the habit, in his early days, of going up to London, I know not how; and there, by the help of some connexions or relations, he would work for a few weeks on a London tailor's shop-board. And having learnt the last metropolitan mysteries of his art, he would return—well primed and loaded—to discharge his duties in his native valley. To these mysteries of his skill the old statesmen owed some of those large decorated coat-sleeves and lapped waistcoats which were many years afterwards worn, in a threadbare state, during Dent's decline, by men who had been brought low through poverty.

"There is no man among my countrymen, or in any neighbouring valley, to match this enterprising old tailor. Many long measures he had; but not one so long as that by which he measured his own standard. He was made by the times in which he lived; and the change of times has made it impossible for us to find a recurrence of his similitude.

"Certainly I have neither room nor time for many biographical notices of my countrymen; but one more name I must mention—that of Blackburne, the barber and wig-maker. To me he was historical, and only known by his works; for he had been called away some years before I was counted among the living inhabitants of Dent. But he was a man famed in his generation through all the neighbouring valleys. From him proceeded the ample full-bottom; and the three-decker (or more rarely the four-decker), so named from its splendid semicircles of white curls that girt the back of the wearer's pericranium; and he made also the humblest of all wigs, the scratch—fitted for a poor man's head. Nor must I forget, in this list of our native artist's works, the formidable tie-wig with a tail like that of a dragon, and with winged curls at the ears. I have heard this wig called, by the school-boys of my day, the flying dragon; and let that be its name, for it well deserved it. All such capital monuments of art were turned out in their glory by the man who with cunning hand and head had built up the crowning decorations of our countrymen. The place of his ancient shop was marked by a great pole, with its symbolical fillet and basin; which I used, in my childhood, to look up to with respectful wonder. But the genius of the place was gone; and I saw only the decayed monuments of the great wig-maker's constructive skill.

"I have not stated such facts as these that I might hold up our ancestors of a former century to ridicule; but in the hope of giving my countrymen a graphic proof of the great change of manners wrought by time; and of a sorrowful change in the fortunes of the inhabitants of Dent, that drove many of them away from their early homes, and sank others into a state of depression against which they knew not how to struggle. I well remember (and I first made the remark in my very childhood) that many of the old-fashioned dresses, seen on a holiday, were the signs of poverty rather than of pride. The coats were threadbare, and worn by men who had seen better days. The looped broad-brim was seen, but as a sign of mourning, like a flag hoisted half-mast high; for it was the half-fallen state of the triple cock (still worn by one or two in the parish) with its three outer surfaces pointing to the sky. And, in the same days, old Blackburne's full-bottoms had lost all their crisp symmetry; and the lower hairs of their great convexity were drooping, as if in sorrow, upon the wearers' necks. The three-deckers showed broken lines and disordered rigging; and as for the flying dragons, they had all, like autumnal swallows, taken themselves away. But there were many exceptions to these mournful signs of decay. There still remained many Dalesmen with old-fashioned dresses, and with cheerful, prosperous looks, among the Sunday congregations at Dent; but the ancient fashions were wearing fast away[1].

"Let me here add a word or two on the domestic state and habits of our countrymen, before their old social isolation had been so much broken in upon by the improved roads and rapid movements of modern times. With the exception of

[1] *Memorial*, pp. 57—64.

certain festive seasons, their habits were simple, primitive, and economical. The cottager had, as his inheritance, the labour of his own hands and that of his wife and children : and, in the good old times, that labour made him quite as independent as one of the smaller statesmen. In manners, habits, and information, there was, in fact, no difference between them. Even in the houses of the clergymen and of the wealthier statesmen, there was kept alive a feeling of fraternal equality; and, although external manners were more formal and respectful than they are now, yet the servants, men or maids, sat down at the dinner-table, and often at the tea-table, with their masters and mistresses.

"The dress of the upper statesman's wife and daughters was perhaps less costly than that of the men who affected fashion; and according to modern taste we should call it stiff and ugly to the last degree, as was the fashion of the day. There was one exception however, both as to cost and beauty: for the statesman's wife often appeared at Church in the winter season in a splendid long cloak of the finest scarlet cloth, having a hood lined with coloured silk. This dress was very becoming, and very costly; but it was carefully preserved; and so it might pass down from mother to daughter. Fortunately, no genius in female decoration (like the Archers and Blackburnes of the other sex) seemed to have brought patches and hoops into vulgar use (as in the preposterous modern case of crinoline).

"Among the old statesmen's daughters hoops did however sometimes appear, as one of the rarer sights of the olden time; and I have heard an aged statesman's daughter tell of her admiration, and perhaps her envy, when she saw a young woman sailing down the Church with a petticoat that stretched almost across the middle aisle. That decoration shut her out from a seat on any of the Church forms; but, by a dexterous flank movement, she won a position among the pews; and then, by a second inexplicable movement, the frame-work became vertical, and found a resting-place by overtopping the pew-door—to the great amazement of the rural congregation.

"All the women, with very rare exceptions, learned to read; and the upper statesmen's daughters could write and keep family accounts. They had their Bibles, and certain good old-fashioned books of devotion; and they had their cookery-books; and they were often well read in ballad poetry, and in one or two of De Foe's novels. And some of the younger and more refined of the statesmen's daughters would form a little *clique*, where they met—during certain years of last century—and wept over Richardson's novels. But this sentimental portion was small in number; and it produced no effect upon the rural manners of the valley; which were fresh and cheerful, and little tinged with any dash of what was sentimental.

"While speaking of the habits and manners of my country-women, I may remark that their industry had then a social character. Their machinery and the material of their fabrics they constantly bore about with them. Hence the knitters of Dent had the reputation of being lively gossips; and they worked together in little clusters—not in din and confinement like that of a modern manufactory—but each one following the leading of her fancy; whether among her friends, or rambling in the sweet scenery of the valley; and they were as notable for their thrifty skill as for their industry. And speaking of both sexes, the manners of our countrymen may have been thought rude and unpolished from lack of commerce with the world; and their prosperity in a former century may sometimes have roused the envy, and the jests and satire, of those who were less handy than themselves; but for many a long year theirs was the winning side.

"Their social habits led them to form little groups of family parties, who assembled together, in rotation, round one blazing fire, during the winter evenings. This was called *ganging a Sitting* to a neighbour's house; and the custom prevailed, though with diminished frequency, during the early years I spent in Dent. Let me try to give a picture of one of these scenes in which I have myself been, not an actor, but a looker-on. A statesman's house in Dent had seldom more than two floors, and the upper floor did not extend to the wall where was the chief fire-place, but was wainscoted off from it. The consequence was, that a part of the ground-floor, near the fire-place, was open to the rafters; which formed a wide pyramidal space, terminating in the principal chimney of the house. It was in this space, chiefly under the open rafters, that the families assembled in the evening. Though something rude to look at, the space gave the advantage of a good ventilation. About the end of the 17th century grates and regular flues began to be erected; but, during Dent's greatest prosperity, they formed the exception, and not the rule.

"Let me next shortly describe the furniture of this space where they held their evening *Sittings*. First there was a blazing fire in a recess of the wall; which in early times was composed of turf and great logs of wood. From one side of the fire-place ran a bench, with a strong and sometimes ornamentally carved back, called a *lang settle*. On the other side of the fire-place was the Patriarch's wooden and well-carved arm-chair; and near the chair was the *sconce*[1] adorned with crockery. Not far off was commonly seen a well-carved cupboard, or cabinet, marked with some date that fell within a period of fifty years after the restoration of Charles the Second[2]; and fixed to the beams of the upper floor was a row of cupboards, called the *Cat-malison* (the cat's curse); because,

---

[1] In north-country dialect, a low partition.

[2] Sedgwick notes: "One or two of the Belgian refugees, who had been driven from London by the great Plague in 1664, are said to have found, for a while, a home in Dent, and there to have practised their art of wood-carving; and one of them is said to have settled in Kirthwaite. The art of wood-carving, at any rate, flourished within the period above indicated; and I remember many good specimens of it in the old statesmen's houses in Dent. The art existed, however, in Dent at an earlier period. For there was, in my Father's time, at the old parsonage, a set of oak bed-stocks, which he had brought from his birth-place. They were vigorously though rudely carved, and had the date of 1532."

from its position, it was secure from poor grimalkin's paw. One or two small tables, together with chairs or benches, gave seats to all the party there assembled. Rude though the room appeared, there was in it no sign of want. It had many signs of rural comfort: for under the rafters were suspended bunches of herbs for cookery, hams, sometimes for export, flitches of bacon, legs of beef, and other articles salted for domestic use.

"They took their seats; and then began the work of the evening; and with a speed that cheated the eye they went on with their respective tasks. Beautiful gloves were thrown off complete; and worsted stockings made good progress[1]. There was no dreary deafening noise of machinery; but there was the merry heart-cheering sound of the human tongue. No one could foretell the current of the evening's talk. They had their ghost-tales; and their love-tales; and their battles of jest and riddles; and their ancient songs of enormous length, yet heard by ears that were never weary. Each in turn was to play its part, according to the humour of the *Sitting*. Or, by way of change, some lassie who was bright and *renable*[2] was asked to read for the amusement of the party. She would sit down; and, apparently without interrupting her work by more than a single stitch, would begin to read—for example, a chapter of *Robinson Crusoe*. In a moment the confusion of sounds ceased; and no sound was heard but the reader's voice, and the click of the knitting needles, while she herself went on knitting: and she would turn over the leaves before her (as a lady does those of her music-book from the stool of her piano), hardly losing a second at each successive leaf, till the chapter was done. Or, at another and graver party, some one, perhaps, would read a chapter from the *Pilgrim's Progress*. It also charmed all

---

[1] A less agreeable picture of this continual knitting is drawn in *A true story of the terrible knitters ʄ Dent*, in Southey's *Doctor*, Inter-chapter xxiv. Vol. vii. p. 78, ed. 1847. Miss Sedgwick remembers that boys and old men knitted as well as women.

[2] Loquacious.

tongues to silence : but, as certainly, led to a grave discussion so soon as the reading ceased.

"In all the turns of life the habits of our countrymen were gregarious. A number of houses within certain distances of one another were said to be in the *lating râ* (the seeking row), and formed a kind of social compact. In joy or sorrow they were expected to attend, and to give help and comfort. To follow this subject out would lead me into details too long for my present purpose. But I may mention how it told upon the customs of Dent, on occasions of great domestic joy. Before the birth of a new inhabitant of the hamlet, all the women of mature life within the *lating râ* had been on the tip-toe of joyful expectation : and the news of the first wailing (the *crying-out*, as called in the tongue of Dent)—the sign of coming life—ran through the home-circle like the fiery cross of the Highlanders ; and, were it night or day, calm sunshine, or howling storm, away ran the matrons to the house of promise, and there, with cordials, and creature comforts, and blessings, and gossip, and happy omens, and with no fear of coming evil—for the women of the valley were lively, like the women in the land of Goshen—they waited till the infant statesman was brought into this world of joy and sorrow, in as much publicity as if he were the heir to the throne of an empire. This custom was upheld with full tenacity during the younger years of my life[1].

"There were in ancient times, few observances in the conduct of a funeral, which are not known at the present day. Formerly, however, they kept a watch in the house, with burning lights in the room of death. This passed under the name of the *Lyk-wake:* but the custom had become very rare, and I believe entirely went out before the end of the last century ; and at no period of our history were there hired professional 'mourning women, skilful in lamentation,' as among the Jews of old, to give effect to the wailings of sorrow. As a prevailing custom, many were 'bidden to the

[1] *Memorial*, pp. 68—73.

funeral'; and there was a peculiar refreshment called the *arval*, offered even at a poor man's funeral, before they went with the coffin to the church ; and, after the interment, if the mourning family belonged to the better class of statesmen, those who had been bidden to the funeral had a dinner provided at one of the inns, which the immediate mourners did not attend. The fact is nothing new to my countrymen ; and I only mention it now, because I have many times heard it sneered at, and shamefully misrepresented. I never knew a single case in which this truly kind and hospitable mode of celebrating a funeral led to intemperance or abuse. It may be better now to conduct a funeral with more quiet simplicity. But so long as there was a large gathering of those who had been the neighbours and friends of the deceased, there was nothing unseemly in giving a poor man a dinner, for which he was thankful, or in offering refreshment to friends who had come from afar, and stood in need of it.

"The festivities of Christmas, and other holiday seasons, were kept up among our countrymen with long-sustained, and sometimes, I fear, intemperate activity. They had their morris-dances; their rapier-dances; and their mask-dances. These grotesque and barbarous usages of a former age disappeared a considerable time before the end of the last century. I believe I saw the end of them full eighty years since, while I was in my nurse's arms. Dent was long famous for its Galloway ponies; and its race-course had its celebrity in former centuries. I believe I saw, in my very early boyhood, the last race ever run upon the old course. Since then, the old ground has been so cut up and changed, that, happily, it would be impossible to re-open it as a race-course, were the old taste to come to life again.

"I should think myself ill-employed were I to dwell long upon the by-gone vices and follies of my countrymen : but I should be disloyal to the cause of truth were I only to hold up to the light of day the fairer and brighter side of their character. Among the vulgar sports of England, especially

during Shrovetide, were matches of game-cocks, which for
centuries had kept their place. Nowhere did this vile and
cruel sport flourish more than among the Dales of the north
of England. Men of character joined in it without compunc-
tion; and so thoroughly was it ingrained among the habits of
society, that the Masters of the chartered grammar-schools
received a Shrovetide fee from their scholars; and in return
gave game-cocks to the boys, to be matched for the honour of
the school! This fee (known by the boys as the *cock-penny*) is
given to the present day[1]; and I have paid it myself many
times. But, for about a century and a half, the Master has
ceased to give any return beyond an acknowledgement of
thanks. I have been present during some of these matches
as a looker-on in my early days (what school-boy will not get
into mischief if he can?); and I have witnessed their fruits;
which were reaped in gambling, quarrels, drunken riots, and
bellowings of blasphemy. Thank God, they have gone from
sight; and will never again, I trust, defile the light of day.
So far as Dent is concerned, this form of cruel sport died
away in the unhappy years that closed the last century.

"In conclusion, I will add a few words more upon the
social decline of my countrymen, which no ingenuity on their
part could have averted; for the gigantic progress of mechanical and manufacturing skill utterly crushed and swept away
the little fabric of industry that had been reared in Dent.
Many of the inhabitants gradually sunk into comparative
poverty. The silken threads that had held society together
began to fail; and lawless manners followed[2].

"Through all the years of my boyhood and early manhood,
the Magistrate, nearest to Dent, who acted for the West
Riding of Yorkshire, lived at Steeton (in the valley of the
Aire), which was about forty-two miles from Dent, and forty-
seven or forty-eight from Sedbergh. In those years several
well-educated men of ample fortune lived within the parish

[1] This was written, it must be remembered, in 1868.
[2] *Memorial*, pp. 74—77.

of Sedbergh, but not one of them was in the Commission of the Peace. Whether from want of patriotism, or love of ease, or a too modest estimate of their own powers, they refused the office. My Father, through all the vigorous years of his very long life, refused to act as a Magistrate, believing its duties inconsistent with those of a Parish Priest. My brother John thought differently, and obtained his Commission soon after he became Vicar of Dent, to the real benefit of the country. For he knew the people well; knew how to temper justice with mercy; and, without flinching from his duty in its sometimes painful exercise, he was honoured, trusted, and beloved; and to the end of life was called the poor man's friend—a character engraven on his monument by those who knew him well[1]. The cost and trouble of seeking justice put law for awhile in abeyance; or, if a check were put upon coarse manners and a disorderly life, it was sometimes done in the way of lynch-law, like that which on occasion has reigned in the back-settlements of America. I could tell some tales of this kind that might raise a laugh; but in very truth they ought to be called tales of sorrow[2].

"I remember, one Sunday evening, when I was a young school-boy, seeing a man in a brutal state of drunkenness, tumbling and bellowing like a maniac among the graves and tombstones of the church-yard, and challenging any one in Dent to fight him. He was a man of very great strength, and of considerable pugilistic skill, which he had gained in London, where he had resided for some time with a relation, but had been sent back to Dent for insubordination and intemperance. When sober he was a good-tempered cheerful man, and a (so-called) 'good companion'; but he had not one grain of principle. He had learnt to regard sin as life's jest, and good

---

[1] *Supplement*, p. 30, *note.* The words of his epitaph, in Dent Church, here referred to, are: "His manners and temper were gentle and kind, And endeared him to all who were under his pastoral care: Nor did he forfeit their love while Faithfully discharging the duties of a magistrate, For he was merciful, as well as just."

[2] *Memorial*, p. 77.

manners as a mask or mockery, put on to serve a purpose. When under excitement he became fierce and dangerous, and for several years he was the terror of the Dale.

"On the occasion just alluded to, the constable of the parish came with a pair of handcuffs and one or two assistants to secure the drunken maniac. After looking at the formidable brute for a few seconds, the constable said, 'If I fix these things on, I dâre not tak 'em off without ganging to th' Justice, and that will cost the parish I knâ not what. He is oër drunk to be dangerous, and I'll give him a good basting.' So he laid down the handcuffs upon a tombstone, and being himself a man of activity and great strength, and no mean artist, he had, in less than a minute, so pounded the maniac's face that it lost all semblance of humanity, and the monster, for a while, had got his *quietus*. The constable then walked home with his handcuffs, cheered and thanked by his neighbours for his cheap way of doing justice.

"Of this strange scene, acted in our quiet village on a Sunday evening, I was a witness. And on another occasion, at one of our annual Fairs, I saw the same drunkard put in handcuffs by the same constable. The day following, the constable and his prisoner, and an assistant, each well mounted, began their journey towards Steeton. The horses required food; the men regarded such excursions as a kind of holiday-keeping, and lived well; and the party could not return before the third day. This was not called cheap justice. The magistrates hated dealing with country brawls, and often quashed the cases with the cheap benefit of some good advice. And, if the case led to the prisoner's committal, there were two more very long journeys for the parish officer, and more cost for the parish[1]."

"The great French Revolution seemed to shake the whole fabric of society to its foundation; and the shock was felt even in the retired valleys of the north of England. But the inhabitants of Dent, though sorely lowered in position, had

[1] *Supplement*, p. 31, *note.*

learnt no lesson of disloyalty. They burnt Tom Paine in effigy—a kind of fact sure to fasten itself upon the memory of a boy; and one of the statesmen, who had inherited a fortune far above any previously known in the valley, engaged the parish singers, and others with lungs that were lusty and loyal, to make nocturnal parades about the parish, singing melodies like *Rule Britannia* and *Hearts of Oak;* and when the parade was over, they were allowed to crown the day with squibs, crackers, loud cheers, and deep potations. Such fooleries could do no good; and they did much harm to those who acted in them.

"The war that followed brought new taxes, and increased poor-rates; and no new gleam of reviving hope shone upon our countrymen. I was still living at the Parsonage at the end of last century; and I well remember the two years of terrible suffering, when the necessaries of life were almost at a famine price, and when many of the farmers and land-owners—before that time hardly able to hold up their heads —had to pay poor-rates that were literally more than ten times the weight of what they had been in former years. It was indeed a time of sorrow and great suffering. But I will not end with notes of such a dismal sound.

"Dent has again revived, and taken a new position. Emigration has relieved the burthen of the five hamlets. Education has made good progress. Roads are greatly improved[1]. I remember some roads in Dent so narrow that there was barely room for one of the little country carts to pass along them; and they were so little cared for, that, in the language of the country, the way was as 'rough as the beck staëns.' I remember too when the carts and the carriages were of the rudest character; moving on wheels which did not revolve about their axle; but the wheels and their axle were so joined as to revolve together. Four strong pegs of wood, fixed in a cross-beam under the cart, embraced the axle-tree, which revolved between the pegs, as the cart

[1] *Memorial,* pp. 77, 78.

was dragged on, with a horrible amount of friction that produced a creaking noise, in the expressive language of the Dales called *Jyking*. The friction was partially relieved by frequent doses of tar, administered to the pegs from a ram's horn which hung behind the cart. Horrible were the creakings and *Jykings* which set all teeth on edge while the turf-carts or coal-carts were dragged from the mountains to the houses of the dalesmen in the hamlets below. Such were the carts that brought the turf and the coals to the vicarage, during all the early days of my boyhood. But now there is not a young person in the valley who perhaps has so much as seen one of these clog-wheels, as they were called; and our power of transport, to be more perfect, only wants a better line of road, that might easily be made to avoid those steep inclines, which are now a grievous injury to the traffic of the valley. But, with all our modern advantages of transport, Dent has lost the picturesque effect of its trains of pack-horses : and many times, on a Sunday morning, I have regretted that I could no longer see the old statesman riding along the rough and rugged road, with his wife behind him mounted upon a gorgeous family pillion ; and his daughters walking briskly at his side, in their long flowing scarlet cloaks with silken hoods[1]."

The ancestors of Adam Sedgwick have been statesmen of Dent for more than three centuries, but their origin, the orthography of their name, and its etymology, have occasioned many rival theories. In 1379 the name is spelt *Syggheiswyk, Seghewyk, Segheswyk*[2]; in 1563 *Seeggeswyke*; between 1611 and 1619 the parish registers of Dent give *Sidgsweeke;* in 1624, *Siddgswicke;* between 1645 and 1696, *Sidgwick* or *Sidgswick*. Between 1700 and 1737 the name is entered

---

[1] *Supplement*, p. 41.

[2] From the rolls of the collectors of the subsidy granted to King Richard II. in the West Riding of Yorkshire 1378—79, printed in *The Yorkshire Archæological and Topographical Journal*, Vol. v. pp. 1—51. These and the succeeding references to the spelling of the name are due to the kindness of Arthur Sidgwick, M.A., Fellow of Corpus Christi College, Oxford.

thirty-six times. Of these entries, two in 1701, and one in 1736, give *Sidgwick;* all the rest *Sidgswick.* The earliest *Sedgwick* at Dent appears on the tomb of the Rev. James Sedgwick, great-uncle to Adam Sedgwick, who died in 1780; but in his register of baptism, 30 Sept. 1715, he is entered as son of John *Sidgswick.* Adam Sedgwick maintained that the spelling of the family name was deliberately changed by this James Sedgwick, at the suggestion of the then Master of Sedbergh school. On the other hand, a branch of the family who had settled at Wisbech in the Isle of Ely called themselves *Sedgewick* at the beginning of the seventeenth century[1], and they are said to have adopted a characteristic crest, a bundle of sedge bound up in a form like that of a wheat-sheaf. Hence it became natural to seek for the origin of the family in 'a village built on fenny ground, with an abundance of the water plant called sedge'; and Sedgwick in Westmoreland, near the head of Morecambe Bay, was fixed upon as the birth-place of the clan, from the similarity of sound, though it does not fulfil the other conditions. Adam Sedgwick rejects these theories for the following reasons :

" 1. Because the word sedge is, I think, unknown in the dialect of the northern Dales.

" 2. Because the well-known village, Sedgwick, is built upon a high and dry soil that is washed by the beautiful waters of the Kent, a stream that runs brawling over the rocks.

" 3. Because the word *Sedgwick* does not give the sound of the name as it was uttered among the ancient inhabitants of the mountains ; nor does it come near to the spelling used in former centuries. The name is at this time commonly pronounced *Sigswick* by the natives of the Dales[2]."

The same information is cast in a more humorous form in the following letter :

[1] *Visitation of Cambridgeshire,* made 1619 by Henry St George, Richmond herald; printed by Sir Tho. Phillips, Middle Hill Press, 1840.

[2] *Supplement,* p. 18. Miss Sedgwick informs me that the name is still commonly pronounced *Sigsick.*

S. I. 3

"The Sidgwick you mention is of the Dent stock. His great-grandfather was brother to my great-grandfather. It may be one step higher; for I am a sorry genealogist. In the old Parish Register the spelling was always with two *i*'s. My father's uncle altered the spelling, and adopted the *Cyclopic* form (at the foolish suggestion of an old pedant of Sedbergh School), when he was a boy. He afterwards educated my father and sent him to College; and so it came to pass that all my dear old father's brood were born with the one *i*; or at least were so dockited on all high-ways, and in all post-towns.

"The name is still pronounced (except where children's tongues have been doctored by 'pupil teachers') *Siggswick*, and has nothing to do with *Sedge*. Neither the name nor the plant are known among my native hills[1]."

The etymology of the word Sedgwick has been most kindly investigated by Professor Skeat, with the following result. "There can be no doubt," he says, "that Sedgwick was at first a place-name, and then a personal name. 'Wick' is not a true Anglo-Saxon word, but simply borrowed from the Latin *uicus*, a town, or village. 'Sedge," or 'Sedj' or 'Sedg' is simply the later spelling of the Anglo-Saxon *secg*. Two distinct words were spelt thus: (1) the modern 'sedge'; (2) a word which has now become quite obsolete, but was once in common use, like *uir* in Latin. It is a derivative from *secg-an*, to say, and meant *say-er, speak-er, orator*, and generally *man, hero, warrior*. It could easily be used as a personal name, as 'man' is now; and *Secg-wic* is therefore a town built by Mr Secg or Mr Mann, as we should say at the present day[2]."

When Sedgwick was making a geological tour in Saxony, he met a gentleman who was both a geologist and an antiquary. They fell into conversation about the etymology

---

[1] To Archdeacon Musgrave, 13 March, 1862.

[2] In the same way Sedgeberrow, in Worcestershire, is *Secges-bearuwe*, i.e. *Secg's-grove;* and Sedgeleigh, in Hampshire, is *Secges-leah*, which means *Secg's-lea*, not a sedge-lea.

of his name, and it was decided that it might originally have been *sieges-wick*, 'village of victory'; whereupon, taking into account the position of the village of Sedgwick, they amused themselves by inventing the following story :

"Soon after the abandonment of England by the Romans, the Anglo-Saxons invaded the valley of the Kent, and settled there after they had driven out the ancient Britons. Then came successive crews of new invaders, Danes and Norsemen; and, during a lawless period, there were many conflicts between the earliest settlers and the piratical crews, which landed and were engaged in the highly exciting work of burning, plundering, and cattle-lifting. On one of these occasions, the plundering sea-rovers were repulsed by the older Anglo-Saxon inhabitants, in a battle fought on the banks of the Kent; and the victory was commemorated, at first perhaps by a heap of stones, and then by a village built near the spot, which took the name of *Siegeswick*, or village of victory[1]."

Professor Skeat rejects this etymology, and Sedgwick did not advance it seriously; but, even if it be erroneous, it may still be admitted that the hero who gave his name to the village may have established his reputation by an achievement not so very different from that which Sedgwick and his friend invented; and, in fact, there is still to be seen a large cairn or tumulus near the village in question, under which those who fell in some such raid may have been interred. Moreover the presence of northern invaders in old days is amply attested by traces of their language still to be met with in the dales.

Here, however, we must leave these interesting speculations for the surer ground of legal documents still in the possession of the Sedgwick family. From these it can be ascertained that towards the end of the reign of Queen Elizabeth a Sedgwick was in possession of an estate in Dent called Bankland. His son Leonard, who owned, besides

[1] *Supplement*, p. 19, *note.*

Bankland, a second property in Dent, called Gibshall, had
twin sons. The elder of these inherited Bankland, the
younger Gibshall, of which, at the end of the seventeenth
century, a John Sedgwick was in possession. He was a man
of energy; and, having a grove of oak trees on his property,
turned tanner, and realised a handsome fortune out of their
bark. He had two sons, Thomas, born 1705, who followed
his father's trade, and James, born 1716, who took Holy
Orders, and became Master of the endowed Grammar-school
at Horton in Ribblesdale. The former, grandfather to
Adam Sedgwick, had an only child, Richard, born 7 March,
1736. He was educated at Dent Grammar School until the
age of fifteen or thereabouts, when he was sent to his uncle at
Horton for more advanced instruction. After spending some
months at Horton, he was removed to the school at Sedbergh,
at the suggestion of his uncle, who has the further credit of
having persuaded his brother to give his son the benefit of
a University education. But, as he had not enjoyed that
privilege himself, while Mr Wynne Bateman, then Master of
the school, had graduated at St John's College, Cambridge,
in 1734, it is surely more probable that the step was taken
in consequence of his advice. Be that as it may, Richard
Sedgwick, after a few months instruction in mathematics
from Mr John Dawson of Garsdale (of whom more below),
matriculated at Cambridge in 1756 as a sizar of St Catharine's
College, or, as it was then called, Catharine Hall. It has
been recorded that before starting he bought a horse, rode
it to Cambridge, and there sold it at a profit. There was
nothing unusual in this at that time. His cousin, another
Richard Sedgwick, went up to St John's College, Cambridge,
in the following year in the same manner; and when
Mr Paley, Master of Giggleswick School, took his son to
Cambridge in 1758, he rode on horseback, with the boy on a
pony beside him. The future archdeacon was a bad rider,
and at first got a good many falls. His father paid but little
attention to his misfortunes, merely turning his head and

exclaiming: "What, William, off again! Take care of thy money, lad!"

Richard Sedgwick proceeded to the degree of Bachelor of Arts in 1760, when he was placed seventh in the second class of the Mathematical Tripos. He was ordained in that year or the next, and from 1761 to 1768 held the curacy of Amwell near Hoddesdon in Hertfordshire. He was also assistant-master in a boarding-shool at Hoddesdon, kept by Dr James Bennet, a gentleman of some distinction in the literary world, for he not only published an edition of Roger Ascham's English works, but obtained the collaboration of Dr Samuel Johnson, who is said to have written for him a life of the author and a dedication to the Earl of Shaftesbury. In 1766 Sedgwick married Dr Bennet's daughter Catherine, and in 1768 was presented to the living of Dent by the patrons, twenty-four of the leading statesmen of the Dale—a strong proof of the popularity of the family, for they could have known little or nothing of him personally. His young wife accompanied her husband to Dent, where she died before the end of the summer (31 July, 1768), leaving one child, who survived her mother for some years. After her death (28 June, 1777) her father married a distant cousin, Margaret Sturgis, by whom he had seven children: Margaret (1782), Thomas (1783), Adam (1785), Isabel (1787), Ann (1789), John (1791), James (1794).

The names of Adam Sedgwick's brothers and sisters will occur so frequently in the course of our narrative that it will be best to mention in this place the leading events in the lives of each of them. Margaret married late in life a distant cousin, the Rev. John Mason, Vicar of Bothamsall in Nottinghamshire, and chaplain to the Duke of Newcastle. She became a widow, 29 October, 1844, and returned to Dent, where she resided till her death, 13 January, 1856. Thomas passed his whole life in Dent, and died, unmarried, 19 September, 1873, aged 90. Isabel died unmarried, 18 January, 1823. Ann married, 22 September, 1820, Mr William Westall,

an artist of considerable reputation in his day, and died in
1862. John was admitted a sizar of St John's College in 1810,
and proceeded to the degree of Bachelor of Arts in 1814, but
without obtaining Honours. He was ordained immediately
afterwards, and became curate of Stowe in Lincolnshire.
Ultimately, in 1822, he succeeded his father as Vicar of Dent,
an office which he held till his death, 9 February, 1859.
James, like his brother, was admitted a sizar of St John's
College, Cambridge in 1813, and proceeded to the degree of
Bachelor of Arts in 1817. He was not deficient in ability, and
it was hoped that he might take a good degree, and perhaps
be elected to one of the Fellowships then appropriated to
Sedbergh School. But he lacked industry; was placed no
higher in the Tripos than sixth in the third class, and subsequently failed in the Fellowship Examination. Soon after
taking his degree he was ordained, and in March, 1818,
became curate of Freshwater in the Isle of Wight, where he
remained until June, 1839. He then removed to Downham
Market in Norfolk, and in 1840 was presented by the Dean
and Chapter of Norwich to the vicarage of Scalby near
Scarborough. He died, 28 August, 1869.

No son ever spoke or wrote of his father with greater love
and respect than Adam Sedgwick; and, when it is remembered that the old man lived till 1828, when his son was
forty-three years of age, everything that he says about his
public character, his management of the parish, and his
influence in the dale, may be accepted as the deliberate
judgment of one man by another. Numerous allusions to
him, and anecdotes illustrating the fervid religious feeling,
tempered by sound common-sense, for which he was so
remarkable will be found scattered through the letters printed
below. For the present we will select the following passages
from the *Memorial* and its *Supplement*. They are not only
interesting in themselves, but reveal the source of many of
the convictions which were most deeply engrained in his son's
character.

"When in my childhood I saw, on a Sunday morning, the ample convexity of my father's well-dressed and well-powdered wig, I thought it one of the most beautiful sights in the world. I remember too, as he went, with his usual light step, towards the church, and saluted his friends who were come to join in the sacred services of the day, that each head was uncovered as he passed. They loved my Father, because by birth he was one of themselves, and because of his kindness and purity of life. They were proud of him too, because he was a graduate of the University of Cambridge, and had been living in good literary society some years before he fixed his home in Dent. Part of his influence arose, also, from the reputation of his skill in athletic exercises; and from a principle of action which he carried out through his long life—never to allow his conception of his sacred duties to come, on questions of moral indifference, into a rude collision with the habits and prejudices of the valley. The consequence was that he held an almost unbounded influence over his flock. Of this I will mention one example; for it deserves notice as a fact of which it would be in vain to look for a match in the present condition of the Church of England.

"Some years before I ever saw the light there was an unexpected contest for the county of York. Mr Wilberforce, a young man of bright presence and great eloquence, was then first named as a candidate; and he had even then become famous as an enthusiastic advocate for the abolition of the Slave Trade. This fact set every chord of my father's heart in motion. He consulted his early friends, the good old Quakers in Kirthwaite, and his other friends in all the five hamlets; and he personally canvassed the valley from house to house. Then at least the inhabitants of the valley formed an united Christian brotherhood. At that time the freeholders abounded, and every vote was pledged for Mr Wilberforce. Soon afterwards came a solicitor to canvass on the other side; but he soon left his canvass, finding himself

unable to advance a single step. For wherever he asked for
a vote the reply was, *Naë use Sir, we o' here gang wi' th'
Parson.* So the solicitor left the field ; mounted his horse at
the door of the Sun Inn; and, uttering an anathema, cried
out aloud, that Dent was "the —— Priest-ridden hole in
England[1]."

"Great injustice should I do to the memory of my Father,
were I to describe him as turning his influence as a Parish
Priest to serve the purposes of a political movement. The
Slave Trade he regarded as a foul national sin, which (however deep its roots might be struck into the policy of the
State) every man, who believed in the over-ruling Providence
of God, was bound, by all lawful means within his reach, to
root out and trample under-foot. The influence he had over
the minds of his flock rested on his humble teaching of
Gospel truth; on the cheerful simplicity of his life; and on
his readiness, at every turn and difficulty, to be in true
Christian love an adviser and a peace-maker.

"Were then the inhabitants of Dent, in any high sense,
religious men during the old times of their prosperous
industry? They were honourable in their dealings, active in
their daily work, steady in the external observances of the
Church Services, and without the bitterness of controversial
spirit. They had an ancient custom which I may mention
here, (and many times when I have thought of it I have
felt sorrow that it had ever been abandoned), of assembling
and holding a Communion in the Church at a very early
hour on Easter Sunday morning. The custom had come
down to them from ancient times,—probably before the
Reformation. There was nothing superstitious in such an
observance; and it was well fitted to touch the conscience

---

[1] This story refers to the general election of April 1784, when Parliament
had been dissolved on the recommendation of Mr Pitt. Mr Wilberforce, then a
young man of 25, successfully contested the county of York, with Mr Duncombe,
in opposition to Mr Weddell and Mr Foljambe, the nominees of the great Whig
families. *Life of Wilberforce*, by his sons. 8vo. London, 1838: i. 50—64.

of any one who believed in his heart that his Saviour had, at an early hour, as on that day, triumphed over the grave, and opened to the race of fallen man the gate of everlasting life.

"They had some customs that raised in their hearts no reproach of conscience, but which in our day would, by many, be thought inconsistent with the conduct of a man who professed to be leading a Christian life. I will mention one notorious example. It had been a custom, dating from a period, I believe, long before the time of James the First, for the young men of Dent to assemble after Sunday Evening Service, and finish the day by a match at foot-ball. My father might perhaps have put down this ancient custom; but he did not interfere, because he thought the contest, if carried on in good-will, tended to health and cheerfulness: and he knew well that it was not thought sinful or indecorous by the old inhabitants of Dent. He dreaded, too, the acts of intemperance and drunkenness which might arise out of the sudden suppression of a generous and healthy exercise in the open field.

"There was often at the old Parsonage, on a Sunday evening, a small tea-party for those whose homes were distant from the Church; and, later in the evening, my Father read, to a small assembled circle, from some serious book (it might be an extract from one of Bishop Wilson's sermons); and the little service ended with a short family prayer. Now it was by no means unusual for one, who had been contending robustly in the foot-ball match, to come and join in the grave and quiet Sunday evening Service at the Parsonage; and the only kind of question the old Pastor ever asked was one which expressed his trust that the game had gone on in cheerfulness and good-will[1]."

"Athletic sports were held in rivalry by different parishes, and were conducted with great spirit. Matches at leaping, foot-racing, wrestling, and foot-ball, were all in fashion among

[1] *Memorial*, pp. 64—68.

the Dalesmen. But the victory of the foot-ball match was regarded as the crowning glory of the rural festival. My father never opposed such games, because he thought they promoted health, temperance, and good social temper. The spirit of parochial rivalry sometimes, however, led to mischief; and in some rare instances the games were carried on with a savage energy.

"I remember an occasion, in my very early life, when one of the old statesmen, John Mason of Shoolbred, came in great haste and out of breath into the Vicarage, and wished to see my father. 'I hope you will kindly come and help us,' he said, 'or there will be mischief at the field-sports in the Great Holm. At a late parochial meeting there was a sad accident, which led to mutual charges of foul-dealing. Several of us have been asking them to pledge their word, as true men, that all shall be done fairly and kindly; but their blood is up, and they refused with scorn, till one of the men cried out, 'We will play fairly if Mr Sedgwick will come and be the umpire of the foot-ball match.' 'I will go with all my heart,' said my Father, 'that I may be a peace-maker; and I should like to see the game. Come, Adam, take my hand, and you shall walk with me to the foot-ball match.' I right willingly obeyed the order; and though more than eighty years have passed away since that day, yet I remember standing on the high embankment by the river-side, and my father's figure at this moment seems to be living before my mind's eye. I remember his cheerful coun-tenance, beaming with kindness, and lighted by the flush of health; his broad-brimmed hat, looped at the sides in a way that told of a former fashion; his full-bottomed wig, well-dressed and powdered; and his large silver shoe-buckles; all of them objects of my childish admiration. But what I wish most to notice was the respectful manner of the crowd. Many of them came to thank my Father, and each one spoke with uncovered head. Harmony and good-will were restored to the excited combatants, and the great foot-ball match went on and ended in joyful temper and mutual good-will[1]."

To these reminiscences may be added another, told by Sedgwick in the presence of Miss Lucy Brightwell,.daughter of Mr Brightwell of Norwich, and written down by her shortly afterwards.

"During one of the last visits the Professor paid to my dear father, then nearly blind from cataract, he was led to mention his own father, who had been similarly afflicted. He told us how the good old man was honoured and loved in his old age; and concluded with the following story.

One of his parishioners, an ungodly-minded man who had no faith in the Scriptures, called one day at the Vicarage; and, being directed to find his way to the study, came unawares upon the aged pastor, whom he heard (as he supposed) conversing. He waited and listened, and found that the converse was in truth Prayer. 'My father,' said Professor Sedgwick, 'being absorbed in feeling, was unconsciously uttering aloud the breathings of his soul before his Maker. The man remained spell-bound for some minutes, and then went away, without saying a word. But he had heard what convinced him of the reality of religion—he had found true and genuine faith, and from that moment ceased to be an unbeliever.' The Professor, as he told this, was weeping, and we were ready to weep with him."

During the last twenty years of his life Richard Sedgwick was afflicted with total blindness, and was in consequence obliged to keep a curate. But he knew the different services of the church by heart, and generally took part in them. Those who remember him record that it was especially touching to see and hear him when conducting the service for the Burial of the Dead. Led by one of his sons, or by a friend, he would meet the mourners at the gate of the church-yard, and precede them into the church. He had a clear voice, very sweet in its tones, and pronounced the Lesson in a tone of triumphant exultation, which never failed to make a deep impression on the congregation; and, when they came to the grave, the prayers were not read, but prayed. "My father," Sedgwick wrote in 1829, "was a vèry happy old man, and over and over again said that his blindness was a

[1] *Supplement*, pp. 42, 43.

blessing, as it made him more religious and more fit to die. In the last years of his life he was, out of all comparison, the most perfect moral creature I have ever had the happiness of knowing[1]."

In 1822 he resigned the Living, and went, with his eldest son and daughter, to live at Flintergill, a small house at no great distance from the Parsonage, where he died, 14 May, 1828.   His son has recorded that "he was cheerful and happy to his last day, and died as quietly as a child goes to sleep in a cradle[2]."

[1] To Rev. W. Ainger, 7 February, 1829.
[2] To Miss Kate Malcolm, 16 April, 1849.

CHAPTER II.

(1785—1804.)

BIRTH OF ADAM SEDGWICK. CHILDHOOD AND BOYHOOD AT DENT.
LOVE OF HIS NATIVE DALE. HIS SISTER ISABELLA. GOES
TO SCHOOL AT SEDBERGH. HIS MASTER AND SCHOOLFELLOWS.
SELECTION OF A COLLEGE AT CAMBRIDGE. READS MATHE-
MATICS WITH MR DAWSON. ACCOUNT OF HIM.

ADAM SEDGWICK was born at the vicarage of Dent, early
in the morning of the 22nd March, 1785. The surgeon who
attended his mother was Mr John Dawson of Sedbergh,
already celebrated as a mathematician, as mentioned in the
previous chapter, but still compelled by poverty to follow the
more lucrative profession of a general medical practitioner.
Sedgwick always took a singular pleasure in recording, both
in conversation and in writing, the circumstances under which
he came into the world, and his letters contain several versions
of them. Of these by far the most graphic and humorous
occurs at the end of a letter to Mr Charles Lyell of Kinnordy,
father of the geologist. He had been one of Mr Dawson's
pupils, and had written to Sedgwick urging him to draw up
a memoir of him, and asking for some information respecting
his life and writings[1]. "A fourth sheet! I had no notion
that I had finished a third till I tried to find a blank page.
Well! as I have a blank page, I will tell you of my very first

[1] The rest of the letter, dated 26 January, 1847, is printed below, pp. 61—69
A few details have been added from a letter to Miss F. Hicks, 28 March, 1841.

acquaintance with our old Master, and I will tell it you as nearly as I can in his own words: 'On the 21st of March, 1785, I was called to attend your mother. The night was tempestuous, and I had much difficulty in making my way to Dent through the thick snow; and when I got to the old vicarage I found that my difficulties were not over. The moment was critical; and though you seemed anxious to show your face in the world, you were for doing it in a strange preposterous way.' Here he referred me laughing to an early page of *Tristram Shandy*. 'So I sent' said he 'your Father's servant to knock up old Margaret Burton to help me to keep you in order.' She was a celebrated midwife, of firmer nerves than the old mathematician. Between them the work was done, and by hook or by crook, I was ushered into the world at about 2 o'clock, a.m. on March 22nd, 1785. I was then carried downstairs, in old Margaret's apron, to the little back-parlour, where my Father was sitting in some anxiety, as he had been told that his youthful son was beginning life badly, and not likely to take good ways. Margaret threw back the corners of her apron and cried out: 'Give you joy, Sir, give you joy! a fine boy, Sir, as like you Sir, as one pea is to another.' My Father looked earnestly at me for a moment or two, kissed me, and then, turning to the old midwife exclaimed: 'Like me do you say, Margaret, why he is as black as a toad!' 'Oh! Sir! don't speak ill of your own flesh and blood, if I have any eyes in my head he is as white as a lily,' she replied, much shocked, while old Mr Dawson shook his sides, as much, I dare say, as he did when he told me the story. To stumble on the threshold was of old counted a bad sign, and what I have told you may be the reason why I am sticking as a Senior Fellow without getting on in the world."

Sedgwick's complexion, though not so black as to justify his father's comparison, was still extremely dark. He inherited it from his mother. Mrs Burton acted as nurse to young Adam for a short time, but she died when he was still quite a boy, and he lost sight of her family for many years. In 1840, however, an accident made him acquainted with her great-granddaughter, and they soon became sworn friends on the strength of their common obligations to the skill of the same person[1].

As a child, Adam was active and merry, fond of play, and given to tearing his books rather than reading them. When he was five years old, his godfather Mr Parker, then Master of the Grammar School at Dent, gave him a new and handsomely bound spelling-book, which the little fellow thought too good to be torn. So he set to work, and soon learnt to read it. Like his father before him, he was sent, probably at a very early age, to the public grammar-school, to be educated with the other boys of the Dale. Mr Parker was much attached to his godson; and in the summer of 1793, when Adam was eight years old, he took him, riding behind him on horseback, to visit his friends at Hesket-Newmarket in Cumberland. The events of this summer made an indelible impression upon Sedgwick's memory. He refers to it again and again in his letters, and always with pleasure. He was shown Carlisle, and remembered that at that time the old walls were standing, and that he had walked all round the city on the ramparts; and he spent a few days at Scaleby Castle, an old place of strength about six miles from Carlisle[2], the occupier of which was Mr Fawcett, a distant cousin. But the most delightful reminiscences of all occur in a letter which he wrote in 1853 for the amusement of one of his nieces, who had just been revisiting her birthplace, the Isle of Wight. "We all delight," he says, "to revisit the scenes of our childhood. Such visits produce emotions, some cheerful and some sorrowful, that do our hearts good, and they ought to teach us healthy lessons. I almost envy you the pleasure of your visit to the sweet Island of your child-

---

[1] To Rev. W. Ainger, 23 March, 1840.

[2] To Miss Fanny Hicks, 18 April, 1849. The walls of Carlisle were pulled down in 1811. *Life of Isaac Milner*, 8vo. Cambridge, 1843, p. 453.

1793.
Æt. 8.

life. When I was a child of eight years old I spent the summer vacation at Hesket, a small village between Penrith and Carlisle. My schoolmaster took me home with him during his summer holidays, and the Grand Turk does not think himself half so great a man as I then thought myself. Exactly thirty years afterwards, I landed one Saturday evening at Hesket, during my geological tour in 1823. The old man who kept the Inn when I was a child was still living; and I remembered him by this token, namely, that he had pulled my ears because I had, with a lad older than myself, caught his turkey-cock, and pulled some beautiful feathers out of its tail. I quite rejoiced to see the old man. I forgave the ear-pulling which I had deserved, and he had quite forgotten my theft of the feathers. Next day I went to church. A woman opened the door of a pew, and when I looked round I saw that it was the very seat I used to sit in during my visit in 1793—the year the king of France was beheaded, you know. Of course I knew no one, but I observed that the well-looking middle-aged woman who had received me into her pew went to the house where my old school-master lived. So I followed her, and she proved to be the sister of my master—a young woman who in 1793 was very kind to me. I was very sentimental all that day, and in the evening drove to Penrith[1]."

When they got home to Dent, Mr Parker told Mr Sedgwick how much everybody had been struck by his son's powers of observation.

Soon after this, Mr Parker left Dent[2], and Mr Sedgwick was solicited by the Governors of the school to undertake the Mastership, with one or more under-masters to assist him. After some hesitation he consented, and taught his own boys along with his other scholars. About Adam's progress tradition is silent; but as, in after years, his knowledge of Greek

[1] To the same, 27 July, 1853.
[2] In 1812 Sedgwick went to stay with him "near Macclesfield". To Rev. W. Ainger, 14 February, 1812.

and Latin was superior to that of most men as distinguished in mathematics as himself, it may be safely assumed that he worked hard; and that he had been well grounded, in the first instance by his godfather, who was a clever man, and afterwards by his father, who is said to have been a very good scholar, and whose early experience under Dr Bennet would now stand him in good stead.

It is an old story that the boy is father to the man; and it is interesting to find that the portrait which family tradition has drawn of Adam Sedgwick as a boy would be true had it been drawn of him when he was grown up. He did not give any special promise of future intellectual power, but he was remarkable, we are told, for a frank, genial disposition; he was full of fun and high spirits; he delighted in rambling over the fells and climbing the hills which bound his native dale; his powers of observation were great; and he had a plentiful share of sound practical common-sense. He was also distinguished for undeviating truthfulness in all that he said and did. Among his brothers and sisters and school-fellows, if Adam said a thing was so, there was no further question about the matter. " I almost lived out of doors,"— he said in conversation with Mr J. W. Salter[1]—" at fourteen years old I was trusted with a gun, and coursed over the heathy moors the whole autumn day. I believe I was a tolerably good shot. I was a fisherman too at this age, and was particularly careful to obtain the exact feathers which were considered the most killing flies for trout, grayling, etc. Nor, though I ought to confess it with some reluctance,— save that I never had an unworthy selfish thought in the matter beyond the joy of sport—was I quite free from the

[1] Mr Salter accompanied Sedgwick to North Wales in 1842 and 1843, and afterwards prepared under his direction *A Catalogue of the collection of Cambrian and Silurian Fossils contained in the Geological Museum of the University of Cambridge*, 4to. Cambridge, 1873. While engaged upon this work he persuaded Sedgwick to dictate to him a few reminiscences of his early life, apparently with the intention of preparing a complete biography; but ill-health and other en-gagements prevented the completion of the task.

crime of poaching. Old —— and I were great friends, and
many a night have I met him by appointment to try our
flies, and our snares for rabbits, hares, and pheasants. I
believe he always had the game; the sport was quite enough
for me. But to this day I like to hear the click of a fowling-
piece; and, as I pass a mountain burn, I can scarcely help
speculating in what holes the trout may be lying.

"But I did not quite forget the rocks and the fossils.
One of my early employments on a half-holiday when nutting
in Dent woods, was, as I well remember, collecting the conspicuous fossils of the mountain-limestone on either side of
the valley. It was not till many years afterwards that I
understood its structure, but these early rambles no doubt
aided to establish a taste for out-door observations."

These geological reminiscences must be received with
a certain caution, partly because Mr Salter was not sufficiently well acquainted with the district to follow Sedgwick's
descriptions of it, partly because even the most truthful of
men cannot help imparting to events which happened in their
early years the colouring derived from fuller study at a later
period; still, as Sedgwick could hardly have drawn upon his
imagination, even in extreme old age, for the whole story, the
record has been thought worth preserving in this place.

As might be expected from the recollections of Dent in
former days which have been printed in the previous chapter,
he loved to talk with the old statesmen, and to hear their
stories of the Scotch incursions, and of the rebellion of 1745.
Some of them, his father among the number, had been at
Kendal in that eventful year, and seen the disorderly retreat
of the remnant of the Pretender's army. Two incidents of
the '45 can fortunately be reproduced almost in the words
in which Sedgwick used to tell them.

"When I was a boy there was living at Dent a certain
Matthew Potts, whom we lads looked up to as a hero, because,
when he was an apprentice, he and another lad ran away
from their masters, and followed the fortunes of the Scottish

army on their retreat northward. They were witnesses of the battle of Clifton Moor near Penrith, which Scott has described in *Waverley*, and Potts carried off from the battle-field a broadsword, a target, and a tortoise-shell comb stolen from the body of a young Highland officer. The young scamps lived by milking the cows they found upon the moors; and they grew so familiar with scenes of battle that they boasted of having set a dead Highlander astride of a stone-wall with a pebble in his mouth to keep it open."

"When the Highlanders marched through Kendal they were very badly shod, and laid hands on all the boots and shoes they could find. One of them went into the stables of the King's Arms and appropriated a pair of boots belonging to the ostler, who was out of the way. As the thief was lagging behind his comrades to put on his ill-gotten spoils, he was overtaken by the owner of the boots, a truculent fellow, who ran him through with a pitchfork and killed him on the spot. My father, who was a boy of ten years old at the time, knew the ostler well, and had often heard the story. I think the histories record that a soldier of the Duke of Perth's regiment met with his death at Kendal, but say nothing of the manner of it[1]."

The old men of the village were delighted to have the bright boy as a listener, and used to speak of him afterwards to his father as 'a lad aboon common'. We can readily imagine that "Adam o' the Parson's," as he was called in the dialect of Dent, soon became a leader of the lads of his own age; but the only records of his youthful prowess that have come down to us are both connected with bonfires. The first of these incidents befel in 1798, when Sedgwick was thirteen. "I well remember," he wrote in 1863, "the day that brought to my native valley the news of the battle of the Nile, and we contrived (for the schoolmaster gave us a holiday) to pile up such a heap of turf, sticks, and tar-barrels,

---

[1] These stories were told by Sedgwick, about a fortnight before his death, to W. Aldis Wright, M.A. now Fellow and Vice-Master of Trinity College.

that we had nearly set the village of Dent on fire[1]." The
other was an annual celebration. In 1871, having occasion to
write a letter to Canon Selwyn on the fifth of November, he
began as follows: " This was, in my younger days at Dent, a
day of joy—a holiday, a hunt, and a bonfire. I was busy,
with my schoolfellows, a good part of the day in gathering
logs of wood, collecting turf, and breaking hedges, with
loyalty to King George, and detestation of the Pope and
Jesuits. We clubbed from our poor purses, and sturdily
begged of those who had a little to spare, to raise a stock to
purchase tar-barrels—each of which cost about eight-pence—
and finally used to put old Dent into such a blaze that there
was a serious risk of our turning the village itself into a
bonfire[2]."

An interesting anecdote respecting this period has been
preserved by family tradition. When Adam was about
twelve years old, the conveyance of a farm in Dent to the
trustees of a charity had to be executed by some gentlemen
who resided in the north of the county of Durham. Thereupon a formidable difficulty presented itself. By what means
were the indispensable signatures to be obtained, and the
deeds brought back to Dent in safety? Said one of the
trustees : " If anybody can be depended upon, it is Adam
Sedgwick; and I, for one, am quite willing to entrust the
deeds to his care, if his father will let him go." Mr Sedgwick,
to the boy's great delight, made no objection ; and mounted
on his father's mare " Bet", Adam rode off all alone, with
the precious deeds secured in saddle-bags. He came back at
the end of a week with the deeds in safety, all properly signed
and sealed.

Such an expedition as this, taken together with the sense
of importance conferred by it, would have gratified any boy
of spirit and intelligence. But, keenly as Adam admired
new scenes and fresh objects of interest, we do not hear that

[1] To Dean Trench, 21 April, 1863.
[2] To Canon Selwyn, 5 November, 1871.

he was fired with any special ambition to leave his native 1793 to 1801.
dale, and try his fortune in the great world beyond it. A few Æt. 8—16.
years later, when he had tasted the sweets of University life,
he was apt to find Dent somewhat dull ; and in middle life he
had other objects of interest, and perhaps thought less of his
home than of those who lived there. But, when he became
an old man, all his boyish fondness for the place returned to
him, and he has left many a charming description of its
scenery, and of his own feelings towards it. Here is one,
written from Dent to Lady Augusta Stanley, 3 July, 1865,
which recalls his young days there with singular vividness.

" I wish I could, for a minute or two, transport you to this
place. Not so much that you might look at my rugged old
face, as that you might gladden your eyes by gazing over the
sweet scenery of this rich pastoral valley. The home scenery
is delicious ; and glowing at this moment (6.30 a.m.) with the
richest light of heaven ; and from the door of this old home
of my childhood I can look down the valley, and see, blue in
the distance, the crests of the Lake Mountains which rear
their heads near the top of Windermere. All around me is
endeared by the sweet remembrances of early life. For here
I spent my childhood and my early boyhood, when my
father and mother and my three sisters and three brothers
were all living in this old house. Our home was humble ;
but we were a merry crew; and we were rich in health, and
rich in brotherly love. Forgive me for going on at this rate.
I think you are a lover of rural scenery ; and indeed when I
return to Dent I feel again as if I were in my true home—and
I quite naturally talk and think of the joys of boyhood, and
begin ' to babble of green fields '."

Fond as Sedgwick was of his brothers and sisters collectively, his chosen friend and companion was his sister Isabella,
who, it will be remembered, was just two years younger than
himself. She died in 1823, and when we come to that part of
his life, we shall see how passionately he mourned her loss.
Nor did his sorrow diminish as time went on. Throughout

his life he cherished her memory with the tenderest affection, and his letters of sympathy to friends in trouble frequently contain allusions to her, and to their early days together. For instance, in 1849, twenty-six years after her death, he wrote to a lady who had just lost her own sister : " I can feel for your great sorrow. I once lost a sister, the dearest of all my sisters, and the darling companion of all my early years. She was a woman of most placid temper, yet of great personal courage. We had our little squabbles about our toys when children ; but after we reached our teens I think I never heard so much as a word from her lips that was not spoken in kindness. Her death was a grievous blow to me, and I never visit my native hills without being reminded of her at every turn[1]." And again, writing in 1864 to the mother of one of his numerous godchildren: "My *especial* love to my dear god-child. I rejoice to know that she is a jolly tomboy. That shows that she has good spirits, good health, and the right use of her limbs. I had a sister, not quite a year younger than myself[2]. When a young girl she was my never-failing companion; and I taught her, in our wild valley of Dent, all sorts of boys' tricks. For example, she would run like a monkey up a tree to peep into a magpie's nest. The effect of this training gave her excellent robust health—and it matured her sweet natural temper—and when she grew up she was a mild, gently feminine, unselfish person, beloved by everyone. May my dear god-daughter have health and temper like that of my beloved tomboy and my darling sister, and may God bless her with a longer life[3]."

When Sedgwick was sixteen he was sent to the Grammar-School at Sedbergh, the mastership of which was then in the gift of St John's College, Cambridge. At that time the post was held by the Rev. William Stevens, a former fellow of that

[1] To Mrs Horner, 14 March, 1849.
[2] The account of Sedgwick's brothers and sisters given in the previous chapter, which has been verified from the parish-registers of Dent, shows that his memory was slightly at fault here.
[3] To Mrs Martin, 26 February, 1864.

The farmhouse near Sedbergh where Sedgwick boarded when at school : from a water-colour drawing by Mr William Westall, taken about 1820.

house, who had been fifteenth wrangler in 1791. As far as mathematics went he must therefore have possessed sufficient knowledge for his position; but the school did not flourish under his rule. Indeed it has been whispered that he neglected his duties, and that "for years together he had the school locked up, teaching in his own house a few boys, hardly ever amounting to ten[1]." Mr Sedgwick, however, must have had confidence in his abilities, for he sent all his sons, one after another, to be taught by him, and the two families were on terms of close intimacy. Adam describes him as "an excellent scholar, and a good domestic and social man[2]." Perhaps he took more pains with him than he did with others; perhaps his social qualities endeared him to his pupils. For it should be mentioned that he had been for a short time a chaplain in the Royal Navy, and had been present at Lord Howe's victory off Cape St Vincent in 1794. In those days of martial enthusiasm such a man would appear little short of a hero in the eyes of boys, especially if he entertained them with stories of his naval experiences. Indeed it is not improbable that Sedgwick's lifelong interest in all things naval may be traced to the influence of his old schoolmaster.

Discipline was not strict at Sedbergh School. The boys boarded at the neighbouring farm-houses, and when their work was done, few inquiries were made as to the employment of their time. Half holidays were generally spent in fishing in the Rawthey or the Lune, or in rambles over the mountains. The farm-house called "the Hill", where Sedgwick boarded, along with three other boys, stands a little to the north of the town, just under Winder. As the woodcut shows, it has been but slightly altered from the aspect it bore in his day. The farmer was a quaker named Edmund Foster, a near connexion of the Fosters of Hebblethwaite Hall, who were great friends

1801 to 1804. Æt. 16— 19.

---

[1] *History of the Parish and Grammar School of Sedbergh, Yorkshire*, by A. E. Platt. 8vo. Lond. 1876, p. 159.

[2] *Supplement*, p. 65. Writing to Ainger, 14 Feb. 1812, Sedgwick speaks of Mr Stevens' "zeal for the interest of his pupils"; but adds "there is something about him which I neither like nor understand".

of the Sedgwicks. "We were treated", Adam has recorded, "with infinite kindness by the family, and our happy freedom made us the envy of our schoolfellows[1]." The lodgers were of course obliged to accommodate their hours to those of their host; and it was to this training that Adam attributed the habit of early rising which he never abandoned[2].

The schoolhouse, erected in the last century, has not

Door of the school-house at Sedbergh.

been materially altered since Sedgwick's time. It is an oblong stone building in two floors, about sixty feet long by twenty-two feet wide. The west door, long disused, and

---

[1] *Supplement,* p. 55.
[2] To Dr Livingstone, 1865.

blocked by slabs of stone, is a picturesque example of the classical style prevalent in the early part of the last century[1]. The date, 1716, probably indicates the completion of the existing building. The lower floor is now fitted up as a chapel, and the upper floor as a library and reading-room. The words A. SEDG-WICK, 1803, are cut on a stone at the south-west ex-

ternal corner. The inscription was no doubt intended to commemorate his leaving the school, but whether it was cut by himself, or by some other person, is not now known.

Of Sedgwick's schoolfellows at Sedbergh two at least must be specially commemorated, William Ainger, and Miles Bland. The former, who in University standing was his senior by one year, was the chief friend of his early life. They lived in close intimacy at Cambridge, and afterwards, though they met but seldom, corresponded with tolerable regularity. Mr Ainger, after holding several pieces of preferment, became Principal of the Theological College at St Bees, and Canon of Chester. He died rather suddenly in 1840, attended by his old friend. Mr Bland became a Fellow of St John's College, and during his residence at Cambridge, he and Sedgwick probably saw a good deal of each other. On the occasion of the death of their old Master, Mr Stevens, in 1819[2], they raised, by their joint exertions, a large sum of money for the benefit of his widow and children, for whom he had neglected to make any provision. But their friendship was never very close, and after 1823, when Mr Bland accepted the living of Lilley, in Hertfordshire, and left the University, it ceased altogether.

Adam Sedgwick's school days were prolonged, we do not know why, until a later period than was usual at that time, for he was not sent to Cambridge until 1804, when he was

1801 to 1804.
Æt. 16—19.

---

[1] The beautiful drawing from which the woodcut was taken is due to the kindness of Edward G. Paley, Esq., of Lancaster, Architect.

[2] Mr Stevens died 9 November, 1819, aged 50.

well on his way to twenty. In the autumn of the previous year a council was held to determine the particular college at which he should be entered. It was composed of his father, Mr Stevens, and their common friend the Rev. D. M. Peacock, then Vicar of Sedbergh, who had been Senior Wrangler in 1791, and subsequently a Fellow of Trinity College. Boys educated at Sedbergh were usually entered at St John's College, because on that foundation there were then three Fellowships and ten Scholarships appropriated to the School; and Mr Stevens urged that his two cleverest boys, Bland and Sedgwick, should both be entered there. But Mr Peacock held a different opinion. "Bland", he said to Mr Sedgwick, "is a better mathematician than your son. He will always "beat him in examinations, and, afterwards, if it comes to a "question of a Sedbergh Fellowship, your son will not be the "successful candidate." Mr Sedgwick assented, and suggested his own college, Catharine Hall. "No!" said Mr Peacock, "Adam is a clever lad, let him go to my college, Trinity, and take his chance[1]." And so it came to pass that he was entered as a sizar at Trinity College, 18 November, 1803, under the popular tutor Mr Jones[2].

During the summer-months of 1804 he became a pupil of Mr John Dawson, the surgeon-mathematician of Sedbergh, who had given similar instruction to his father just forty-eight

[1] This account has been communicated by Miss Isabella Sedgwick, and therefore represents family tradition. Sedgwick himself gave a different version of the reason why he was not entered at St Catharine's College, which we find stated as follows in a diary kept by the Rev. George Elwes Corrie, B.D., then Fellow and Tutor of that College, and afterwards Master of Jesus College : "16 *June*, 1843. I met Professor Sedgwick in the Library, who, among other things, told me that he was anxious to have been admitted to our College, in consequence of his Father having been of the Society, but that owing to the disputes in College at that time the Master did not allow any admissions." A reference to the *Cambridge Calendar* for 1803 shows that the offices of Tutor, Bursar, Dean, Lecturer, and Steward were all vacant, and that there was only a single undergraduate on the boards. The Master was the Rev. Joseph Procter, D.D. (1799—1845).

[2] "18 Nov. 1803. Admissus est sizator Adamus filius Ricardi Sedgwick de Dent in Com. Ebor. e Scholâ apud Sedbergh in eodem Com. sub præsidio Mag. Stevens, ann. nat. 19. Mag. Jones tutore." Admission Book of Trinity College.

years before. It has been already related that Mr Dawson had been called upon to use his skill as a surgeon to usher Adam Sedgwick into the world; and now, by a singular coincidence, he was to use his skill in another department to fit him for his entrance into the University.

This seems to be the fitting place to give some account of this remarkable man, whose career affords one of the most striking instances on record of self-help, of indomitable perseverance triumphing over difficulties which could hardly have been surmounted by anyone, unless he had belonged, as Sedgwick said, "to the very highest order of intellectual greatness."

Sedgwick had a warm affection for Mr Dawson, and has left two accounts of him: one in the letter to Mr Lyell, from which a single passage has been already quoted[1]; the other in the *Supplement*[2] to the *Memorial*. These accounts, as might be expected, travel over much of the same ground, especially when describing Mr Dawson's early struggles. An attempt has therefore been made, in the following pages, to combine them into a continuous narrative, of which the letter to Mr Lyell forms the basis.

"The outline of Mr Dawson's early life I will try to give you in a very few words. He was the son of a very poor statesman in Garsdale[3], with perhaps not more than £10 or £12 a year, and his vocation was to look after the paternal flock of sheep on the mountains. In this situation he remained, I believe, till he was more than twenty years of age. But he was a forgetful shepherd, and was living in an ideal world of his own. He had no money, no friend, and could not therefore buy books. But he begged them, or borrowed them, or *invented* them, for he actually worked a system of Conic Sections out of his own brain. Some small sums of money he gained by teaching, and in 1756 three young men went to

---

[1] See above, p. 45.   [2] *Supplement*, pp. 49—54.

[3] He was born at Rangill Farm in Garsdale, and, as he was baptized 25 February, 1734, was probably born in January.

read with him before they entered the University. My father, who often spoke of the Garsdale summer as one of very great happiness and profit, was one of them; Dr Haygarth[1], afterwards a physician at Leeds, and in his day a man well known, was one of this happy number; the third took his degree at St John's College, Cambridge, and afterwards had a living in Leicestershire. This was perhaps the first dawn of the youthful shepherd's fortunes[2].

"Soon after this triumvirate went to the University, young Dawson was taken as a kind of assistant—he was too old to be a regular apprentice—by Mr Bracker an eminent surgeon of Lancaster, a man of science and good sense, who had a name among the northern worthies of last century. There he remained several years, compounding medicines, solving crusty problems on an inverted mortar, and learning the duties of his profession as a surgeon. His condition was, during this time, much improved. He had now no lack of books, or want of sympathy; and he rapidly made that great and generous progress which marks an intellect of first-rate power, when urged onwards in its work by a never-tiring will. Before long he was capable of holding consultations with good professional men, and of measuring weapons with mathematical analysts of the highest name in England. But he had only stolen hours for his favourite studies, as his master —I think I may say his master—was in great and wide practice. As soon as he thought himself fit for the duties

[1] John Haygarth, M.D. of Leeds, was born in Garsdale in 1740. He was distantly related to Dawson.

[2] Professor Pryme, who read with Dawson during the summer of 1799, gives a somewhat different account of his early years: "He was the son of a yeoman in Garsdale. He had availed himself of some books belonging to his brother, an excise officer, to gain a knowledge of arithmetic, and of the rudiments of mathematics, and then became an *itinerant* schoolmaster; for many parts of that mountainous district were not sufficiently peopled to maintain one capable of teaching anything beyond mere reading and writing. He stayed two or three months at a time in one house, by arrangement, teaching the children of the family and neighbourhood, and then removing to another. In the meantime he was pursuing his own mathematical studies." *Autobiographic Recollections of George Pryme.* 8vo. Lond. 1870, p. 29.

of his profession, before he had attended any regular academic medical course, and without any medical diploma, he became surgeon and apothecary at Sedbergh. There he remained about a year—living on a crust, and saving every farthing he could scrape together, till he was possessed of nearly £100. This sum he 'stitched up in the lining of his waistcoat'—I am telling you *his own words*, which I have more than once heard from his mouth when I was a schoolboy at Sedbergh. Off he set, staff in hand, and walked to Edinburgh. There he lived, in a cheap garret, upon a sum so incredibly small, that I dare not trust my memory with the mention of it.

"While making way with the medical cycle, and encountering a formidable range of severer studies, he lived with the sternest self-denial. But no economy could save his funds from wasting; and the external sinews of his movements were on the very point of failing, when once again he packed up his whole stock, took his good staff in hand, and strode back to Sedbergh. He had then no difficulty in meeting the common wants of life. The country was longing for his return, and professional practice flowed in upon him. But he still lived with great self-denial, and by using all means within his power, both of head and of hand, he amassed a sum I believe about three times as great as that with which he walked to Edinburgh.

"Again he left his native mountains, and went, partly on foot, partly I believe in a stage-waggon, to London, with all his gold stitched, as before, in small parcels under his waistcoat. 'But I could not', said he, 'live as I had done at Edinburgh.' Neither could he live in the same retirement, for the sound of his name had passed beyond the Dales, and several men of science sought his personal acquaintance. Among the rest he mentioned, I think, the late celebrated Dr Waring, Lucasian Professor at Cambridge[1]. 'My money

---

[1] Edward Waring was Lucasian Professor of Mathematics from 1760 to his death in 1798. He proceeded M.D. in 1767.

1804.
Æt. 19.

went from me', said he, 'faster than I wished, but I got through one good course of surgical and medical lectures.' In short, he became a regular member of his profession; obtained a diploma; and then returned on foot to Sedbergh. His course was now clear, and he had won a good position for himself. He married, and had one daughter[1], whom you must well remember; settled in his own house, and had the command of the best medical practice in all the neighbouring Dales; and sometimes, to his sorrow, his duties carried him far beyond them.

" He still went on with his favourite studies, and his mind hardly seemed to have a pause. It was said of him, perhaps in jest, that he could solve a problem better when riding up the Dales on his saddle than when sitting at his private desk. At any rate he made himself master of every standard mathematical work known to the scientific literature of this country, and was counted among the very first analysts of his day. This was not the mere admiring gossip of a country town; its truth is proved by various profound essays on contested points of Physical Astronomy, into which he was led, not through vain-glory, but from the simple love of truth.

"The rest you know, I think, as well as I. After several years of honourable, and most successful, practice, in a wild and poor country, and amidst a thousand interruptions to his severer studies—he gained a small fortune and a great reputation. Cambridge undergraduates flocked to him every summer. About 1790 he entirely relinquished his practice as a surgeon, and devoted himself exclusively to mathematical teaching[2]. Between 1781 and 1794 he counted eight Senior

[1] The Sedbergh Registers show that John Dawson married Ann Thirnbeck, 3 March, 1767. Their daughter Mary was born 15 January, 1768. Mrs Dawson died 21 January 1812; her daughter, unmarried, 18 July, 1843.

[2] The expense of attending Mr Dawson's class was not great. We have seen a letter written by Dr Butler, Headmaster of Harrow and Dean of Peterborough, who became his pupil in 1792. The letter is dated 17 June in that year. The writer, after describing his journey from London to Sedbergh (which began on a Tuesday at 7 a.m. and ended on the following Saturday at 11 p.m.), mentions that Mr Dawson charged 5s. per week for instruction; and that he would have to pay

Wranglers among his pupils. In 1797, 1798, 1800, and 1807
the Senior Wranglers were also Dawsonians[1]. I only knew
him in his decline; but he continued, I believe, to have pupils
till about 1810[2].

"I became a pupil of our late Master in 1804, a few months
before I entered at Cambridge; and at that time he was full
70 years of age. His intellect was then as grand as it had
ever been, but his memory had begun to fail. I remember a
singular example of this in the vacation of. 1806, when I was
reading with him some of the later sections of the first book
of the *Principia*. You remember his 'Peripatetic' mode of
teaching. He came to the back of my chair. 'Here is a
proposition', said I, 'but no proof, and I think it is by no
means self-evident.' He looked at the proposition for a
minute, and then took his pencil and worked out a kind of
mongrel proof, half analytical, and half geometrical, on a slate.
This was done in about ten minutes. Judge of my surprise!
On turning over the leaf I found Newton's proof on the next
page! I called Mr Dawson back, and showed it to him. He
was as much surprised as myself, and said, with an expression
of sorrow, that he had gone over this section many hundred
times, and ought not to have forgotten the proposition. 'But',
he added, 'I am beginning to have an old man's memory.'

at the King's Arms Inn "the best in the town" 1*s*. 6*d*. per week for an excellent
room; for dinner 10*d*. a day, and for breakfast 2*d*. "Dinner" he says "consisted
of a leg of mutton and potatoes, both hot; ham and tongue, gooseberry tarts,
cheese, butter, and bread: pretty well for 10*d*." The letter is printed in *The
Sedberghian* for December, 1881.

[1] The Senior Wranglers of these four years were: 1797, John Hudson, Trinity,
afterwards Fellow and Tutor; 1798, Thomas Sowerby, Trinity, afterwards Fellow
and Tutor of Queens' College; 1800, James Inman, St John's, afterwards Mathematical Professor in the Royal Naval College; 1807, Henry Gipps, St John's.
The previous eight cannot all be identified, but the following may be safely claimed
for Dawson: 1786, John Bell, Trinity; 1792, John Palmer, Professor of Arabic
on Sir T. Adams' foundation; 1793, Thomas Harrison, Queens'; 1794, George
Butler, Sidney.

[2] There is a slight error here. Miss Ann Sedgwick writes to her brother 25
July, 1812: "James is attending Mr Dawson during Mr Stevens' vacation. It is
his last opportunity, as Mr Dawson takes no more pupils after this summer. He
has at present fourteen."

S. I. 5

"About twelve or thirteen years afterwards, I called on him during one of my visits to my native valley of Dent. He seemed delighted to see me, roused himself, talked of Cambridge, of his own early studies, of our change of system, of some new work of analysis, and especially of the *Calculus of Variations*, with which, he said, he had just begun to be acquainted when his mind began to fail, and he was by old age made incapable of any new and severe study. 'I have sometimes grieved,' he said, 'but perhaps it is ungrateful of me, that I did not know this powerful implement of discovery in early life. I thought that I might have grasped it, and then tried my hand with some of the great problems of physical astronomy; but now I am a feeble old man, and my days are nearly numbered.' While this conversation was going on, with no small animation on his part—to the utter amazement of his daughter and myself, for she told me afterwards that she had hardly heard him talk collectedly for many weeks, and I expected only to see a venerable intellectual ruin, and bid him a sorrowful adieu for ever—in came some neighbours and interrupted him. He became silent at once. They went away in ten minutes, and I then tried again to arouse him, and lead him back to the subjects on which he had dwelt with so much energy and apparent delight. But all in vain. He had forgotten every syllable of our conversation, and was become again a mere dotard. I left him with most oppressive feelings of admiration and sorrow, and I never saw his face again.

"Such is my outline, but I have given it you in more words than I first intended. He published one or two good mathematical tracts in the *Manchester Memoirs*[1], which, as you well know, have become famous as the vehicle of Dalton's[2]

---

[1] This is a mistake. Dawson wrote no papers in the *Manchester Memoirs*. Sedgwick was probably thinking of some letters by him in Hutton's *Miscellanea Mathematica*, signed "Wadson," in which he attacked a theory advanced by the Rev. C. Wildbore, *On the velocity of water issuing from a vessel in motion*.

[2] John Dalton, of Manchester, F.R.S., Chemist and natural philosopher, born 1766, died 1844.

great chemical generalisations. He was engaged in a contro- 1804.
versy with that strange but clever mathematical bear Emerson[1]. Æt. 19.
Thomas Simpson[2] had shown, in an analytical investigation,
that Newton had made a slip in his great problem of *Precession.*
Spite of this blunder it is one of Newton's very greatest
triumphs. Emerson worshipped Newton, and abused every
other son of man. So on principle he opened a foul-mouthed
volley on poor Thomas Simpson. Dawson replied by sending
an entirely independent analytical investigation of the same
great problem, which gave a result identical with that of
Simpson, at the same time modestly pointing out the slip in
Newton's demonstration. This produced a volley of abuse
and nothing better, from his opponent, and so the matter
dropped.

"At a later period Mr Dawson wrote an excellent memoir
on certain problems in the Lunar Theory, in reply to a
geometrical essay on the same subject by Matthew Stewart
(father of the celebrated Dugald Stewart), of Edinburgh.
Our old tutor proved to demonstration that purely geometrical
methods must in many approximations necessarily fail, and in
nearly all must be unsatisfactory, inasmuch as they do not
give us any certain data as to the value of the terms left out
in the approximation. This controversy was carried on in the
pure love of truth, and in a most excellent spirit. Dawson
had the entire victory, and his tract procured him the personal
friendship of some of the leading Edinburgh philosophers[3].

---

[1] William Emerson, mathematician, born 1701, died 1781.

[2] Thomas Simpson, F.R.S., mathematician, born 1710, died 1761.

[3] In 1763 Mr Stewart published an essay on the "Sun's Distance", which he
made out to be far in excess of previous estimations of it. Dr Chalmers (s. v.
Stewart) says that "even among astronomers, it was not every one who could
judge in a matter of such difficult discussion. Accordingly, it was not till about
five years after the publication of the 'Sun's Distance' that there appeared a
pamphlet, under the title of *Four Propositions*, intended to point out certain errors
in Dr Stewart's investigation, which had given a result much greater than the
truth." After describing the nature of these errors, Dr Chalmers adds: "And it
is but justice to acknowledge that, besides being just in the points already mentioned,
they [the *Four Propositions*] are very ingenious, and written with much modesty
and good temper. The author, who at first concealed his name, but afterwards

1804.
Æt. 19.

After this, he was visited at Sedbergh by Playfair[1], Lord Webb Seymour, and Lord Brougham.

"The pamphlet you inquire after was written, I believe, sometime between 1780 and 1790, in reply to some of the published doctrines of Priestley in his tract on *Philosophical Necessity.* Dawson thought the doctrine of immoral tendency, and not true. He attacked it with firmness, but with his usual calm temper. Priestley replied, also in a good spirit, and ever afterwards spoke of Dawson in terms of respect and admiration[2].

"Our old master was a firm believer, and a good sober practical Christian of the old school. His moral influence was felt by all near him, but he loved a good story, and told his rough adventures among his country patients with infinite humour. I have, however, no time to tell you any of them now. His sphere of usefulness was limited. The wonder is that he could do anything, circumstanced as he was, to make himself a name. On the whole I think him a man belonging to the very highest order of intellectual greatness.

"I knew him well in his honoured old age, for I was his

consented to its being made public, was Mr Dawson, a surgeon at Sedbergh in Yorkshire, and one of the most ingenious mathematicians and philosophers which this country at that time possessed."

[1] John Playfair, mathematician and natural philosopher, of Edinburgh: born 1749, died 1819.

[2] Thomas Priestley published, in 1777, a pamphlet entitled *The Doctrine of Philosophical Necessity.* Dawson's reply was first published anonymously in 1781: *The Doctrine of Philosophical Necessity briefly invalidated,* 8vo. 1781. An answer appeared in *The Monthly Review or Literary Journal* for July, 1781 (p. 66), but Dawson's name is not mentioned. Subsequently Dawson published a second edition: *The Doctrine of Philosophical Necessity briefly invalidated,* by John Dawson of Sedbergh. Second edition, to which is now added an Appendix. "And binding Nature fast in fate, Left free the human will." Pope. 12mo. Lond. 1803.

It should be mentioned that Dawson had paid a good deal of attention to metaphysics and theology, as shown by his correspondence with the Rev. Thomas Wilson, who had been his pupil in early life. The quotations in these letters prove that he had at least a respectable knowledge of Latin and Greek, though he laments his inability to read the Fathers in the original. *Selections from the Poems and Correspondence of the Rev. T. Wilson,* Cheetham Society, 1857, pp. 106—125.

pupil during three successive summers of my undergraduate 1804.
life, but it is hard for me to do full justice to the head and Æt. 19.
heart of my dear old master. Simple in manners, cheerful
and mirthful in temper, with a dress approaching that of the
higher class of the venerable old Quakers of the Dales, without
any stiffness or affectation of superiority, yet did he bear at
first sight a very commanding presence, and it was impossible
to glance at him for a moment without feeling that we were
before one to whom God had given gifts above those of a
common man. His powerful projecting forehead and well-
chiselled features told of much thought; and might have
implied severity, had not a soft radiant benevolence played
over his fine old face, which inspired his friends, of whatever
age or rank, with confidence and love.

"Happy were the days, both to young and old, when the
genial-hearted philosopher walked over the hills, which he did
frequently, to spend a few hours at the Vicarage of Dent!
Whenever he and my father met, their hearts seemed to be
warmed with the spirits of two schoolboys meeting on a
holiday. And well might they be happy in the sweet remembrances of God's mercies so long vouchsafed to them, and in
those firm unflinching Christian hopes that gave a bright
colour to the days of their old age[1]."

Our portrait of Mr Dawson is reduced from a beautiful
water-colour drawing by William Westall—who married
Sedgwick's sister Ann—dated 1817, three years before
Dawson's death. It represents him therefore "in his honoured

---

[1] Mr Dawson died 19 September, 1820. A few years afterwards a monument
was erected to his memory, on the south side of the central aisle of Sedbergh
Church, by some of his former pupils. It consists of a black marble niche,
enclosing a bust, inscribed "Sievier sculp^t. Aug. 1825." Beneath, is a
white marble tablet bearing the following inscription: "In memory of John
Dawson, of Sedbergh, who died on the 19th Sept. 1820, aged 86 years.
Distinguished by his profound knowledge of mathematics, beloved for his
amiable simplicity of character, and revered for his exemplary discharge of
every moral and religious duty. This monument was erected by his grateful pupils,
as their last tribute of affection and esteem." The inscription was written by
Mr John Bell, the distinguished leader at the Chancery Bar, who had been one
of Mr Dawson's Senior Wranglers, as mentioned above, p. 65, *note*.

old age", when advancing years had probably increased the natural severity of his features. An earlier portrait, by the Rev. D. M. Peacock, taken in or about 1809, represents him standing beside a pupil, who is seated at a table, working out a mathematical problem. In this picture the face is thoughtful, but far less severe than in the later work[1].

---

[1] In 1809 Mr Joseph Allen, a portrait-painter, having been commissioned to paint Dawson's picture, Mr Peacock, then vicar of Sedbergh, invited him to stay at his house. When the picture was finished, Mr Peacock was not satisfied with the likeness, and being himself an artist of no mean capacity, painted a replica of Allen's work, with the advantage of fresh sittings from Mr Dawson. This picture is now at Ripon in the possession of Miss Cust (Mr Peacock changed his name to Cust on inheriting an estate near Northallerton). Allen's portrait cannot now be traced. Sedgwick gives the following history of it. "It was the property of Mr Leigh of Leeds. This gentleman was unfortunately killed during a furious election riot about the time the Reform Bill was passing through Parliament, and I heard some time afterwards that the portrait was knocked down at a public auction in Leeds, but to whom I never learnt." Fortunately, however, it was engraved; and a comparison of the engraving with Mr Peacock's portrait shows that the two works are absolutely identical. The engraving is lettered "Painted by Joseph Allen. Engraved by W. W. Burney. To the Friends and Pupils of Mr Dawson, of Sedbergh, this Engraving, from an original Picture in the possession of R. H. Leigh, Esq. is respectfully Inscribed by Joseph Allen." The picture may be thus described: "Canvass, 17½ in. broad, by 21¼ high. Dawson is standing up, seen to waist, three-quarters face to the spectator. His right hand, holding spectacles, rests on the back of a chair; his left on an open folio volume lying on a table covered with a green cloth. The fingers are closed, except the fore-finger, which points to a mathematical figure on the right page. Dawson's figure is tall and spare; face rather severe, with deep lines on it. Complexion clear and ruddy; eyes grey; short grey hair. White neckcloth twisted round his neck; high black waistcoat; chocolate coat, with row of five buttons. A pupil is seated at the table, his back turned to the spectator. He is bending over his work; right elbow on table; head resting on right hand. Light hair, light brown coat, white neckcloth like Dawson's." The name of the pupil represented is said to have been Litterdale.

JOHN DAWSON, OF SEDBERGH.

Born 1734; *died* 1820.

From a water-colour drawing by William Westall, 1817.

To face page 70, *Vol. I.*

CHAPTER III.

(1804—1810.)

BEGINS RESIDENCE AT TRINITY COLLEGE, CAMBRIDGE (1804). THE REV. T. JONES. CHRISTMAS AT WHITTLESEA. COLLEGE EXAMINATION (1805). SUMMER AT DENT. FEVER. LIFE AND FRIENDS AT CAMBRIDGE (1806). PREPARATION FOR DEGREE (1807). SYSTEM OF ACTS AND OPPONENCIES. SEDGWICK'S FIRST ACT. UNIVERSITY ELECTION FOR M.P. HIS FATHER'S ACCOUNT OF THE YORKSHIRE ELECTION. SCHOLARSHIP AT TRINITY COLLEGE. BLINDNESS OF HIS FATHER. SENATE-HOUSE EXAMINATION (1808). REVISITS DENT. READING PARTY AT DITTON. CLASSICAL WORK (1809). FELLOWSHIP (1810).

SEDGWICK went up to Cambridge, accompanied by his friend Bland, in October, 1804. They left Dent, or Sedbergh, on Saturday, 29 September; spent Sunday with a friend at Kirkby Stephen, and on Monday joined the "Paul Jones" coach at Brough, in which delightful vehicle, "an old six-inside," they passed three days and two dismal nights[1].

1804.
Æt. 19.

The fear of a French invasion was at that time uppermost in the minds of Englishmen. As Sedgwick often said 'England looked like a great camp'[1]; and the occupation of arming and drilling, which had penetrated even to the University, together with the paralysis of trade, and the high prices of

---

[1] To Miss Sedgwick (U.S.), 12 October, 1853. To Miss Isabella Sedgwick, 1 October, 1852.

the common necessaries of life, had seriously diminished the number of young men who could afford a University education. In 1804 only 128 presented themselves for matriculation, the smallest number since 1775, when it had sunk as low as 121[1].

More than one of Sedgwick's contemporaries has been heard to say that when he first made his appearance in Trinity College he was thought uncouth, and that some time elapsed before he was properly appreciated. This unfavourable judgment need not surprise us. It must be remembered that though he was nearly twenty years of age, he had never quitted his home, or its immediate neighbourhood, for more than a few days at a time. Carlisle was probably the furthest point he had reached in any of his brief excursions. Nor, in that remote corner of England, could he have had any opportunity of associating with men of the world. When we have mentioned his own father, the vicar of Sedbergh, Mr Dawson, and two or three of the neighbouring country-gentlemen, the list of his older friends is exhausted. From the young men of his own age, whose ideas of amusement were confined to sport, wakes, and drinking-bouts[2], he could have learnt nothing but tastes and customs 'more honoured in the breach than the observance', and that he never gave way to such himself, to any serious extent, is a proof either of his father's influence, or of his own strength of will. On

---

[1] In the first six years of the present century, 1800—1805, the average (neglecting fractions) was 150: in the last six years of the previous century it had been 156. Between 1806 and 1811 it rose to 232 ; and between 1812 and 1817 to 278.

[2] A friend—a clergyman—writing to Sedgwick 28 July, 1807, records the following experiences: "I had fully determined to write to you immediately after my return from Yorkshire; but a number of Feasts crowded upon me in such quick succession that I found myself unable to devote a single hour to correspondents. Mr ——'s house was filled with company on Sunday, Monday, Tuesday, and Thursday, many of whom never thought of *flinching* till three or four o'clock each morning; they then staggered home to bed....We met with the usual hospitality in Dent....They even kicked up a dance for us in the old school-room on Whit Friday. Your Brother John and I unluckily dined with Thomas Fawcett on that day, where we met a set of rook-shooting Bloods from Kendal. They all drank very freely, and tho' John and I left early in the afternoon, yet we were far from sober. This however did not prevent our joining in the dance, and frisking away as well as the best of them."

a young man brought up as he had been, the University 1804. would produce the sense of wonder and bewilderment which Æt. 19. we associate with the first sight of a crowded capital. He must have felt out of his element among scenes for which he would be less prepared than most undergraduates even at that time, when journeys were not lightly undertaken, and when most people could say, with the Vicar of Wakefield, that their longest migrations were from the blue bed to the brown. His dress, his manners, and his bearing would bespeak him a plain unsophisticated Dalesman[1].

Sedgwick's tutor at Trinity College, as mentioned in the previous chapter, was the Rev. Thomas Jones. It would be difficult, if not impossible, to name any one man, who by force of personal character, vigour of intellect, and unwearied devotion to his duties, was enabled to effect a more enduring influence for good on the moral and intellectual life of his college. He had been Senior Wrangler and first Smith's Prizeman in 1779, and was elected Fellow in 1781. A few years after his election he took part in a movement of such far-reaching importance, that we must briefly notice it. For some years the examination for scholarships and fellowships had been conducted in private—candidates "went their rounds to the electing seniors[2]". This system could never have been a good one, and of late years it had been grossly abused, inasmuch as it was notorious that some of the Seniors had voted at elections without having examined, and others had been personally influenced in favour of particular candidates. In October, 1786, an election to Fellowships, made " exactly in the most improper, as well as the most unpopular, manner possible[3]", brought matters to a crisis. Ten of the Junior

---

[1] Unfortunately none of the numerous letters which Sedgwick is known to have written to the home-circle at Dent have been preserved.

[2] *Memoirs of Richard Cumberland*, 4to, London, 1806, p. 106. The author gives a detailed account of the way in which the whole examination was then conducted, which well deserves careful perusal.

[3] Mr John Baynes to Mr Romilly, in *Memoirs of the Life of Sir S. Romilly*, ed. 1841, i. 253. The writer, one of the Junior Fellows who signed the Memorial,

1804.
Æt. 19.

Fellows—among whom was Jones—addressed to the Master and Seniors a remonstrance, which alleged that the practice was in opposition to the statutes, and tended to destroy the objects of the foundation. The Master and Seniors, greatly incensed, cautioned the memorialists to behave with greater deference to their superiors, and entered an admonition to that effect in the Conclusion Book. From this sentence the Junior Fellows appealed to the Visitor. The appeal was heard by Lord Chancellor Thurlow on behalf of the Crown. His sentence, while condemning the form which the action of the memorialists had taken, condemned the practice complained of more strongly still. The result was eminently satisfactory. Not only were the remonstrance and the admonition both withdrawn, but the Master insisted ever after on each elector being a *bonâ fide* examiner; and, in the following year (1787), notwithstanding the share which Jones had taken in the memorial, he was elected Senior Tutor—an office which he held until his death in 1807, at the comparatively early age of fifty-one[1].

It must be recollected that in those days private tutors were not commonly resorted to, and indeed, during the two years which preceded the examinations for degree, were forbidden by Grace of the Senate[2]. In consequence of this, and of the comparatively small number of undergraduates, the college tutors were brought into closer relations with their pupils than is possible at present, for they were not

takes credit to himself for having originated it. As he and Miles Popple, another of the Memorialists, were parties to the appeal to the Visitor, it is probable that they were the prime movers in the whole matter.

[1] Cooper's *Annals*, iv. 424; Gunning's *Reminiscences*, ed. 1855, ii. 100; Monk's *Life of Bentley*, ed. 1833, ii. 423.

[2] Grace, 25 February, 1781. It is printed at length in Whewell, *Of a Liberal Education*, Part I., p. 220. The time at the end of a student's career during which reading with a private tutor was prohibited, was gradually diminished. By Grace, 9 April, 1807, it was reduced to a year and a half; by a subsequent Grace, 3 July, 1815, to one year. Ibid. p. 221. The Grace of 1781, it should be observed, limits its prohibition to tutors engaged within the precincts of the University (*intra Academiam*). A student was at liberty to read with whom he pleased during the vacations.

only advisers, but instructors. In both capacities Mr Jones 1804. was preeminent. His friend and biographer, the Rev. Æt. 19. Herbert Marsh, has recorded that in his duties as College Tutor

"he displayed an ability which was rarely equalled, with an integrity which never was surpassed. They only, who have had the benefit of attending his lectures, are able to estimate their value. Being perfect master of his subjects, he always placed them in the clearest point of view; and by his manner of treating them, he made them interesting even to those who had otherwise no relish for mathematical inquiries.

"As a companion, he was highly convivial: he possessed a vein of humour peculiar to himself; and no one told a story with more effect. His manners were mild and unassuming, and his gentleness was equalled only by his firmness. As a friend, he had no other limit to his kindness than his ability to serve. Indeed his whole life was a life of benevolence, and he wasted his strength in exerting himself for others. The benefits which he conferred were frequently so great, and the persons who subsisted by his bounty were so numerous, that he was often distressed in the midst of affluence. And though he was Head Tutor of Trinity College almost twenty years, with more pupils than any of his predecessors, he never acquired a sufficient capital to enable him to retire from office and still continue his accustomed beneficence[1]."

To this sober prose may be added a few lines from an elegy by Robert Dealtry, LL.D., written after Jones' death. The verse is poor, but the sentiments have the true ring of sincerity:

> "The wild unbroken boy he led, not drove,
> And changed coercion for paternal love.
> By mildness won, youth found resistance vain,
> Bound in a silken, yet a snapless chain.
> Around his sacred tomb th' ingenuous band
> Of sorrowing pupils oft shall pensive stand,
> Shall hail the Tutor faithful to his trust,
> Revere his memory, and bedew his dust."

It is not likely that Sedgwick made many new friends during his first term of residence. His two school-friends Ainger and Bland being both members of St John's College

---

[1] *Memoir of the late Rev. Thomas Jones.* Signed, Herbert Marsh, Cambridge, Feb. 19, 1808.

he probably was more frequently in their society than at
Trinity; and we shall find by and bye that he made
several friendships in their college.

Dent was far too distant to be visited in one of the short
vacations; so, when the Michaelmas Term was well over,
Sedgwick went, on December the 17th, 1804 (a date he was
fond of recalling), to spend Christmas with Ainger at his
father's house at Whittlesea. From this memorable visit
dates his lifelong friendship with the whole Ainger family.
He never forgot the simple pleasures which he there enjoyed.
In 1851, forty-seven years afterwards, he could still write to
Mr James Ainger, his friend's brother, that "when Christmas
came round, I remembered the happy, genial, joyful Christmas
I spent in your father's house in 1804". It was on this occasion
that he made the acquaintance of Henry Smith, son to the
surgeon of Whittlesea, then a boy of sixteen, who afterwards
became a distinguished Indian general, and is known to fame
as Sir Harry Smith. Sedgwick watched his career with
affectionate interest, and his name will frequently recur in our
narrative.

In the College examination held in June 1805, Sedgwick's
name appears in the first class, in company with only six
others—a distinction which shows, to those familiar with the
practice of Trinity College at that time, that he must have
got up the classical subjects with thoroughness, as well as
the mathematical. The ordeal was in those days specially
formidable, for a *vivâ voce* examination used to be held in the
Hall in the presence of the Master, who sat in the centre of
the dais, with the Seniors to his right and left, as Byron has
recorded:

> " High in the midst, surrounded by his peers,
> Magnus his ample front sublime uprears:
> Placed on his chair of state he seems a god,
> While Sophs and Freshmen tremble at his nod[1]."

[1] *Thoughts suggested by a College Examination.* The lines were printed in
Hours of Idleness, published in 1807, only two years after Sedgwick's first
examination.

The persons to be examined stood in front of this awe-inspiring assemblage, and questions were passed down the line from one to another by the presiding examiner, the Master occasionally interposing a word of praise or reproof—more frequently the latter. Men have been known to faint with apprehension even before it had come to their turn to be questioned. When the examination was over the names of those only were published whom the examiners thought specially worthy of commendation[1]. A place in the first class was therefore a considerable distinction.

In the summer Sedgwick went down to Dent, accompanied by his friend John Carr, an undergraduate of Trinity College, one year senior to himself. Carr was a distinguished mathematician, who became second Wrangler and second Smith's Prizeman in 1807, and subsequently obtained a Fellowship. Sedgwick's intimacy with him indicates that he was by this time making his way in the College, and had got into the society of men older than himself—at all times a sure indication of popularity. The vacation was spent, as recorded in the last chapter, in reading mathematics with Mr Dawson.

Soon after Sedgwick's return to Cambridge, he was attacked by a typhoid fever which nearly proved fatal. His medical attendants despaired of his life. They had in fact left his rooms, and were walking up and down on the pavement beside the Chapel, waiting to hear the last news before they left College. The news, however, did not come, and it was at last suggested that they should go back and look at their patient again. To their surprise they found him not only not dead, but apparently somewhat stronger than when they had left him. One of them, Sir Busick Harwood, said to the other: "This is a very strong young man, let us try if we can do anything more for him." Accordingly a blister was suggested. The poor patient shrunk from the anticipated suffering, and asked what effect the application would have

[1] Pryme's *Recollections*, p. 90.

1805.
Æt. 20.

upon his flesh. To this very natural question he is said to have received the somewhat brutal answer: "Oh! — the flesh, if we can only save the life!" The blister was applied, and the patient survived, to tell the story almost as it is here related[1]. He was nursed through his illness with unremitting diligence by his friend Ainger, and he used to say that he owed his life more to him than to his physicians.

The date of this fever is fixed exactly by an interesting circumstance which Sedgwick recollected when the funeral of the Duke of Wellington reminded him of the death of Lord Nelson. News of the success or failure of his operations against the French fleet must have been waited for with an anxiety of which we, in these peaceful times, can form no conception. If he failed, England would almost certainly be invaded; if he were successful her safety was secured. Intelligence of the battle of Trafalgar reached London on Wednesday, 6 November, 1805, and Cambridge on the following day. The volunteers assembled in the market-place, and fired three volleys; the bells of Great St Mary's Church rang a dumb peal; and in the evening the town was illuminated[2]. Sedgwick resided at that time in a set of garrets between the Chapel and the Great Gate[3]; and, half-delirious as he was, he insisted, as soon as he heard the bells, on being carried to the window, that he might see the illuminations[4]. In those days the Sun Inn was opposite to the College, and would no doubt be conspicuous for its patriotism.

Sedgwick's recovery was slow, and it was not till the

---

[1] This story is taken from an excellent article headed " Adam Sedgwick," by the Lord Bishop of Carlisle, in *Macmillan's Magazine* for April 1880.

[2] *Cambridge Chronicle*, Saturday, November 9, 1805. "At noon yesterday [the column is headed ' Cambridge, November 8'], the Cambridge Volunteers were drawn up in the Market Place, and fired three *feux-de-joie*. A dumb peal was rung at Great St Mary's, as a testimony of respect to the memory of the brave Admiral, and in the evening a general illumination took place."

[3] Sedgwick told his niece, Miss Isabella Sedgwick, that his first rooms were in this part of the College.

[4] To Mrs Richard Sedgwick, 21 November, 1852.

early spring of the following year (1806) that he was able   1806.
to leave his rooms. He recollected this in 1871, when fine   Æt. 21.
weather on the last day of the year had enabled him to take
a short walk under the shelter of the Chapel : " The bright
sunshine ", he wrote, " has tempted me out, and I have had a
turn on the flags before our Chapel. I tried the bowling-
green, but it was in the shade, and did not suit me. While
walking I could not but think of the early weeks of the year
1806, when I crawled out of my rooms, one bright sunny day
(about February I think) after my long confinement from
typhus fever. I had great difficulty in getting back. But
then I was young, and my rate of recovery was astonishing.
I am not strong now, but I mounted my staircase quite
briskly on my return today[1]."

The rest of the year 1806 is singularly barren of information. At the examination in June Sedgwick was again placed
in the first class—which shows that he had by this time
entirely recovered from the effects of his fever—and when
term was over he went down to Yorkshire, where he spent
the summer in reading mathematics with Mr Dawson. We
get, however, a glimpse of his Cambridge life and interests,
from the following letter, which, though written sixty-one
years afterwards, belongs, by its subject-matter, to 1806.

To Rev. H. C. G. Moule, Trinity College.

FAKENHAM, *Dec.* 16, 1868.
My dear Sir,

I wish it were in my power to give you more
information respecting the two very remarkable persons (H.
Kirke White and Robert Hall), than I have at my command.
Both of them, in their way, were men of great genius. I did
not know Kirke White till a little while before his death—
that is, I never met him and conversed with him during his
first academic year. But whenever I met him in the streets
I was impressed by his look and bearing. He was a tall

---

[1] To Miss Isabella Sedgwick, 31 December, 1871.

1806.   thoughtful-looking young man, with fine features, and with a
Æt. 21. complexion that seemed to indicate a life of severe study.
In his second year, a month or two before his death, I several
times met him in society. His manners well matched his
character. They were simple, earnest, winning, and unaffected.
He had the look of a man of genius. So far as regards his
features, Chantrey's medallion gives, I think, a good *general*
notion of them, so far as can be given by a profile likeness in
low relief[1].

Robert Hall had ceased to live in Cambridge before my
Freshman's year[2]; and of his manners in society I have no
right to speak, as I do not remember to have ever exchanged
a sentence with him; though, on public occasions, I have
once or twice met him. But he occasionally revisited Cambridge; and then he always preached at the Baptist Meeting-
House in St Andrew's Street; and whenever I could secure a
seat on such occasions, I always attended the Meeting. He
always began with a prayer (sometimes of considerable
length), uttered with great earnestness and simplicity, but
injured in effective power from an apparent asthmatical
difficulty of articulation. There was the same constitutional,
or organic, difficulty in the commencement of his sermons.
But the breathing of his sentences became more easy as he
advanced, and before long there was a moral grandeur in his
delivery which triumphed over all organic defect or physical
weakness. While he rolled out his beautiful and purely
constructed sentences one felt as if under the training of a
higher nature. In occasional flights of imagination, in discussions of metaphysical subtlety, we were for a while amazed,

[1] Henry Kirke White commenced residence at St John's College in October
1805; but he did not matriculate (as a Sizar) until 26 May, 1806. He died
19 October in the same year. The medallion by Chantrey is in All Saints' Church,
Cambridge.

[2] Mr Hall resigned his duties at Cambridge, owing to ill-health, in November,
1804: returned in April, 1805; and resigned finally 4 March, 1806. In 1803 he
had been advised to reside at Shelford, and his biographer ascribes the mental
affection of 1804 to the loss of society occasioned by this removal. Works of
Robert Hall, 8vo. Lond. 1843. v. 443, vi. 73—75.

and almost in fear for the Preacher. And then he would
come down, with an eagle's swoop, upon the matter he had in
hand, and enforce it with a power of eloquence such as I
never felt or witnessed in the speaking of any other man.
Such is my feeling now. Many a long year has passed away
since I last heard Robert Hall. I have listened with admiration to many orators in the two Houses of Parliament, and to
many good and heart-moving preachers; but I never heard
one who was, in my mind, on the same level with Robert
Hall.

I am at a friend's house, and I have very little spare time
on my hands; but I have stolen away from the party for a
few minutes, that I might do my best in answering your letter
which has just reached me. Alas! I know too well that this
letter is not worth your reading; but I have, at any rate, endeavoured to shew my goodwill to a brother-Fellow.

I remain,

Very faithfully yours,

A. SEDGWICK.

Sedgwick's recollections of Henry Kirke White, who was
of St John's College, confirm what was mentioned above, viz.:
that his friendship with Ainger, his senior in University
standing by one year, would readily introduce him to
Johnian society. Two other men of Ainger's year and
college became intimate with him, Robert Bayne Armstrong,
and James Tobias Cook, who were afterwards both Fellows
of the House. In his own college he knew Charles James
Blomfield, afterwards Lord Bishop of London, William Clark,
afterwards Professor of Anatomy, Richard Ward, and Edward
Peacock. These were all of his own year, and distinguished
for their hard-reading and academical success. Of those
senior to him, besides Carr, may be mentioned George
Pryme, afterwards Professor of Political Economy, who had
obtained his degree in 1803, and his Fellowship in 1805. He

had been a pupil of Mr Dawson, and while at Sedbergh had played cricket with Sedgwick[1], but they could hardly have seen much of each other at Cambridge before 1808, when Pryme gave up the pursuit of the law in London, and returned to College. They then became very intimate friends.

At the beginning of 1807 Sedgwick began to prepare in earnest for his degree. In those days the University required no proof of a student's proficiency until his third year, in the course of which he had to keep two Acts and two Opponencies, as they were called. Of the former the first took place in the Lent Term, the second in the Michaelmas Term. This system has been so long completely obsolete that a description of the way in which it was managed is almost indispensable. We will try to make this as brief as possible.

At the beginning of January in a given year the Moderators obtained from the Tutor of each college a list of his pupils who aspired to Honours in January of the ensuing year. Out of these lists—each name on which was noted for the Moderators' guidance, *reading, non-reading, hard-reading,* as the case might be—a complete list was formed, and transcribed into a book. On the second Monday in the Lent Term the Moderator for the week—(the two Moderators divided the term between them)—sent a written notice to one of the students on his list—who apparently was selected quite arbitrarily—to the effect that he was to appear in the Schools on that day fortnight as a disputant. Shortly after receiving this summons, the student, now called a *Respondent,* waited on the Moderator with three Propositions, usually termed Questions, the truth of which he was prepared to maintain against any three students of the same year, whom the Moderator chose to nominate, and who were called *Opponents.* The first question was generally taken from Newton's *Principia;* the second from some other writer on Mathematics or Natural Philosophy; the third was called the *Moral Question,* and, in connexion with it, Locke, Hume,

---

[1] *Recollections,* p. 30.

Butler, Clarke, Hartley, Paley, were alternately attacked and defended. During the fortnight's preparation, it was usual (at least at the time we are considering) for the Respondent to invite the Opponents to wine, or tea, or breakfast, in order to compare arguments, and generally to rehearse the performance[1]. On the day appointed the Moderator entered

1807.
Æt. 22.

The Moderator's seat, from a photograph[2].

[1] *Alma Mater*, 8vo. Lond. 1827, ii. 35—38. The system here sketched was abolished in 1839, and for about sixty years previous the disputations had lost much of their original vitality. Compare *The Origin and History of the Mathematical Tripos*, by W. W. Rouse Ball, M.A., 8vo. Camb. 1880; and *The Mathematical Tripos*, by J. W. L. Glaisher, Sc.D., F.R.S., 8vo. Lond. 1886.

[2] The seat here figured is that in the Law School, which was somewhat larger than that in the Arts School, but in all other respects resembled it exactly. The photograph was taken in January 1886, just before the Law School was fitted up for Library purposes, as authorised by Grace of the Senate, 17 December, 1885 (*Cambridge University Reporter*, p. 297).

the Arts School at 1 p.m., ascended the rostrum on the west side, and said: *Ascendat Dominus respondens.* Thus summoned, the Respondent ascended the rostrum on the opposite side of the school, and read a Latin dissertation, which generally occupied from ten to fifteen minutes, on any one of the three questions. As soon as he had finished, the Moderator called upon the first Opponent to begin (*Ascendat opponentium primus*). He ascended a rostrum beneath that of the Moderator, and propounded his arguments in the form of syllogisms, which the Respondent answered as best he could. When the first Opponent had finished, his place was taken by the second Opponent, and so on. The first Opponent was obliged to bring forward eight arguments, the second five, and the last three. "When the exercise has for some time been carried on according to the strict rules of Logic"—says the authority from whom we have borrowed this account—"the Disputation insensibly slides into free and unconfined debate, the Moderator, in the mean time, explaining the argument of the Opponent, when necessary; restraining both parties from wandering from the subject; and frequently adding, at the close of each argument, his own determination upon the point in dispute. The three Opponents, having, in their turns, exhausted their whole stock of arguments, are dismissed by the Moderator in their order, with such a compliment, as, in his estimation, they deserve (*Domine opponens, bene disputasti—optime disputasti—optime quidem disputasti*); and the Exercise closes with the dismission of the Respondent in a similar manner[1]."

At the close of the Act, the Moderator assigned to the Disputant a certain number of marks, which he set down opposite to his name in his book; and, when all the Acts had been kept, the two Moderators conjointly formed the students into classes according to the number of their marks.

---

[1] *Remarks upon the present Mode of Education in the University of Cambridge.* By the Rev. John Jebb, M.A., Ed. iii. 8vo. Camb. 1773, pp. 18—20. Wordsworth, *Scholæ Academicæ*, pp. 32—43: 368—376.

The disputations took place on five days in each week, and each occupied about two hours. The language used was Latin, which, by the beginning of the present century, had degenerated into a strange jargon ; and the logic was little better than the language. Still, as a man's place in the classes depended upon the impression he made upon the Moderator when he kept his Act, it was imperative (at least in 1807) to take pains. Moreover, at that period, when the distractions of University life were few, Acts still excited considerable interest, and were well attended, both by graduates and undergraduates. The disputant was therefore on his mettle, and anxious to distinguish himself at a performance which he felt would have a great share in determining his academical reputation[1]. The Acts were often performed with great spirit, and tradition still preserves the names of some of those who were specially successful for their ingenuity of attack or defence. The system was also useful in a social way, by making men of the same year, but of different colleges, known to each other. By this means Sedgwick "became well acquainted with" Henry Bickersteth of Gonville and Caius College, who was Senior Wrangler in his year. He mentions this fact in a letter which is printed in Bickersteth's *Life*, and adds: " He did not quite do himself justice in his first public Act, in the Lent Term of 1807; but his Act in the October Term of the same year was the most triumphant I ever witnessed. He literally seemed to trample his opponents under foot[2]." Their acquaintance, thus begun, ripened into a friendship which was maintained after the one had become Master of the Rolls and Lord Langdale, and the other Woodwardian Professor.

Sedgwick kept his first Act in February, 1807, and soon

---

[1] *Of a Liberal Education in General*, Part I. By W. Whewell. Ed. ii. 1850, p. 183.

[2] *Memoirs of the Right Honourable Henry Lord Langdale.* By T. D. Hardy. 8vo. Lond. 1852. i. 232.

1807.
Æt. 22. afterwards wrote to Ainger to record his success, with other fragments of University intelligence[1]:

"......I came off better than I expected in the schools. I perhaps might fairly say that I came off better than I could have previously wished. The honour I received from Woodhouse[2] was: *"omni tuo officio multâ cum laude perfunctus es."*

"Since you left Cambridge we have had a most eloquent sermon from Dr Milner; it was delivered with that peculiar emphasis which you might expect from a man of his powers, conscious at the same time of the truth and importance of what he was delivering[3]. We have had four sermons from *Lemma* Vince[4], the most strange things ever preached in pulpit. The first Sunday he took us thro' the three laws of motion and Wood's chapter on projectiles[5]. The 2nd Sunday he got into his complete system of astronomy[6]. The 3rd he took us through the 11th section of Newton, and to conclude gave us a dissertation on optical glasses. His pulpit lectures have now ended......"

The Act over, Sedgwick evidently thought of nothing but preparation for the Senate House; and is rallied by his friend on his devotion to his studies:

"......How possibly can you, deeply immersed as you are in all the sublimities of Mathematical Science, take any interest in the grovling concerns of one who, since he left you, has merely been scampering about the Fens in order to get rid of time? In truth,

---

[1] To William Ainger, 23 February, 1807.

[2] Robert Woodhouse, M.A., Fellow of Gonville and Caius College; Lucasian Professor 1820—22; Plumian Professor 1822—28.

[3] Isaac Milner, D.D., President of Queens' College, 1788—1820. In after-life Sedgwick did not speak of Dr Milner in such complimentary language. The sermon referred to must be the one preached at Great St Mary's Church, 30 January, 1807, against the Emancipation of the Roman Catholics, which produced a great sensation in the University. Life of Milner, 8vo. 1842, p. 344. It is printed in Milner's *Sermons*, 2 vols., 8vo. Lond. 1820, i. 1.

[4] Samuel Vince, M.A., Plumian Professor 1796—1822. He was Select Preacher in February, 1807. The Sundays in February 1807 fell on the 1st, 8th, 15th, 22nd.

[5] An allusion to a then popular mathematical work: *The Principles of Mechanics; designed for the use of Students in the University*, by James Wood, B.D. 1796.

[6] *A Complete System of Astronomy*, by Rev. Samuel Vince, 3 vols. 4to. 1797.

Sedgwick, had I anything more important to acquaint you with, I would not presume to inform you that last Tuesday sen'night I was capering at Wisbeach to the sound of a Fiddle, and that the deepest speculation in which I have engaged, has been an attempt to learn the character of an eccentric girl whom you may recollect I once mentioned as the only female likely to make an impression on your iron heart. Positively I think her as great an oddity as yourself; and surely this is saying enough to excite any one's curiosity who is not so much infected with the Mathematical Mania as to scorn everything·which is lower than the stars...."[1]   1807. Æt. 22.

We do not know what became of this damsel, or whether she was ever aware of the honour destined for her; but the next letter shows that the self-denying student could at any rate find time to take interest in University politics :

" Since you left us we have had a dead calm in Cambridge, which in all probability would have been of long duration if the King of his great goodness had not caused a dissolution of parliament. We are now in confusion and uproar. The Johnians are exerting themselves to the utmost in grunting out the praises of their brother Palmerston. Lord H. Petty's interest is considerably diminished in consequence of Mr Jones' illness. We were under the greatest apprehension for the life of our old Tutor, but the last reports from London were more favourable. He is now there, under the care of an eminent physician, attended by his friend Professor Marsh....

" As my sister has now left Cambridge, I can begin a system of close reading to which I hope to adhere. I have finished the first volume of Newton, and just begun to look at the second. In making an attempt, last night, upon the philosophy of sound, I got so completely fast, that after retiring to rest I was disturbed with the most horrid dreams you have the power of conceiving. I intend to rise at five all this summer; if you will do the same I can promise you great advantage from it....[2]"

Sedgwick had been taught by his father to abhor the

---

[1] From William Ainger, 11 April, 1807.
[2] To William Ainger, 4 May, 1807.

slave-trade, and he told Bishop Wilberforce in 1848 that he
had signed a petition against it as soon as he "had learnt to
scrawl his name in childish characters". He had also imbibed,
probably from the same source, a wholesome horror of the
Church of Rome, and had not yet become a sufficiently
decided Liberal to see the justice of removing the civic
disabilities of Roman Catholics, as he did a few years later.
He would therefore take a peculiar interest in the University
elections of 1806 and 1807, of which the former turned on
Abolition, the latter, at least to some extent, on Catholic
Emancipation. After the death of Mr Pitt, Lord Henry
Petty of Trinity College, afterwards Marquis of Lansdowne,
then Chancellor of the Exchequer in the cabinet of Lord
Grenville, had come forward, as an abolitionist, at the suggestion of Mr Wilberforce[1], who had personally canvassed
on his behalf. He was elected by an enormous majority,
his competitors, Viscount Althorp and Viscount Palmerston,
polling together only 273 votes, while 331 were recorded for
him[2]. At the same time many of those who usually agreed
with Mr Wilberforce were not a little scandalised at his
support of one who had opposed Mr Pitt, and some, among
whom was Dr Milner, voted against his candidate[3]. At the
general election of 1807 the feelings of the constituency were
no longer the same. The Abolition Bill had received the
Royal Assent on the 25th March, and therefore the slavery
question had passed out of sight for the moment. Lord
Grenville and his colleagues—better known as the ministry of
"All the Talents"—had been dismissed by the King on a
point connected with the relief of the Roman Catholics; and
their successors, the Duke of Portland and Mr Perceval, had
dissolved Parliament. The question placed before the consti-
tuencies was partly the relief of the Roman Catholics—partly
the vindication of the King's conduct, who had demanded

---

[1] *Life of Wilberforce, ut supra,* iii. 255, 256.
[2] Cooper's *Annals,* iv. 484.
[3] *Life of Milner, ut supra,* pp. 316—320.

a pledge from his ministers that they would never again, under any circumstances, offer him advice on the Catholic question. Protestantism and loyalty—both equally unreasonable—were roused to passionate enthusiasm in the country and in the University. There were four candidates for the two seats : Lord Euston, who had been Mr Pitt's colleague ever since his first election in 1784, and Lord Henry Petty (whigs); Sir Vicary Gibbs, Attorney General, and Viscount Palmerston (tories). "We are all in a flame for Church and King" wrote Dr Milner. "Most seriously, I do think that the greatest constitutional question, by far, that has happened in my time is now at issue; and if the 'outs' were to get the better, I think that the royal prerogative would be in imminent danger." He concludes by urging his correspondent to support "Sir V. Gibbs and Lord Palmerston, who, at present, represent the constitutional side, against Lord Euston and Lord Henry Petty, the friends of the ex-ministers[1]."

The polling, then limited to one day, took place on May the 8th. At the end of the morning's voting Gibbs and Palmerston were considerably ahead, Lord Euston was third, and Lord Henry Petty "hopelessly at the bottom". Thereupon Mr Pryme and about forty of Lord Henry's supporters, who had promised to vote for him only, called a meeting of his committee, and with his consent divided their votes between him and Lord Euston[2]. In consequence the latter headed the poll with 324 votes, and Sir V. Gibbs was second with 312, just beating Viscount Palmerston, who polled 310. Lord Henry Petty polled only 265[3]—a curious instance of the fickleness of fortune, when compared with his great majority only fourteen months previously. Sedgwick was probably right in attributing his failure to the loss of the active assistance of so influential a member of his own college as

---

[1] *Life of Milner, ut supra,* p. 349.
[2] Pryme's *Recollections, ut supra,* p. 79.
[3] Cooper's *Annals,* iv. 487.

1807.   Mr Jones, who must have supported him on general grounds,
Æt. 22.   for he was the pupil of another tutor, Mr Porter.

An interesting glimpse of an election contest of those times is afforded by a letter written to Sedgwick by his father from Dent, 11 June 1807. The candidates for the county of York were Mr Wilberforce, who had been member since the memorable contest of 1784, Mr Lascelles, and Viscount Milton. The polling, held at York, lasted for fifteen days, at the end of which Mr Wilberforce and Viscount Milton were declared to have been elected[1].

"You will naturally conjecture what a bustle and ferment this County has been put into by the contested election. Wilberforce had every freeholder's vote in this parish except Tattersell's, who gave a plumper for Lord Milton....Dr Dawson and I with 12 others from Dent and Sedbergh set forward from Kirkby Lonsdale at the same time, but were divided at Skipton for want of immediate conveyances. From thence Mr John Fawcett and Hen. Hodgson were my sharers in a chaise to York and home again...

"Mr Leigh Bland's friend was a warm advocate for Lascelles, but very careful about spending money for liquor. Lord Milton's agents were just the reverse, they spared no expence to get a single voter. Most of his freeholders returned home with about five guineas clear of all expences. The other candidates were before him in canvassing; and Milton would have got few here had not a report prevailed that his Committee allowed very liberally for expences. Indeed we were all carried almost free from expence by stopping at houses that were open for accommodation."

The same letter reports that Mr Dawson is anxious to know "in what Book or Section of Newton your mathematical question was which you kept your Act upon in the Schools: he also wishes to know upon what you mean to keep your next Act[2], if you have already determined that point." This inquiry shows that the old mathematician was watching with interest the steps of his pupil's career at the University. Mr Sedgwick, whose knowledge of the higher mathematics had grown somewhat rusty, contented himself with general advice: "You have been engaged in lectures

---

[1] *Life of Wilberforce, ut supra*, iii. pp. 315—337.

[2] No reference to Sedgwick's second Act, or to either of his Opponencies, occurs in his correspondence.

out of my sphere. I wish you to apply regularly to your
studies, but at the same time not so closely as to injure your
health; take also regular exercise when you can[1]."

In the Easter vacation Sedgwick was elected to a scholarship in his college. As this was his only opportunity of
competing for a distinction without which, according to the
rules then in force, he would not have been allowed to sit
for a Fellowship, success was a matter of vital importance
to him, and his chances probably caused him the gravest
anxiety. The election over, he allowed himself a short pedestrian excursion by way of holiday, in which he had invited
one of his earliest Yorkshire friends and schoolfellows to join
him. The answer indicates how acutely he had suffered from
apprehension as to the result of the examination:

"It would have given me heartfelt pleasure to have been with
you in the delightful walk you mention, that I might have congratu-
lated you, while your mind was filled with the most agreeable ideas
from the honour which you had achieved. As your breast would be
entirely free from those anxieties which you could not but feel in
preparing for that day on which your reputation in a considerable
manner depended, you could not but be in a cheerful mood, and, be
assured, I should have enjoyed your cheerfulness[2]."

The beginning of the Long Vacation of 1807, which
Sedgwick spent in College, preparing for his degree, was
saddened by the death of the respected and beloved tutor,
Mr Jones, which took place on the 18th July. His claims on
the affection of his pupils have been already recorded, and
therefore need not be repeated. That he had obtained it is
proved by a single extract from a letter of one of Sedgwick's
friends: "Poor Jones! I am sure there is not a man one
could have less spared, or one who is more lamented by our
College or by the University in general." Sedgwick no doubt
shared this common sorrow; but he had, in addition, a
private trouble of his own. His father was gradually becoming blind, and a letter from his sister Margaret, after giving
the sad news in detail, had concluded with the ominous words,

[1] From Mr Sedgwick, 18 November, 1806.
[2] From John Brown, 10 April, 1807.

" he fears he shall not be able to teach the school next winter.
My Mother says that if they can get over this winter, you and
Thomas may manage it next summer.   How will you like to
be a Dent's schoolmaster?"   Much as Sedgwick loved Dent
—much as he might have liked such an offer had he never
left it—he now contemplated the possibility of enforced
return with infinite bitterness of spirit.

"My Father's eyes," he wrote to Ainger, "have long failed
him, but he has lately perceived their imperfections so much
that he fears a total blindness; in other respects he is as well
as at his time of life could possibly be expected.   These
accounts for some time produced such a depression in my
spirits that I was prevented from reading; my sorrow,
indeed, was in a good measure selfish (few of our sorrows
are otherwise), for, if my Father's sight should continue to
decline, a fixed residence in Dent must be my inevitable lot.
This situation of all others I should dislike.   Little as I have
seen of the world, I have seen enough to find that to me no
pleasures are to be found in illiterate solitude.   These thoughts
are to me too gloomy to dwell upon......   Pray has Henry
Smith escaped the fate which many of our brave countrymen
have met in Egypt?   I believe his regiment was in the expedition[1]."

His friend, with many apologies for being a bad correspondent, for Sedgwick had begun the above letter with a
page of abuse on that subject, lost no time in replying:

WHITTLESEA, *Aug.* 3, 1807.
My dear Fellow,
        ...Henry Smith, after whom you enquired, did not go
into Egypt, but to Buenos Ayres.   His father had a letter from him
after the engagement.   His captain was killed by his side in the
outset; the command of the company then, of course, devolved to
Henry, who, I believe, acquitted himself very creditably, and did not,
to use his own expression, get a single scratch.   Last week brought
his friends another letter from Monte Video, which acquainted them
that he was then (in April) just recovering from the attack of a fever
which appears, Sedgwick, not to have been less formidable than

[1] To William Ainger, 1 August, 1807.

yours was. He says he has lost all his flesh; but I find he retains all 1807.
his spirit.   Æt., 22.

It is needless to say that I am truly sorry for the accounts which
you have received from Dent. However, you are certainly right in
endeavouring to banish gloomy thoughts. I do not indeed think it
is your character to indulge in them much, and in the present
instance I trust the occasion is not so serious as to justify them. In
the sincere hope, at least, that it may not be found so,

I remain, Dear Sedgwick, your friend,

W. AINGER.

We can imagine that Sedgwick's natural elasticity of spirits
would enable him, before long, to chase away gloomy anticipations, and brace himself to his work and his amusements
as if his future were without a cloud. In 1807 a Long Vacation offered few distractions, with the exception of Sturbridge
Fair, with its varied diversions and excellent theatre, which
even older members of the University did not scruple to
frequent. To these Sedgwick was evidently looking forward,
for Ainger (probably in reply to an invitation to Cambridge)
wishes him "much fun at it", and regrets his own inability to
be present. In the more important matter of work he had
made such good progress that a notion seems to have been
current that he had a chance of being Senior Wrangler. His
friend Carr, who had been Second Wrangler in 1807, and
therefore knew the Senate-House examination well, had
written to him soon after his own degree: "Mind you read
hard, and I have not the least doubt that we shall have the
Senior Wrangler next year. You have my best wishes;" and
Ainger heard by accident at the very end of the year, that
the Fellows had formed a very high opinion of his powers.

ICKLEFORD, *Jan.* 1, 1808.

Dear Sedgwick,

...One morning Mr Professor Lax[1] called here. He
enquired of me about the *great* men who were going out this year.
Your name, of course, was mentioned. He enquired whether you
were to be Senior Wrangler. This was a question which I could not

---

[1] The Rev. William Lax, M.A., Fellow of Trin. Coll. and Lowndean Professor
of Astronomy and Geometry 1795—1836. It was probable that, in virtue of his
office, he might be appealed to to settle the position of the candidates in the brackets.

1808.
Æt. 23.

positively answer; but I took the liberty of saying that you, *without doubt*, would be among those whom he would have the pleasure of examining. He said he had heard a splendid account of you froḿ some of the fellows of Trinity, among whom, I think, he specified Hudson[1].

From a man of your celebrity, Mr Sedgwick, I certainly cannot expect the honour of a letter 'till your expectant brows are finally crowned with the laurels which at present hang over them ; and, by that time, I may, probably, be again at Cambridge. I shall assuredly be most happy to make one in the train of your triumph.

I conclude, Dear Sedgwick, by repeating my wish of a happy new year to you. If benedictions are of any service, may mine avail you in the Senate House. Seriously, however, Sedgwick, may success attend you! but, whether successful or unsuccessful, rely equally on the sincere esteem and unaffected friendship of yours truly,

W. Ainger.

The Senate-House examination began on Monday, 18 January. It was at that time conducted partly *vivâ voce*, partly by printed papers, each class[2] being seated at a separate table. At the conclusion of the first three days examination a new classification was published of those who had passed with the greatest credit. This consisted of a series of brackets, arranged in order of merit, the names in each being placed alphabetically. These brackets (which were hung up on the pillars of the Senate House at 8 a.m. on the Thursday morning) were regarded as a first approximation to the final list, " and men who were joined together in the same bracket had the opportunity of fighting the battle out under the direction of some Master of Arts appointed for the purpose. Sedgwick was in the first bracket, and the battle was fought out under the direction of the Rev. George Barnes, then Tutor of Queens' College, who said that he found no reason to alter the order in which the names came to him; that the men were so different in their reading that he could have put them in almost any order by a special choice of questions; but that the man who impressed him most as possessing inherent power was Sedgwick[3]." The result of

---

[1] The Rev. John Hudson, M.A. succeeded Mr Jones as senior tutor.
[2] These classes were arranged by the Moderators. See above, p. 84.
[3] *Macmillan's Magazine, ut supra*, p. 477.

this last trial was published on the Friday morning, when
the successful candidates were admitted to the degree of
Bachelor of Arts. Sedgwick's name stood fifth in the first
class, or Wranglers. Those above him were: Bickersteth, of
Gonville and Caius College (Senior Wrangler); Bland, of
St John's College, his old friend and schoolfellow, who had
always, as we have seen, been considered certain to beat him;
Blomfield, of Trinity College, who used to say that Sedgwick
was a much better mathematician than himself[1]; and White,
of Gonville and Caius College.

The next two years were spent by Sedgwick in preparing
for the Fellowship examination at Trinity College. His own
wish was to read for the Bar, to which he was stimulated by
the example of his friend Bickersteth; but he was deterred
by the consideration that his father's health was failing, that

The Vicarage of Dent, from a water-colour drawing by Mr William Westall.

[1] *Memoir of C. J. Blomfield*, 8vo. 1863, i. p. 9.

1808.   his two younger brothers had to be educated, and that it
Æt. 23. was therefore his duty to create an independence for himself
as soon as possible[1].   Some relaxation, after the strain of the
preparation for the degree, was, however, indispensable, and
in January, 1808, he went down for a few months to Dent,
which he had not visited since October, 1806.   The change
proved the reverse of agreeable, and his letters betray a good
deal of disappointment, not to say ill-humour.

                               DENT, *February* 19, 1808.

   Dear Ainger,

         I hope you have executed the commission which I
left with you, viz. 'to see my box well fortified with a wrapper,
cord &c., and set on the right road for finding its way into
the North.'   The said box contains all the valuable property
of your humble servant.   You need not doubt therefore that he
is extremely anxious for its arrival.   I am anxious on another
account; it contains some books of which I already begin to
feel the want.   Indeed all the last week I have thought myself
a fish out of water.   I rise about 9 in the morning ; come down
stairs in all due form, and commence breakfast, which consists
of a large mess of oat-meal porridge, to which I drink about a
quart of excellent milk.   This is by far the greatest animal
comfort which I enjoy, for I no sooner have finished breakfast
than I become miserable for want of employment.   The weather
is so bad that to walk is impossible.   I have therefore nothing
to do all morning but amuse myself with my own pleasant reflections, surrounded and perplexed with all the clamour of
domestic music.   I hope next week to find more *rational* pleasures, for I have procured an excellent grey-hound.   You may
depend on it that dogs are the best company a man can have
with him in the country.   Your pleasures, Ainger, I know are in
some measure different from mine ; marriage may be all well
enough when a man is on his last legs, but you may depend
on it that to be linked to a wife is to be linked to misery.

   [1] This statement is made on the authority of Miss Isabella Sedgwick.

From the horrid estate of matrimony I hope long to be de-   1808.
livered. I arrived at Dent without any incident during my   Æt. 23.
journey which is worth mentioning. I found my friends at
Dent better than I expected ; my father, though his eyes have
so far failed as to be of no use to him in doing the Church
service, still keeps up a good flow of spirits, and his general
health does not seem materially impaired....

Believe me, dear Ainger,

Yours truly,

A. SEDGWICK.

DENT, *April* 23, 1808.

Dear Ainger,

I merely write this to be out of your debt, for there
is nothing here which I can tell you *totidem litteris*, but Bland
will be able to tell you much better *totidem verbis*. As soon
therefore as you may think convenient after the sight of this,
write to me, write to me about anything but love and friendship. Indeed a plain matter-of-fact letter will be most agreeable, inasmuch as you are the only person from whom I can
expect to hear what my old friends are doing in Cambridge....
Bland tells me you have been lately on a visit, and that
Miss H—s has been of your party. I think your plan of enjoyment most rational. I wish some blooming damsel could
contrive to kindle a flame in my breast, for then I might
stand some chance of keeping up a proper degree of animal
heat ; without some artificial aid of this kind I am fully confident that my lamp of life will be soon extinguished. Indeed,
Ainger, such is the inclemency of the season, that at the
moment I am writing, most of the farmers in the higher parts
of our valley are busily employed in digging out sheep which
have been covered up in the snow. An "over-drive" at this
time is most unfortunate as many of the sheep are on the
point of bringing forth lambs. So much for the pleasures of
a country life and a crook. I have now completely exhausted

S. I.   7

myself; love naturally led me to talk of sheep, and sheep lead me to talk of I know not what. I shall therefore only add that I am, dear Ainger,

Yours &c. A. SEDGWICK.

Sedgwick returned to Cambridge in May, 1808. The summer was spent on a reading-party with seven pupils. That ingenious device for combining instruction with exercise and pleasure was then in its infancy, and Sedgwick went no further afield than Ditton, a small village on the River Cam about two miles below Cambridge, where he established the head quarters of his colony, as he called it, at Mr Bond's farmhouse. One, at least, of his pupils lodged there with him; another established himself at Horningsea, a village about a mile further down the river, but spent the day at Ditton; the others probably had lodgings in the village. Sedgwick appears to have given himself up to his tutorial duties with unremitting diligence. Vainly did Ainger try to tempt him away to Whittlesea by the attractions of "a day's excursion on the Mere"; "my engagements preclude the possibility of my leaving Ditton" was the stern reply; and in fact the only amusement which he allowed himself was an occasional visit to the theatre at Sturbridge Fair. Still he and his pupils seem to have enjoyed themselves, for more than one of those who read with him refers in after years to his summer at Ditton as a pleasant experience. Sedgwick was nicknamed "the Commissioner", and his authority as "head of the colony" was successfully maintained. All went well, and the party did not return to College until near the end of October.

The names of five out of his seven pupils can be recovered with certainty from his correspondence, viz.: Robert Roberts, and Arthur Savage Wade, of St John's College; Henry Rishton Buck, of Pembroke College; Oliver Hargreave, and John Bayley, of Trinity College. To these should probably be added two other men of the same college, William Robinson Gilby, and Samuel Duckworth. None of these gentlemen were specially distinguished in after-life; unless indeed we except Gilby and Bayley, who became Fellows of Trinity and Emmanuel respectively. Buck went into the army in 1809, as soon as he had obtained his degree, and saw a good deal of active service in Flanders and Holland. Ultimately his regiment took part in the battle of Waterloo, where he was killed, 18 June, 1815. A good many letters passed between him and Sedgwick, who was evidently sincerely attached to him. He carefully preserved all Buck's letters, and on a slip of paper dated 22 September, 1815, he recorded the last events of his friend's career, concluding his brief notice with these words: " Peace to his soul. A man of more cool undaunted courage never existed. If he had lived he would have been an ornament of his profession."

One interesting reminiscence of the autumn of 1808 was frequently recalled by Sedgwick in conversation. He happened to come over from Ditton to Cambridge on Tuesday, 4 October, and on reaching College found that Professor Porson was to be buried on that very day in the chapel. As may be easily imagined, his dress was suited to the country rather than to such an occasion, but, being anxious to honour the memory of so distinguished a scholar, he borrowed a black coat from a friend, and took his place in the procession.

In October, 1809, Sedgwick sat for a Fellowship, but was unsuccessful. There were but two vacancies, to which Charles James Blomfield and William Clark[1] were elected. The former, it will be remembered, was third Wrangler in the same year as Sedgwick, and had also been senior Chancellor's Medallist; but Clark was only seventh wrangler, and had obtained no University distinction in classics. On the supposition therefore that there was no first-rate candidate of the upper year, Sedgwick had a good prospect of success. In the examination, however, which was partly classical, partly mathematical, Clark did extremely well in classics, and gave special satisfaction to the examiners by a translation of a

[1] Professor of Anatomy from 1817 to 1866.

passage from Pindar into English verse. Sedgwick was therefore compelled to wait for a year, which he spent in improving his classical knowledge. "What are you about now?" writes his friend Duckworth, 22 April 1810: "How many vacancies? What number of books of Thucydides, and plays, &c. perused?" Others write in the same strain. On the next occasion, in October, 1810, there were four vacancies, to fill which the Master and Seniors selected Sedgwick; George Edis Webster (8th Wrangler), Edward Peacock (9th Wrangler), and Richard Ward (7th Senior Optime, and second Chancellor's Medallist), all of the same year.

Sedgwick's friends hastened to express their joy at his success. Letters of congratulation have a certain uniformity of style, and those addressed to Sedgwick—or "Sedge" as he was called for brevity's sake—form no exception to the rule. None need be quoted at length—but from that written by Samuel Duckworth, more enthusiastic than the others, we will cite a brief extract. "Escaped from the clutches of x and y; no longer bounded by right lines, superficies, or solids; no longer impelled to move in a diagonal by the joint action of ambition and lucre on the one hand, and of indolence on the other, you may commit yourself entirely to the influence of the latter, enjoy *otium cum dignitate*, and listen to the congratulations of your friends resounding from the banks of the Mersey and Humber to those of the Cam. I for one sincerely congratulate you, my dear Sedge," and so forth. But of all those which he received probably none gave him so much pleasure as that from his old school-master at Sedbergh, who added in a postscript: "Mr Dawson begs to join in congratulations."

CHAPTER IV.

(1810—1818.)

WORK WITH PUPILS. READING FOR ORDINATION (1810). VISIT
TO LONDON. SOCIETY AT CAMBRIDGE. CONTESTED ELECTION
FOR CHANCELLOR AND M.P. FOR UNIVERSITY. INSTALLATION
OF CHANCELLOR. READING PARTY AT BURY ST EDMUNDS.
CHRISTMAS AT DENT (1811). WINTER VISIT TO LAKES.
VISIT TO LONDON. PETITION AGAINST ROMAN CATHOLIC
CLAIMS. READING PARTY AT LOWESTOFT. APPOINTED SUB-
LECTURER AT TRINITY COLLEGE. SECOND PETITION AGAINST
ROMAN CATHOLICS (1812). SERIOUS ILLNESS. SUMMER AT
DENT (1813). SEVERE FROST. VISIT OF MARSHAL BLUCHER.
SUMMER AT DENT. EXCURSION IN YORKSHIRE (1814).
PROJECTED TOUR ON THE CONTINENT. FARISH'S LECTURES.
CAMBRIDGE FEVER. GOES HOME TO DENT. NEWS OF
WATERLOO. ASSISTANT TUTORSHIP AT TRINITY COLLEGE
(1815). TOUR ON CONTINENT (1816). ORDINATION. SUMMER
AT DENT. HARD WORK IN MICHAELMAS TERM (1817).
ELECTED WOODWARDIAN PROFESSOR OF GEOLOGY (1818).

THE long-coveted distinction of a Fellowship at Trinity
College did not bring to Sedgwick all the pleasure he had no
doubt anticipated from it. He had sacrificed his own inclina-
tions, as already mentioned, to his duty towards his father
and his brothers, and he never repented of that decision.
But the course which it compelled him to follow was not the
less distasteful because it was right. He had never taken

any deep interest in mathematics, or done any original work in them; and when it became necessary to teach them, merely for the sake of money, to young men of whom some probably approached their study with unwillingness as great as his own, he regarded both the subject and the pupils with feelings little short of detestation. He worked hard, and evidently did his duty conscientiously and thoroughly[1], but he was at heart profoundly dissatisfied with himself and his surroundings. "Six of these blessed youths I have to feed each day," he exclaims in one of his letters; in others he deplores his wasted life, his inability to find leisure, even in the summer vacation, for the private reading which he is longing to begin; and he looks forward, with eager anticipation, to a future in which he will be able "to have done with the system altogether". Nor were other causes wanting to make his life less pleasant than it had been.

In the first place, his health had become impaired. He had been extremely anxious about the result of the Fellowship examination, and had over-worked himself in preparation for it. In fact, the chronic ill-health from which he suffered during the rest of his life, and which occupied so large a space in his letters and his conversation, may be traced to the mental and physical strain of that period. He broke down completely in 1813, as will be related below; but even at the end of 1810, though he declined to allow that he was ill, his appearance was such that his friends had become solicitous on his behalf. As one of them wrote: "a man who is reduced two or three stone below his standard weight cannot be very well[2]." The state of his purse, however, would not allow him to take the rest of which he stood

---

[1] For instance, when Ainger was coming to Cambridge in the Easter Term, 1812, and had suggested that Sedgwick should meet him on the road from Whittlesea, he received the following reply: "I would have met you with a gig with my whole heart if I could have done it with a good conscience. But consider; I have six pupils, and it is now within three weeks of the examination. An absence of two days would at this time of the year be a great trespass."

[2] From Rev. John Mason, 3 October, 1810.

so much in need, and he was "driving on just as usual[1]". Moreover, his determination to give up the Bar entailed a further, and more important, step, namely, the adoption of the clerical profession, for which he had no very decided inclination. He came, however, to the wise conclusion that it would be best to commence the study of such a subject without delay, and before the end of 1810 he wrote: "I intend to begin my theological labours in about a fortnight. I wish I had a better motive than I have for beginning these labours. I acknowledge the necessity and importance of them, but I feel an indifference to serious subjects which I shall find it difficult to conquer. We must hope for better times." We shall see that the commencement here announced was again and again deferred: but the feeling that he ought to be studying theology was constantly present to his mind, and added to his anxieties. Lastly, he felt acutely the loss of his old set of friends—all of whom, with the exception of Carr, had left Cambridge. It was probably their absence, far more than ill-health or uncongenial work, that made him feel so ill at ease. As usual, he poured his troubles into the sympathising ears of Ainger, now comfortably settled in a curacy at Beccles in Suffolk, who had ended his last letter with these words: "I conclude with earnestly advising you, and the whole set of you, to follow my example, and get away from College as fast as you can."

TRIN. COLL. *February* 11, 1811.
Dear Ainger,

I feel much obliged to you for the circumstantial account you have given me of your situation, your prospects, and your society. You know little of my feelings if you are not convinced that everything relating to yourself must always excite in me the warmest interest. I am frequently gloomy when I consider that in the common course of things I may not hereafter have it in my power to spend many weeks in that society which for the last six years has contributed so

[1] To Rev. W. Ainger, 7 December, 1810.

1811.
Æt. 26.

much to my happiness, I might almost say to my existence. If Carr was away I should consider myself alone at Cambridge.

You will recollect that I made several resolutions to read divinity this vacation. I did begin, but that was all, for I made no progress. Boswell's *Life of Johnson* is the only book I have seen this Christmas. While the men were in the Senate House, I did nothing. The result you will of course have seen. I should otherwise have sent you a tripos in some corner of this sheet. It has been a noble year for Trinity. Armstrong says he fears the glory of St John's is gone for ever[1].

Soon after the degrees were conferred I set off to Town, where I spent a gay and agreeable week; indeed I was engaged out to dinner every day I remained there. Gilby and I one evening left a party early and went to the Opera. I was much more astonished than pleased with the performance. Catalani's powers are certainly transcendent, yet I felt more surprise than delight at her Italian warblings. The music was much too refined for my taste; they sacrifice everything to execution: however I have no right to censure what I do not understand. After taking an early dinner with Harrison[2], I one evening went with Armstrong and Duckworth to Covent Garden. *Cato*[3] and the new pantomime of *Asmodeus* were performed. I am Goth enough to acknowledge myself more pleased with Grimaldi's wry faces than with Addison's declamation.

I returned to Cambridge last Tuesday, and am now leading a life of dull uniformity. I am at present almost prevented from leaving my rooms by a violent cold which has already taken away three of my senses; I can neither hear, smell, nor taste. But while I have one sense left I shall ever remain

Yours sincerely and affectionately,

A. SEDGWICK.

[1] There were 15 Wranglers, of whom Trinity had 7; the 1st, 5th, 6th, 7th, 8th, 9th, 14th: St John's only 2; the 10th and 12th.

[2] Charles Harrison, Trin. B.A. 1810.

[3] A revival of *Cato* with John Kemble as Cato and Charles Kemble as Juba.

In April 1811, after Sedgwick had proceeded to the degree of Master of Arts, he had the right of dining at the High Table in Hall; but these new surroundings gave him but little pleasure, at any rate at first. For a time he enjoyed Carr's society there; but at the end of 1811 or the beginning of 1812 he also left College, and Sedgwick remained alone, to make new friends as best he could. His judgment on his brother Fellows—at least on those senior to himself—was not favourable:

"I find a great want of Carr; for, though I am more at home among our Fellows than I was formerly, I find none amongst them to supply his place. On the whole I have been rather disappointed in the society of Masters of Arts. Many are gloomy and discontented, many impertinent and pedantic; and a still greater number are so eaten up with vanity that they are continually attempting some part which they cannot support[1]."

The University has changed so completely since Sedgwick wrote these words, that a few remarks on University life at the beginning of the present century will be not out of place. In attempting to picture to ourselves what it was, it must be remembered in the first place that foreign travel was impossible, that communication with other parts of England was slow and costly, and that therefore journeys were seldom undertaken. Many Fellows made Cambridge their home, which they rarely left, and died, as they had lived, in their college rooms. Newspapers—such as they were—travelled as slowly as individuals, and the arrival of a letter was a rare event. We have often wondered what the Fellows conversed about at dinner or supper, or at the *symposia* in the Combination Room which in winter filled up the long interval between the two meals. And yet those gatherings were described as cheerful; and Sedgwick himself has been heard to expatiate with delight on the recollection of a certain Christmas, when they had been so fortunate as to secure the company of an

1811. Æt. 26.

---

[1] To Rev. W. Ainger, without date.

Irish Captain, specially famous for his comic songs. Christmas, however, comes but once a year, and the joviality of the twelve days over which its festivities then extended was no doubt enhanced by the dullness of the remaining three hundred and fifty-three.

Again, the refining influence of ladies' society was almost wholly absent. With the exception of the Heads of Colleges, there were very few married men in the University, and the Heads were averse to general society. Dr Mansel, who had daughters to establish, gave a few evening parties ; the other members of the oligarchy thought themselves too important to associate with anybody whose degree was below that of a Doctor, or who had not achieved the dignified position of a Professor. Nor was it the custom for Fellows of Colleges to see anything of the undergraduates. It was rather the fashion to ignore their existence. The old custom of a Fellow sharing his room with one or more undergraduates had died out a century before, and had not been replaced by the frank intercourse which has now become usual, to the common benefit of both. It is evident, from what Sedgwick says of the Fellows, that he knew nothing about them until he was enrolled among their number. There were Fellow-Commoners it is true, who were sometimes numerous ; but the Fellows saw very little of them. From the way in which Professor Pryme speaks of the pleasure which he and some of his friends derived from intercourse with "the most cultivated of the Fellow-Commoners[1]", it is evident that such intercourse was as rare as they found it agreeable.

Few Fellows of Trinity College, except the officials, had any definite occupation. With two exceptions, they were bound to take Holy Orders. Some held small livings in Cambridgeshire, tenable with their fellowships; others, who had been appointed to the office of College Preacher, held more lucrative pieces of preferment. But in neither case was residence compulsory ; parishes were held to be sufficiently

[1] *Recollections*, p. 89.

provided for by the appointment of a curate, and the performance of an occasional Sunday service. With the large
majority of the residents, the fact that they were clergymen
imposed upon them no duties, and effected no difference in
their manners, habits, or language. Men of ambition went
out into the world and boldly courted fortune, as soon as
they had obtained their fellowships—some without even
waiting for that assistance. Those who despaired of success,
or had no energy to strive after it, remained behind. One
resource, and one only, remained to them, the chance of
obtaining a College living; and for this, and for the marriage
which in many cases depended on it, a man would wait, year
after year,

> " Sickening in tedious indolence,
>   Hope long deferr'd, and slow suspense[1];"

till not only had he become unfit for active work in a parish,
but his hopes of domestic happiness had too often ended,
sometimes by mutual consent, sometimes in a sadder way,
by the death of the intended wife.

No wonder that discontent and ill-humour became chronic,
except with those happy dispositions whose natural gaiety
can never be checked; no wonder that those who had to
endure a life which had all the dullness of a monastery
without its austerity or its religious enthusiasm, should
become soured, eccentric, selfish, if not intemperate and
immoral.

We have drawn a gloomy picture, but one which represents, we fear, a painful reality. Before long, however, a
great change took place. The restoration of peace put an
end to the isolation of England, and the Universities, in
common with the rest of the kingdom, shook off their torpor,
and became imbued with new ideas. The Fellows of Trinity
College only a few years junior to Sedgwick were men of
powerful intellect and wide interests; with whom, as it will
be our pleasing task to point out, he made common cause

---

[1] *Ode to Trinity College, Cambridge.* By G. Pryme, 8vo. Lond. 1812, p. 21.

1811.
Æt. 26.

against the dullness of a previous age, and inaugurated the modern development of the University.

At all times, even the most stagnant, politics can rouse the soundest sleeper; and in March, 1811, the University was thrown into an unusual state of excitement by two elections, both of which were contested. The Duke of Grafton, who had been Chancellor for forty-three years, died on the 14th; and the elevation of his son, Lord Euston, to the peerage, vacated one of the seats in Parliament. For the Chancellorship the candidates were Prince William Frederick, Duke of Gloucester, and the Duke of Rutland. The latter had acquired an almost paramount influence in the Town of Cambridge, of which he was High Steward, and on this account was opposed by several prominent tories, and notably by Professor Marsh, who considered that his duties to the Borough would clash with his duties to the University. The Duke of Gloucester, as a staunch Abolitionist, had the support of Mr Wilberforce, and Dean Milner was specially active on his behalf. The Duke of Rutland's supporters, on the other hand, hinted not obscurely that His Royal Highness was in favour of Catholic Emancipation, and called his friends "enemies of the Church". This clever electioneering move was, however, unsuccessful, and the Duke of Gloucester was elected by a majority of 117, 26 March, 1811[1].

For the seat in Parliament the candidates were Viscount Palmerston, who had been Under-Secretary for War since October 1809, and John Henry Smyth, M.A. of Trinity College. The tories made the most of "the ability displayed by Lord Palmerston in the administration of the country," and he obtained a majority of 106 over his opponent, a whig, who had taken no part in public life, and was known only as a good classical scholar[2]. The election—which took place on

---

[1] *The Question Examined, whether the Friends of the Duke of Gloucester in the present contest are Enemies of the Church.* By Herbert Marsh, D.D., 8vo. Camb. 1811. Milner's *Life,* p. 450. Cooper's *Annals,* iv. 495. Wilberforce's *Life,* iii. 502.

[2] Mr Smyth had taken an ordinary degree in 1801, but had obtained Browne's medal for a Greek and Latin Ode in 1799, and for a Greek Ode in 1800.

the day following that of the Chancellor—excited comparatively little interest beyond the walls of Palmerston's own
College. Sedgwick, as the next letter shows, was beginning
to take a keen interest in politics—and he puts the objections
to the Duke of Rutland even more forcibly than Professor
Marsh had done.

1811.
Æt. 26.

TRIN. COLL. Tuesday morning.
[*March*, 1811.]

My dear Ainger,

...The University is already in a ferment; I shall
rejoice to see you whenever you may come, tho' I fear many
of my political lectures have been lost on you. It would
be absurd in any one to wish you to vote against Lord
Palmerston; he is no doubt a highly respectable candidate,
and deserves the support of his college.

The candidates for the Chancellorship are both of our
college. The Johnians in general support the Duke of
Rutland. I am astonished at his impudence in offering himself. If we look to his intellectual attainments we shall find
them beneath contempt. He borrows his influence from that
source which ought to render him infamous. He is one of
the greatest borough-mongers in the kingdom. It is con-
foundedly provoking to the men of our year to be without a
vote. The undergraduates of St John's are all about to be
sent out of College to make room for the Masters of Arts.
As you will be here before next Tuesday, I shall trouble you
no further,

I am Yours most affectionately,

A. SEDGWICK.

You must not tell your friends what I have written. I
shall be prosecuted for a libel if you do.

Notwithstanding Sedgwick's good resolutions, his theological studies made little or no progress. During the Easter
Term he was occupied as usual with pupils, and with preparations for his summer excursion to Bury St Edmunds. When

term was over, he stayed in Cambridge in order to participate in the gaieties of the Chancellor's Installation, and evidently forgot all his cares in the bustle and excitement. "Well do I remember," he wrote in 1864, "the tumult of joy with which I plunged into the festivities of 1811, when the Duke of Gloucester was installed our Chancellor. In those days I was a dancing man, and found it a most happy method of discharging my redundant spirits[1]."

The following letter to Ainger, written at the end of term, opens with such an amusing burst of well-feigned indignation, that we are tempted to regret the loss of the remonstrances against unpunctuality in answering letters which provoked it.

TRIN. COLL. *June* 10, 1811.

Dear Ainger,

Your letter was left in my rooms this morning. When I had read it over, and found your name affixed to it, I could with difficulty persuade myself that I was not deceived by my senses. I am at present a solitary, matter of fact man, little accustomed either to give, or receive, the language of abuse. Your letter is indeed couched in terms of right orthodox scurrility. You have been reading books of religious controversy, I presume. Authors of the description you are now studying I am little acquainted with, and I therefore cannot be expected to express the worst passions of human nature with the same strength and propriety that you do. A momentary irritation was the only effect your impudent scrawl produced ; I resolved to throw it aside and think no more of it. I have, however, allowed my judgment to master my feelings, and resolved to give you an opportunity of explaining away this farrago of false accusations. I have been since endeavouring to account on rational principles for this change in your tone of thinking and of writing. Have you become so far intoxicated by the applause of the gaping crowds at Beccles that you expected to overwhelm me with a torrent of

---

[1] To Mrs Hotson, 5 June, 1864.

eloquent abuse, unaccompanied by reason or truth? You have indeed got some egregious Gospel Trumpeters in your parish. Your sermons make as much noise in the papers as Daffy's Elixir, or Bish's Lottery tickets. I hate such pharisaical blasts; I wonder you have not found means to stop them.

You begin by asserting that I am a sad careless fellow, and that you fear I shall never mend. This from you is too much. On turning over your letters I find that, almost without exception, they begin with apologies for neglect. In one you acknowledge "that the happiness you derive from present objects had filled up every moment of your attention; that you had never found time even to think of those who were absent" (dated Hertfordshire); in another you with all due contrition bewail your offences, calling yourself "a wretched caitiff and a miserable sinner"; in a third you say "that the fulminations in my last letter had roused you as it were from a dream, and at length brought you to your senses". In truth, Ainger, out of the six letters of yours which I have before me I could find more expressions of bitter remorse than you could pick out of all the volumes of the Newgate Calendar.

I should not have brought your own words in judgment against you if you had not set me the example. I still profess myself desirous of continuing a regular correspondence. If you had been guilty of no dereliction of duty, you would have experienced none from me. You say it was my turn to write because you had paid *me* a visit. I recollect no such visit. You did indeed just show your face in Cambridge "to do the devil's dirty work for nothing". But let me assure you that if the taking a coach, and driving across the country to vote for a foolish fox-hunting Duke is hereafter to be considered as an answer to my letters, I shall feel no desire of continuing our correspondence.

I cannot yet propound to you any difficult questions in divinity, because I have not taken the trouble to look for them. I have had five pupils during the whole of last term;

when I was not engaged with them I employed myself principally in reading voyages and travels. I managed to get through ten or twelve quarto volumes. Besides these I have read *Malthus on Population* twice through; he is a delightful author, and has made me a convert to most of his opinions. His maxims are too cold-blooded for a man of your temperament. My more serious moments have been devoted to Xenophon, Tacitus, Virgil, Berkeley's Metaphysical Works, and Paley's Sermons. I am at present engaged exclusively in reading Mathematics by way of preparation for our summer's labours. We shall not remove to Bury till after the Commencement. I have of course regularly seen the morning and evening papers. Such a tide of success has been flowing in upon us, as Perceval says, that even you, dead as you are to all political feeling, must have joined in the general exultation. The reappointment of the Duke of York is a cursed drawback. "The Talents" have behaved infamously; they are nothing better than the vile refuse of a party.

Carr has left Cambridge and all its festivities behind him; he is now in the North, and will continue there during the summer. I almost envy him. At one time I intended to have seen Dent before July, but I could not raise the wind. In truth, Ainger, I scarcely dare appear in the streets; some terrific gaping dun stares me in the face at every corner. Bland is now at Sedbergh. I had some conversation with him before he left us. When speaking of you he seemed quite in a pet, and muttered something between his teeth very much like an oath; he seems resolved never more to think of you, or to write to you, because you never give yourself the trouble of answering his letters. Armstrong has sent me a very long and amusing letter. He was one of the Stewards at the John Port Latin dinner. On returning to Gray's Inn he was obliged to leave a reverend friend of his in the watch-house. Armstrong seems to have made a long speech in defence of this hopeful divine, but his eloquence

produced no effect on the constable of the night. He mentions no names. "The Parson" (he says in one part of his
letter) "has never written to me since he left town." He
complains of you for your indolence. Your friends, you find,
are abusing you with one consent. Don't be so self-com-
placent as to imagine that you are right and they wrong. If
after this prompt and vigorous exertion on my part you
make no suitable return, I shall for the future consider you a
monster of ingratitude.

I am, dear Ainger,

Yours most truly,

A. SEDGWICK.

The reading party assembled at Bury St Edmunds in
July[1], and, let us hope, passed the summer both profitably
and agreeably, but we know neither the names of the pupils,
nor any particulars of their doings, with this exception, that
Sedgwick, as might be expected, made friends there, whom
he visited on subsequent occasions.

Between October 1811 and April 1812 Sedgwick's life
offers no variety. He worked on as usual; denied himself a
holiday in London towards the end of the year—"Five pupils
and an empty purse interpose difficulties not easily got over[2];"
—passed the Christmas vacation at Dent; and finally, after a
Lent term devoted to pupils, and preparation for a summer at
Lowestoft, allowed himself a hasty glimpse of his old set in
London. It is painful to notice that the tone of his letters is
still depressed. The elasticity of youth had passed away, and
had not yet been replaced by the cheerfulness of a man who
is doing work which he enjoys contentedly and resolutely.

TRIN. COLL. February 14, 1812.

My dear Ainger,

I informed you in my last that I should visit Dent
during the Christmas vacation. My anxiety to be off was

---

[1] Miss Margaret Sedgwick directed a letter (31 July) to "Mr Sedgwick, at
Mr Crisp's, Druggist, Bury St Edmunds."

[2] To Rev. W. Ainger, Beccles, Suffolk, 4 December, 1811.

such that I left Cambridge two days before the end of the October term. You will perhaps be surprised when I say that I travelled outside all the way from Alconbury Hill to Ingleton. I had fortified myself with a box-coat of huge dimensions and impenetrable thickness, so that, notwithstanding a keen north wind and hard frost, I found little inconvenience from the weather.

My Father still retains that freshness of complexion, and activity of limb, for which he was remarkable, and, though deprived of his sight, he is not cut off from all communication with books, for my sisters read to him by turns the greater part of every evening. On the whole there are perhaps few men who enjoy a more rational, or a more happy, existence. My mother looks old, though she does not complain of ill health.

At Sedbergh the empire of dulness is firmly fixed. With Stevens I spent some pleasant days; his hospitality, and, above all, his zeal for the interest of his pupils, cannot be too much admired; but yet there is something about him which I neither like nor understand. He has seven daughters and a son, and there is another on the stocks, but of its sex one cannot speak with certainty. Mrs Dawson died about ten days before I left the north. Her dissolution had long been expected. Mr D., if one may judge from his appearance, will not be long in following her. He looks quite cadaverous, and is shrunk into a mere skeleton.

I spent about a week with a man of our college who lives on the borders of Windermere. We took many excursions during my visit to the different Lakes in the neighbourhood. The face of nature is certainly seen to a great disadvantage during this part of the year, yet the excursion—

I was interrupted about 10 this morning by the entrance of a pupil. It is now late, the fumes of wine are in my head, and I am drowsy. It had been my intention to give you a bombastic description of the rugged scenery in the neighbour-

hood of Coniston, but I am now quite disabled. I returned
to Cambridge by Manchester and Leicester. The road is
intolerably bad, and the distance about fifty miles greater
than by Leeds. I was induced to return by this route that I
might call on a clergyman near Macclesfield[1] who formerly
had the delectable office of teaching me A, B, C, and with
whom I spent a month in Cumberland about twenty years
ago. I have only seen him twice since that time and after
long intervals. We soon became as intimate as if our acquaintance had been uninterrupted; and as he is not tormented by that bane of domestic happiness *a wife*, we
contrived, during each of the few nights I was with him,
to keep up the conversation till two or three o'clock in the
morning.

I forgot to tell you that I went out with a gun several
times during the Christmas vacation. I have quite lost the
art of shooting. I had many good shots, and literally killed
nothing. Here I am grinding away with six pupils. Under
such circumstances it is impossible to advance one step. But
I am compelled by circumstances to undergo this drudgery.
When I look back on what I have done since I was elected
Fellow I cannot discover that I have made any proficiency
whatever, or gained one new idea. This is miserable stagna-
tion, but I thank God that I am not yet in the "slough of
despond". I hope for better things. You will undoubtedly
have heard of Carr's appointment to Durham School. I find
a great want of him. He was too valuable a man to be
easily replaced. I think he was in the right to accept
the situation, though at present the emoluments are but
small.

I am, dear Ainger,

Yours most affectionately,

A. SEDGWICK.

---

[1] Sedgwick's godfather, Mr Parker. See above, p. 47.

TRIN. COLL. *Sunday Evening,*
April 19, 1812.

My dear Ainger,

I left Cambridge on Monday week, and arrived here last Thursday, having exhausted my stock of cash and curiosity. I had no time to yawn or flag during the visit, as a succession of delightful engagements presented themselves. Our friends Armstrong, Hargreave, Harrison, Duckworth, Gilby, &c. are all well; with some of these I contrived to breakfast almost every morning during my stay, spent the remainder of the morning in spying farlies[1], and ended at one of the Theatres, the Opera, or the House of Commons, seldom finding my way to bed before two in the morning.

My resolution of spending the summer at Lowestoff[2] is still fixed. We shall probably remain there about sixteen or seventeen weeks; as this is a good long time, perhaps I might engage lodgings cheaper on that account. If a comfortable sitting-room and a bed-room could be procured for a guinea per week I should feel quite happy; if for less so much the better; if for four or five shillings a week more, I should not quarrel with them. Some of the men may perhaps take a house, as you recommend. I should prefer being on my old footing. As you have proposed to look out for me, I feel disposed to accept your offer; but should wish to put you to no inconvenience, especially as more than two months must elapse before I shall think of leaving Cambridge. In regard to the formation of a mess, that will be best done after we arrive at the spot, as two or three of us shall probably leave Cambridge together. At the same time a hint from you may be of good service.

Believe me yours,
A. SEDGWICK.

Sedgwick's interest in the contested election of 1811, and

---

[1] In North Country dialect, a *farley* means a wonder, a strange thing.
[2] Sedgwick always writes Lowestoff, not Lowestoft.

his regret that he had no vote wherewith to oppose the tories, as represented by His Grace of Rutland, have been already mentioned. In the Easter term of 1812 he found an opportunity of exercising his newly obtained rights as a member of the Senate, under political circumstances of more than ordinary interest. Early in the year the propriety of making some concession to the Roman Catholics, in connection with the peace and good government of Ireland, had engaged the attention of both Houses of Parliament. The motions, each of which took the form of a petition for a committee on the state of Ireland, were lost; but it was evident that much of the opposition was directed against the form of the proposal rather than against the matter, and that the question would be brought forward again at no distant date. Under these circumstances the Protestantism of both Universities took fright, and it was resolved to send petitions to Parliament. But it is evident that, at least at Cambridge, those who suggested such a course knew that they were not standing on sure ground. The petition was certain to be opposed, and in fact, as the result proved, ran considerable risk of being rejected altogether. Even the all-powerful Heads of Colleges were not unanimous in its favour. The promoters of it therefore determined to bring it forward as secretly as possible. It was presented to the Senate on Monday, 20 April; but, "it was not till Saturday (18 April) that it was surmised in the University that such a Petition was in contemplation, and it was not till Sunday, a day usually devoted to other concerns, that the promoters of the Petition formally promulgated their purposes." Incredible as this statement sounds, it was made in the House of Lords by the Marquis of Lansdowne, in the terms quoted above, and no one ventured publicly to contradict it. The Earl of Hardwicke (Lord High Steward of the University) had already spoken in the same sense, adding that even the Master of Trinity College had not been trusted with the secret; perhaps because, as Lord Lansdowne "had authority to state", he would have opposed the petition had

he not been accidentally absent.   The document was drawn
up by Dean Milner[1], and, from what we know of his character,
it may be safely assumed that the measures for presenting it
to the Senate were devised by the same person.   The petition
is not in his happiest manner, and the main argument is
curiously fallacious.   It stated, with a specious affectation of
liberality intended of course to disarm opposition, that the
petitioners " have never been adverse to liberty of conscience
in Religious or Ecclesiastical Matters ; that they feel no
uneasiness at the Concession of any comforts or advantages
to their Roman Catholic Brethren" ; but that " the controul
of any foreign Power over the Government of this country
either in Church or State is inconsistent with the first principles of all Civil Government...; that the power of the Pope,
though for various reasons lessened in the public opinion, is
notwithstanding more dangerous to us now than ever, being
itself brought under the control of a foreign and most in-
veterate Enemy."   As Sedgwick's friend Armstrong observed:
" According to the wiseacres who framed that petition, the
Pope's influence increases as all external means of doing
mischief are taken away from him ; if he were only shut up
in a dungeon it follows he would be irresistible."   Notwithstanding the precautions that had been taken, the opponents
mustered in considerable force in both the Houses into which
the Senate was then divided ; and the Grace to affix the
University Seal to the document obtained a majority in the
Regent House of only fourteen, and in the non-Regent House
of only five[2].   That Sedgwick voted in the minority is evident
from the following extract from a letter to Ainger (3 May):

" I had a long letter from Armstrong yesterday in reply to
a much longer of mine.   In truth I had filled to the brim a
large sheet of scribbling paper with abuse of those men who
were instrumental in sending that absurd petition to the two

---

[1] *Life* of Milner, *ut supra*, p. 500.
[2] The numbers were: Regents ; Placet 34, Non-Placet, 20 ; Non-Regents ;
Placet 24. Non-Placet 19.

Houses. Armstrong felt exactly as I did, but, in my opinion,   1812.
acted very absurdly, for he not only shewed my farrago of   Æt. 27.
invectives to our common friends, but, by some means or
other got it conveyed to Mr Whitbread, who in consequence
fired off some bitter invectives in the House against the
patchers of the petition. Fortunately the speech was not
reported ; I say fortunately, because if any circumstances had
transpired by which my letter had been made public in
Cambridge I should have found myself in hot water. After
having committed my assertions to paper, the *onus probandi*
would have rested on me ; and I might have found it devilish
difficult to prove every thing I had asserted. But of all this
mum, mum[1]."

Sedgwick's summer residence at Lowestoft was in every
way successful. He frequently spoke and wrote of it in after-
years, with all the pleasure afforded by a thoroughly agreeable retrospect, and, on several occasions, when prebendary
of Norwich, he spent a few days there in revisiting his old
haunts. Of his nine pupils four at least became his intimate
friends; and some of the resident families, delighted to
welcome a set of cultivated young men to their society,
showed him civilities which he never forgot. Among these
were Dr Smith, afterwards Sir James Edward Smith—the
celebrated botanist, founder of the Linnean Society—and his
accomplished wife. In April, 1865, when all but fifty-three
years had elapsed since his first visit, Sedgwick was at
Lowestoft. Saddened by a drive to several country churches,
where he saw "the monuments of friends of bygone years",
he called on Lady Smith, "the most wonderful woman of her
years that I ever beheld. She is now *ninety-two;* yet her
eyes are bright as diamonds; her face is smooth; there is

---

[1] Armstrong's letter (from which a quotation has been already made) is dated
May, 1812. Sedgwick's letter to him has not been preserved. The petition was
presented to the House of Commons (22 April) by Sir Vicary Gibbs, M.P. : but
no debate is reported. The history of the petition, as stated above, will be found
in *Hansard*, Vol. xxii., pp. 506, 507, 722. It was presented to the House of
Lords (21 April) by the Duke of Gloucester, Chancellor.

a *natural* colour on her cheek ; her voice is full ; her gestures active and firm ; her posture as upright as that of a young woman ; her manner of address happy, kind, and cheerful. She is still very good-looking. When young she was very beautiful, and it was a kind of beauty to last well—somewhat of oriental about it, for when she was a girl she sat to Opie as a gipsy, and it was one of the cleverest pictures he ever painted[1]." The conversation naturally turned on their first acquaintance in 1812, and a few days afterwards she sent him a copy of verses written on the departure of the party by one of her young friends, Miss Ritson. This *jeu d'esprit* —which indicates by the way that their time was not wholly absorbed by the study of mathematics—fortunately records the names of all Sedgwick's pupils.

> Whence comes the deep sigh, whence springs the fond tear?
> Why seems my sad heart now so lonely and drear?
> Why beats it so heavy that once was so gay?
> 'Tis because pleasure flies me. The moralists say
> Did you think it would last?—and I answer them—Nay.
> Yet a sigh of regret will arise in my heart
> When I see that my friends with my pleasures depart.
> At the play, at the ball, in the dance, with the song,
> Our hours have sped gaily and swiftly along.
>
> *　　*　　*　　*　　*　　*　　*　　*
>
> Farewell then to *Belgrave*[2], good wishes attend,
> Good sport in the field, and at home a true friend.
> And farewell to *Sedgwick*, the Mentor who join'd
> With the grave mathematics, the life of the mind,
> Who foremost in whate'er was gay or could please,
> With knowledge join'd mirth, and with study mix'd ease,
> Who so justly the *dulce* and *utile* mingled
> That the harp still was soft and the chords never jingled.

[1] To Mrs Atkinson, 18 April, 1865. Pleasance, Lady Smith, daughter of Robert and Pleasance Reeve, was born 1773, and died 1877, aged 103. Her husband died 1828. The picture was painted 1797, soon after her marriage. It is thus described in *Opie and his Works*, by J. J. Rogers, 8vo. Lond. 1878, p. 161. " Canvas, 29½ in. × 24½ in. Seen to waist, three quarters face to right, dressed as a gipsy, her hat thrown back on her neck, and hanging by a muslin scarf tied in front under her chin ; dishevelled hair about her brow, both hands shown, her right fore-finger resting in the left palm ; an arch smile and pretty face."

[2] William Belgrave, St John's, B.A., 1813.

Farewell too, to *Musgrave*[1], polite and refined,
Like a well-tuned piano the chords of his mind.
Tho' grave never stately, tho' wise n'er pedantic
Tho' devoted to music, yet never romantic.
And farewell to *Peacock*[2], to *Lodge*[3], and to *Case*[4],
Who alike pleased us all, by each good-humoured grace;
And though *Adams*[5] at dancing and Ladies may sneer
Still we'll wish him success in his learned career.
And to *Cook*[6] who to study and books ever true;
To the well-bred, polite, lively *Holder*[7] adieu.
And farewell to *Ingle*[8] of marvellous fame,
By mighty comparisons marking his name,
Who, fond of discussion, would oft raise the smile
And join in the laugh the long hours to beguile.
His mind I for once will attempt to compare
To the great bird of Jove, to the prince of the air,
For ever in *alto* his thoughts will arise
And you must take care, they're not lost in the skies.
Keep his wits 'neath the clouds, for they're monstrously clever,
If they once soar above you, you've lost them for ever.
His mind never free from a thousand vagaries,
You would think he had lived in the age of the Fairies.
But the gay tribes are flying and flitting away
And the brown tints of Autumn no longer may stay.
Stern Winter will come, bid his tempests to roar,
So I'll give a fond tear to the Summer that's o'er,
And sweet retrospection of scenes long gone by
Shall paint, and will raise, or the smile or the sigh,
While poor Lowestoft deserted no gaiety knows
And hears nothing more than the keen blast that blows.
But hope may to future sweet scenes look along
Through the gloom of the winter, whose terrors among
May arrest the sad sigh, make my bosom to swell
And wipe off the tear that accords with FAREWELL.

LOWESTOFT, 1 *October*, 1812.

1812.
Æt. 27.

Of the nine gentlemen here celebrated, Case, Belgrave, Holder, and Adams are unknown to fame; nor do their names occur in after-years in connection with Sedgwick's history. With Cook—who became a Fellow of Christ's

[1] Charles Musgrave, Trinity, B.A. 1814.
[2] George Peacock, Trinity, B.A. 1813.
[3] John Lodge, Trinity, B.A. 1814.
[4] Isham Case, St John's, B.A. 1814.
[5] Richard Newton Adams, Sidney Sussex, B.A. 1814.
[6] Joseph Cook, Christ's, B.A. 1813.
[7] Robert Keyse Holder, St John's, B.A. 1813.
[8] Charles Ingle, Trinity, B.A. 1814.

College—he maintained a correspondence until his death, in 1825, while travelling in the Sinaitic Peninsula. Musgrave (the future Archdeacon), Peacock, Lodge, and Ingle, were all numbered among his most intimate friends. The two former became Fellows of Trinity ; and though Musgrave left College in 1821 to become Vicar of Whitkirk, in Yorkshire, they still contrived to meet frequently. Peacock resided in College until he was made Dean of Ely in 1839. Lodge obtained a Fellowship at Magdalene College in 1818, and was University Librarian from 1822 to 1845. Ingle obtained no University distinction, and left Cambridge as soon as he had taken his degree. But a very warm affection had grown up between him and Sedgwick, and they continued to write letters, and to pay frequent visits, to each other. Ingle was a brilliant, impulsive creature—with a warm heart and a weak head—who clung to Sedgwick's stronger character like ivy to an oak. The friendship, however, was far from being all on one side. Sedgwick had a high opinion of his talents, delighted in his society, and selected him as the confidant of his most private thoughts and feelings. Few lives have opened with brighter promise ; none—as it will be our painful task to tell—have had a more miserable close.

Sedgwick paid another visit to Lowestoft in 1869, and sent a charming description of it to his two American cousins, Mrs Norton and Miss Sedgwick. The letter, dated 8 July, 1869, contains so many references to 1812, that it will be best to quote a considerable portion of it here :

" I have no news to communicate except what relates to myself. I am looking over the broad sea. The sands below my windows are covered by groups of merry children, digging away with their little spades, as lustily as if they thought the fate of England was in their hands. The waves are sparkling in the bright sun, and a great number of vessels are running before a side-wind both north and south, and close in shore, for this is the most eastern point of our Island. I ought to have returned to Cambridge to-day to superintend some important

works going on in my Museum; but I could not resist the
temptation of a run for a few hours to the seaside, that I
might breathe the free air of heaven, and visit some old
friends.

"And I have visited two or three to my joy. Among them
was my dear old friend Lady Smith. She has bright manners,
bright eyes, and clear sight; a face still handsome, and with
healthy colour on her smooth, well-rounded, cheeks. She
hears well, and her voice has a cheerful ring with it. All this
may be said of many English women. But Lady Smith is
one of a million—the wonder of the county, and the charm of
her old friends, for she is now happily wearing her way
through her 97th year! I am old and suffering from the infirmities of old-age; but my friend Lady Smith, to whom I
gave a true-love kiss, is *twelve years older* than myself. Let
not my two saucy American cousins laugh at the thought of
a love-kiss given between two such aged remnants of old
Time's gleanings. Love is the dearest attribute of God.
Like Himself it will last for ever. He may plant it here;
and, if we do our part well, it will have its consummation and
perfection after the wreck of all visible worlds.

"Well! to come back. I am going to have a drive, that
I may revisit the pretty rural spots in this neighbourhood,
where I spent many happy days in my youthful life; and to
which I have often brought my young relations during the
periods of their visits to me at Norwich. I spent 1812 at
this bright little seaport on the Suffolk coast. It was the
last time I ever went out with pupils. The whole summer
and autumn were seasons of intense excitement. No railroads, and no telegrams then. So day by day we went out
to meet the mail-coach, on its first entrance, to catch the
first whispers of news from Spain and Russia. It was you
know the year of Napoleon's invasion of Russia, at the head
of the grandest army that ever mustered within the limits of
Europe. The issue of the contest seemed to involve the very
life and death of old England. And in fits of gloom I some-

times fancied that she must fall before the fortune of the
great conqueror, as so many other powers of Europe had
done before her. But such gloomy visions had one bright
side. I said to myself, if England lose her freedom I will
pack up all I have, and go to settle along with my relations
among the free-men of the United States. We had heard
reports of good news, and I took my stand on a little hill
that overlooks the London road along with my party. Several
hundred of the inhabitants joined us. At length the mail-
coach came in sight, rapidly nearing us. On its top was a
sailor, waving the Union Jack over his head, and gaudy
ribbons were streaming on all sides, the sure signs of victory.
The guard threw down a paper to me, and with it I ran to
the Public Room. There, mounting upon a table, I read to
the assembled crowd the Gazette Extraordinary of the Battle
of Salamanca[1]. The cheers were long and loud, but there
were sobs of sorrow too, for some of us had lost those who
were dear to us. Remembrances of this kind gave a quiet
charm to my sweet drive."

When the party had separated Sedgwick took a short
excursion with his friend Mr Daniel Pettiward[2] to Ipswich
and Bury St Edmunds, and then returned to Cambridge. It
is sad to learn that he still found the place distasteful. At
the beginning of the term, however, he was appointed a sub-
lecturer—an office which, by requiring him to take part in the
College examinations, brought him into closer contact with
the undergraduates, and also with the tutors. From this
period may be dated the commencement of the excellent
understanding which ever after subsisted between him and
the rest of the society.

TRIN. COLL. *Oct. 17th*, 1812.
My dear Ainger,

I met Mr Pettiward at Ipswich as I expected. The
day following we commenced our excursion. We had not

---

[1] Wellington defeated the French at Salamanca, 22 July, 1812.
[2] A member of Trinity College: B.A. 1789, M.A. 1792.

1812.
Æt. 27.

descended more than three or four miles when our vessel stuck fast in the mud. This accident I had from the first expected, as the navigation is very intricate, and the whole crew of the wherry was intoxicated before we started; one of the many blessed effects of a general election. We must have remained in the mud till next high-water, if a very beautiful yacht had not most opportunely made its appearance. One of the gentlemen hailed Mr Pettiward, and, on being informed of our situation, offered to take us on board; we wished good night to our brethren in jeopardy, and most gladly accepted the proposal. Our voyage was delightful in every respect; indeed the accident we had met with gave us a greater zest for enjoyment. If you should ever have an opportunity of making the same excursion you will find yourself amply rewarded for sacrificing some time to it. After taking dinner on board the yacht, and drinking nearly a bottle of Madeira each, we landed at Harwich. Every thing there was in a state of uproar, as the election had taken place in the morning. As we approached the inn we heard sounds of boisterous conviviality. All the gentlemen of the town and neighbourhood were assembled in the upper rooms to celebrate a dinner on the event of the election. Mr Pettiward knew many of the party so well that he did not scruple to introduce me. I flatter myself I chimed in with considerable effect. I never was in better mouth.

The next day was well employed in examining Landguard fort, and many other fortifications in the neighbourhood. The day following we went a short excursion up the Manning-tree river, landed at the other side, and walked to Ipswich, where we arrived late in the evening. I spent a day with Mr P. near Stowmarket, and three at Bury St Edmunds most delightfully, among the friends I acquired there last summer. "Past and to come seem best." This I find confirmed by my own feelings. I am already beginning to complain of Cambridge. This may perhaps be accounted for; few of my friends are arrived, and I am indisposed in consequence of the effects of a severe cold.

Your most affectionately,

A. SEDGWICK.

The term was not many weeks old before the Catholic Question became again prominent. Early in November the ruling party in the University determined to send a second petition to Parliament. The document used in the previous April was slightly altered and enlarged, but the arguments were the same as on the former occasion. It was not, however, proposed to the Senate with the same precautions against discovery; though, when presented to the House of Lords (1 December) Lord Hardwicke complained that "due notice had not been given to the non-resident members of the University of the intention to set on foot such a petition." It appears, however, that a notice of six days, instead of the usual notice of three days, had on this occasion unquestionably been given. The excitement among non-residents, especially among junior Masters of Arts in London, was prodigious; and Sedgwick became the corresponding member of "a confederacy," as one of them called it, established among members of Gray's Inn and the Temple, prominent among whom were Gilby and Armstrong, for the purpose of opposing the petition. He was to send information to four specified persons, who were to inform the rest. By this means, no matter how late the notice of congregation might be issued, it was expected that twenty voters would reach Cambridge in time. These elaborate precautions were rendered unnecessary by the length of notice ultimately given, but a number of non-residents did eventually come up. The voting took place on Wednesday, 18 November, when the Grace to seal the petition was passed in the Regent House by eighteen votes, and in the Non-Regent House by eleven—a result which must have

been singularly mortifying to the enthusiastic partisans of 1813.
toleration[1]. Æt. 28.

At the beginning of May, 1813, Sedgwick broke a blood-vessel in the course of an excursion on the river. One of his usual colds—as usual neglected—had ended in a violent cough; and finally inflammation of the lungs, or something very like it, had supervened. The attack must have been serious, from the way in which his father wrote on receiving the news:

"Notwithstanding your caution to the contrary we were all a good deal alarmed by your first letter. However your last letter, in which you seem so much recovered, has made us in better spirits. I hope by this time you are nearly well, and beg that you will take the earliest opportunity after you receive this of letting us know how you go on, if you have not, before it arrives, sent a letter off.

Your mother joins me in requesting that you will set off for the North as soon as you think you are able, and it is safe for you to undertake the journey. Perhaps the journey and change of air may be of service to you."

That the health of one apparently so strong should break down so completely was a subject of great surprise, and sincere regret, to all his friends; and Ainger with his usual solicitude hastened to him. His place as examiner in the next College Examination was taken by Mr Pryme[2], and towards the end of May he was well enough to leave Cambridge. Indeed he felt so strong that before going away he accepted the laborious office of Moderator for the ensuing year. On his way to Dent, however, while staying at the house of his cousin Mr Mason, in Nottinghamshire, he had a relapse which compelled him to give up all thoughts of the Moderatorship. When he got home his friends were greatly shocked at his appearance, and despaired of his recovery. It was thought that he would become consumptive. Complete idleness, however, did wonders. He provided himself with a

[1] The numbers were: Regents; Placet 52, Non-Placet 34; Non-Regents; Placet 53, Non Placet 42. The above account is derived from the letters addressed to Sedgwick by Gilby and Armstrong; from *Hansard*, Vol. xxiv., pp. 111, 134, 218; and from the *Cambridge Chronicle*, 20 November 1812.

[2] *Recollections, ut supra*, p. 91.

horse (which he named *Caliban* from his ugliness) warranted
to "walk on the edge of a precipice or descend a mountain
with any horse in Christendom", a gun, and fishing-tackle,
and lived in the open air. Still he was subject to relapses,
and it was not till near the middle of July that he was well
enough to take an excursion to the Lakes. As he frequently
visited that district in after years as a geologist, it is interesting to note his remarks upon it when he was simply in search
of the picturesque and wonderful.

"About five weeks since, I left Dent on an excursion
which afforded me the most unmixed delight. My head-
quarters were at Hawkshead, in the centre of the Lakes.
The lady whom I visited allowed me to pursue every scheme
of pleasure my fancy could suggest, so that during the first
fortnight not a single day elapsed without my going out on
some expedition from which I generally returned in the
evening fatigued without being exhausted. My appetite,
strength, and spirits, every moment improved; my mind
therefore was in a state above all others suited for receiving
pleasing impressions. Objects combining everything to
delight and astonish were each day presented to my view.
I always left them with reluctance, and started the day
following with increased avidity.

During the first fortnight I not only visited all the principal
Lakes in the neighbourhood, but I also spent an afternoon
among the ruins of Furness Abbey, descended into the iron
mines in that neighbourhood, and explored, I think, more
than a quarter of a mile under ground, the copper-mines near
Coniston. After the celebration of the regatta at Windermere I left Hawkshead, and commenced my last and longest
journey. After passing for twenty-five miles through a most
rugged and desolate country I arrived at the foot of Wastdale.
The day following I ascended about three miles by the side
of a noble expanse of water. On the opposite side a mountain
entirely covered with rock rose abruptly, to the height of
2000 feet, from the margin of the lake; when I arrived at the

top I found myself in an amphitheatre of pyramidal mountains, all of which terminated in sharp rocks. Rudge makes the perpendicular height of the ridge considerably more than 3000 feet. I was this day quite alone, and felt an elevation of spirits infinitely greater than when I was surveying the sweet country on the shores of Windermere. I sometimes thought that the lovely scenery in that neighbourhood made me melancholy. In Wastdale everything was rugged and sublime; about a dozen farm-houses seemed to make the desolation more visible. From this place I pursued my way through a pass in the mountains by a most frightful road, which literally winds among masses of rock which have fallen from the precipice above. I remained on horseback till I became quite giddy ; then I dismounted, and led the way to the top along a road in many places not more than a foot and a half wide. A false step would have been destruction to my beast.

August 19. I was prevented from coming to a conclusion on Tuesday, and am now so busy that I shall be compelled to take the remaining part of my tour for granted. Suffice it to say that I arrived from Wastdale-head to Borrowdale without breaking my neck. I proceeded through the most romantic vale in England to Keswick, where I met with a Cambridge man with whom I prosecuted my journey for the next four days. We walked round Bassenthwaite, ascended Borrowdale, explored the lead-mines, perhaps a quarter of a mile from the surface of the earth : and descended to Buttermere through a fine pass in the mountains. The celebrated Mary has now no personal charms to recommend her, but I must recollect her with gratitude, for I was wet to the skin, and she spliced me with a dry shirt, and her husband's breeches. We scraped an acquaintance with the parson, who introduced us to a party of ladies who were going on Crummock Lake on a fishing excursion. We caught few fish, but had lots of conversation. We returned to Keswick, and went together to Ambleside, where we parted, and where I considered my tour

as terminating. I remained one day near Windermere, which
I spent in sailing, and had I imagine very nearly been upset,
and in the evening I dined with Wilson, the author of *The
Isle of Palms*[1]. He is a clever convivial man, much superior
to what I should have expected from his poetry[2]".

After this he accompanied his sisters to Gordale, near
Settle, and ended the vacation with a visit to his friend Carr,
whose appointment to the headmastership of Durham School
has been already mentioned[3]. In October he was well enough
to return to Cambridge[4], whither he was accompanied by
his brother James, who was entered as a sizar at St John's
College.

For the moment Sedgwick had recovered, but he felt the
effects of this illness throughout his life. " I have been liable
to attacks of congestion ever since 1813," he wrote in 1864 ;
and he told Dr Hooker that he had become "unfit for
sedentary labour after 1813." Fresh air and regular exercise
became indispensable to him; and frequent attacks of ill-
health warned him that he must seek for a profession which
would keep him out of doors for several months in each year.

---

[1] John Wilson, better known as Christopher North, published *The Isle of
Palms, and other Poems*, in 1812.

[2] To Rev. W. Ainger, 17 August, 1813.

[3] The Rev. John Carr died in November, 1833. Mr Robert Surtees, in a
letter dated 15 November of that year, thus sums up his character: "He was
eminently distinguished as a mathematician, and was, perhaps, not less distinguished as a classical scholar. He peculiarly excelled in pure Latin composition, but his private character was to me his chief recommendation. Kind,
unobtrusive, gentle ; most pure, most blameless, wrapped up in domestic feeling,
and neither meddling with nor caring for the world, I firmly believe he had not an
enemy....There was a quiet, unobtrusive independence about him, which I never,
perhaps, saw equalled ; a purity and delicacy of mind and manners arising from
the union of a complete education and the most perfect sense of honour, united to
the most unaffected simplicity of manner. As a schoolmaster, he never looked
like one, but he sent good scholars to Cambridge. No boy ever left Durham
without loving him." A short time before his death Carr had been made
Professor of Mathematics in the newly founded University of Durham. A
monument, by Rickman, was erected to him in Durham Cathedral. Taylor's
Memoir of Robert Surtees, ed. Raine, p. 439.

[4] Many of the details respecting this illness have been supplied by Miss
Sedgwick.

This conviction, more than any other consideration, determined him to become a candidate for the Professorship of
Geology in 1818; and, after forty-nine years devoted to that
science, he could say with thankfulness: " Geology has been
a hard task-mistress, but she has paid me nobly in giving me
health, which I had utterly lost before I put myself under her
robust training[1]."

The immediate effects of Sedgwick's illness are painfully
apparent in the listlessness and want of energy from which he
suffered during the next two years. He was evidently obliged
to take constant care of himself, and felt indisposed for
any intellectual exertion that was not absolutely indispensable. He performed his duties as College examiner; tried to
make his brother James work; and did some desultory reading, partly theological, on his own account. But the old
animation was gone, and the letters which formerly he wrote
so frequently, ceased altogether. And yet events took place
which under more favourable circumstances would have
furnished him with subjects for long and entertaining narratives. There was the great frost of January, 1814, when no
coal-barges could get up the river, and he was obliged, as he
has been often heard to say, to burn his gun-case and some of
his chairs[2]; and in July the dinner in Hall to the Chancellor
and Marshal Blucher, on leaving which he saw the old soldier
snatch up an attractive damsel who was pressing forward to
get a good look at him, and give her a kiss. But on these
trifles, and on graver matters, he is equally silent. In the
summer months a curacy was suggested to him, at a small
place in Northamptonshire, where the church, he was assured,
would not be too much for his lungs[3]; but he had not energy
enough to submit to even the small amount of work then

---

[1] To Mrs Norton, 27 August, 1867.

[2] Professor Pryme records (*Recollections*, p. 113), that the scarcity of coal
was so great and the cold so severe, that some of the trees in the grounds of
St John's College were cut down for fuel, and at all the colleges men sat two or
three together in one room.

[3] From Rev. W. Ainger, 2 May, 1814.

required of candidates for ordination, and so the proposal was declined. In the spring of 1815 he attended the lectures of William Farish, Jacksonian Professor of Natural and Experimental Philosophy, probably more for the sake of obtaining intellectual gratification without thought or trouble, than for any more serious reason. But, listless as he was at Cambridge, the moment he got to Dent he became a different man. In September 1814 Ainger and his brother James paid him a visit there, and they took a pedestrian excursion together. The letter in which he directs James Ainger how to get to Dent shows how difficult it was in those days to reach remote parts of England.

"The distance of our village from Ingleton is not more than eight miles. If the weather should be very favourable, I should be happy to meet you there, and walk home with you over the mountains. On horseback we perhaps might break our necks. If the weather should be bad it will be better to go on to Kirkby Lonsdale, which is about eleven miles from Dent. The road from thence is passable by a carriage, though some parts of it are as steep as any house-roof in Whittlesea. We could easily contrive to meet you there with a horse. At all events it will be necessary that your portmanteau should be conveyed to Kirkby Lonsdale or Kendal, and from one of those places to Dent by a carriage. If your Reverend brother come along with you it will be best to post it from Kirkby Lonsdale, and then you will both arrive bag and baggage."

During the expedition Sedgwick was evidently better able to bear fatigue than either of his friends. Their melancholy experiences evidently made a deep impression on him; for, writing to one of them thirty-seven years afterwards, he laments that he had then become "a poor walker compared with what I was when I took you (I think in 1814) to the Ulverston slate-quarries, and witnessed on your face an expression of anything but comfort[1]."

When his friends had left him, Sedgwick made a short

[1] To James Ainger, Esq., 29 December, 1851.

tour in Yorkshire. He went by way of Leeds to York, which he had evidently never visited before, and was profoundly impressed by the "vastness, harmony of proportion, and rich- ness of execution of the cathedral," though the exterior was "miserably spoiled by a set of dirty houses which press so close upon it that no situation can be found from which the whole pile can be seen at once." Next morning he went on to Hull, taking Beverley by the way. "The minster is a beautiful specimen of the lighter gothic, though in some places much injured by certain modern improvers. The west front is exquisite, infinitely superior to that of Westminster Abbey, which was intended to be an imitation of it. Hull is a fine town of the kind, though, I should think, not a very comfortable residence. The docks are beautiful, and the Humber is superb. As we were crossing it the merchant vessels seemed almost to cover its surface. In our passage we met a steamboat, working its way at a prodigious rate against wind and tide." From Hull he travelled to Cambridge by Peterborough, where he thought the cathedral "very fine; but whoever compares it with that at York should be stoned for blasphemy[1]."

As soon as the Peace of 1814 had been concluded, English- men, who had been shut up within the narrow limits of their own island for more than a quarter of a century, felt a natural anxiety to explore the Continent, and above all France, with whose Government we had been so long at war. Several of Sedgwick's intimate friends had taken the first opportunity to go abroad—Bland to Switzerland, Charles Musgrave to France, whence he wrote exceedingly interesting letters. He notes the richness and the beauty of the country, the cheapness of living, even to a foreigner, and the efforts that were being made, at the sea-ports and elsewhere, to recover from the disastrous isolation to which France, like England, had been so long subjected. The effects of the war were still painfully apparent: "In passing through the country I have

[1] To Rev. W. Ainger, 31 October, 1814.

seen few men in the fields compared to the number of women. The effective male population of France seems to have been completely drained by the war." English troops were still marching through France on their way home from Spain, and the sight of them did not increase the respect for the restored Government, which was on all grounds unpopular, especially with the army : "No one endeavours to conceal that the military are highly disaffected to the King, and are panting for a change." Nor were traces of the Revolution wanting : " Every town on the Loire presents a monastery in ruins, and the chapels of the convents have been principally converted into Diligence-offices and stables. The men who have conducted us down the Loire from Nantes remember to have seen boats crowded with Royalists sent to the bottom, while the Revolutionists fired at them from the shore if they raised their heads above the surface of the water[1]." These and other letters, similar in tone, though less graphically written, were not without their effect on Sedgwick, and in the early spring of 1815 he was planning not only a tour in France, but a first essay in authorship. By way of preparation he spent part of the Lent Term in learning the French language, and in reading French books. His projects, however, were overset by Napoleon's escape from Elba; and he was obliged to exchange Paris for Bury St Edmunds, and French literature for divinity.

"My French journey is quite hopeless. If that were the only ill effect to be expected from Bonaparte's visit it would not, perhaps, be much lamented. So many books of travels have been written that my little volume could not have contributed much to the general stock of information.

"On Monday I shall begin to read divinity. I have got about twenty folios out of our library. By the way, if I should go to Bury I cannot take them with me. I must therefore begin the Monday following. No matter[2]."

---

[1] From Charles Musgrave, dated " Paimbœuf, 18 July, 1814."
[2] To Rev. W. Ainger, 30 March, 1815. Mr Ainger was now curate of Hackney, London.

In the course of the Lent Term Cambridge was visited by an unusually severe epidemic of typhoid fever. At first little or no attention was paid to it, but the deaths of several members of the University, and the serious illness of others, created so great a panic, that on the 3rd May—in deference to public opinion rather than from any conviction of the necessity of such a measure—the Senate agreed to allow the term then commencing to all undergraduates, who, having kept the previous term, chose to absent themselves. Such a permission was of course largely taken advantage of, and the public life of the University practically ended on the day the above Grace passed[1]. Even the Commencement was celebrated with maimed rites ; "not a single fiddle[2]" was heard, and the non-resident Masters of Arts who had journeyed to Cambridge in search of pleasure went back to their chambers and their parishes sorely disappointed. Sedgwick, warned by his own experience of "the Cambridge fever" in 1804, had beaten a hasty retreat at a somewhat earlier period, and gone down to Dent with his brother James, and his friends Lodge[3] and Sheepshanks[4].

1815.
Æt. 30.

DENT, *May* 22, 1815.

Dear Ainger,

The escape of Bonaparte from Elba was not more sudden and unexpected than my flight from Cambridge. I had resolved to remain at all events ; and as I knew that my brother James was not likely to read much at home (for God

---

[1] The exact words of the Grace are worth quotation : " Cum opinio, quamvis vana forsan sit, latè pervagata est, et multorum animis insedit, eam esse hoc tempore loci hujusce gravitatem, ut Juvenes Academici non sine vitæ periculo in eo commorari valeant ; Placeat vobis, quo Parentum potius medeamur anxietati, quam quòd rei necessitas ita postulare videatur, ut scholares in quacunque facultate, vel absentes, hunc terminum complevisse censeantur, ea tamen lege ut nemini qui superiore termino abfuerit, hæc Gratia sit profutura."

[2] From Rev. W. Ainger, 29 July, 1815.

[3] John Lodge, Trin. Coll. B.A. 1814 ; afterwards Fellow of Magdalene Coll. and University Librarian 1822—1845.

[4] Richard Sheepshanks, Trin. Coll. B.A. 1816, a distinguished mathematician, afterwards Fellow, Secretary to the Astronomical Society, and Founder of the Sheepshanks' Fund and Exhibition.

knows his literary zeal is not very great in any place), I was resolved still more on his account to remain in College. The death of six or seven members of the University during the latter part of the Lent term excited so much apprehension, that the walls of many colleges were quite deserted in the vacation. About the beginning of this term two men, one of Christ's and the other of Emmanuel, died the same morning. The fever also began to make its appearance in our College, which till then had escaped all contagion. These melancholy appearances excited so much alarm that many members of our College, and among the rest your present correspondent, were persuaded to scamper off, without having time to give notice of their departure. At the time of our departure we expected to be called back to the University to keep the latter part of the term. The death, however, of another member of Emmanuel College, and one or two new cases of fever, induced the Senate about the beginning of this month to give the term altogether.

Since my arrival I have as usual been engaged in a variety of employments. I have now read the whole of *Gil Blas* twice over, and am reading for the second time certain parts of *Telemaque*. A short analysis of certain chapters in Beausobre's *Introduction* has employed part of my time, but the task is a confounded dry one. Old Carr had the kindness to lend me a very elaborate commentary on St Luke, of which I purpose to make myself master. I hope also to read the Bishop of Lincoln and Burnet on the Articles in course of the summer.

My sisters received five or six weeks since a second package of books for the use of their Sunday Schools[1]. Upwards of 40 children attend, and many of them have made a highly satisfactory progress ; but more of this in my next.

Yours ever,

A. SEDGWICK.

---

[1] The school had been established in, or about, 1813. *Memorial*, p. xi.

1815.
Æt. 30.

This quiet life was interrupted by one great excitement—the news of the victory of Waterloo—which Sedgwick had the pleasure of bringing himself to Dent. How this came to pass can fortunately be told in his own words.

"At that time we had a post three days a week, and each of those days, to the great comfort of the aged postman, I rode over to Sedbergh to bring back the newspapers and the letters to my countrymen. Gloomy reports had reached us of a battle and a retreat; but another and greater battle was at hand; and on one of my anxious journeys, just as I passed over the Riggs, I heard the sound of the Sedbergh bells. Could it be, I said, the news of a victory? No! it was a full hour before the time of the postman's arrival. A minute afterwards I saw a countryman returning hastily from Sedbergh. "Pray what means that ringing?" I said. "News, Sir, sich as niver was heard before: I knâ lile about it; but the Kendal postman has just come an hour before his time. He was all covered with ribbons, and his horse was all covered with froth." Hearing this, I spurred my horse to the Kendal postman's speed; and it was my joyful fortune to reach Sedbergh not many minutes after the arrival of the Gazette Extraordinary which told us of the great victory of Waterloo.

"After joining in the cheers and congratulations of my friends at Sedbergh, I returned to Dent with what speed I could; and such was the anxiety of the day that many scores of my brother Dalesmen met me on the way, and no time was lost in our return to the market-place of Dent. They ran by my side as I urged on my horse; and then, mounting on the great blocks of black marble, from the top of which my countrymen have so often heard the voice of the auctioneer and the town-crier, I read, at the highest pitch of my voice, the news from the Gazette Extraordinary to the anxious crowd which pressed round me. After the tumultuous cheers had somewhat subsided, I said: ' Let us thank God for this great victory, and let the six bells give us a merry peal.' As I spoke these words an old weather-beaten soldier who stood

1815.   under me said : ' It is great news, and it is good news, if it
Æt. 30. bring us peace. Yes, let the six bells ring merrily ; but it has
been a fearful struggle, and how many aching hearts will
there be when the list of killed and wounded becomes known
to the mothers, wives, and daughters of those who fought and
bled for us ! But the news is good, and let the six bells ring
merrily !¹' "

Just a month after this stirring scene had been enacted
before the public of Dent, private news came to Sedgwick
which must have caused a nearly equal excitement in the
home-circle at the vicarage. Hudson, who had succeeded
Jones as Senior Tutor of Trinity College, had accepted the
vicarage of Kendal, and the approaching vacancy of the tutorship—which could not be delayed, at the farthest, beyond the
expiration of the year of grace—had been taken advantage of
by the Master and Seniors to extend the number of tutors to
three—a step rendered necessary by the increase in the number
of students. Two of the assistant-tutors, John Brown, and
James Henry Monk², were promoted, as was usual, to be tutors,
and the offices thus vacated had to be filled up. The selection
of assistant-tutors was at that time left to the tutors under
whom they had to work, and Monk, with the consent of his
colleague, invited Sedgwick to accept one of the vacant posts.

From Rev. J. H. Monk.

CAMBRIDGE, *July* 15, 1815.
Dear Sedgwick,

You are probably aware that J. Brown and myself have
been appointed to succeed Mr Hudson as joint Tutors. I now
write to solicit your aid as mathematical assistant tutor, and to
express the earnest hope of my colleague and myself that you may
find it consistent with your views and your feelings to accept that
office. We are well aware of the great advantage which would
accrue to the College from your assistance in the tuition, and are
persuaded that the appointment would be in the highest degree
gratifying to every person who is interested for the prosperity of our
society.

It is, I believe, Brown's intention to take the higher mathematical
departments himself, though not exactly in the same arrangement as

¹ *Supplement,* p. 38.   ² Professor of Greek 1808—1823.

Hudson has done. But I am certain that in this, as well as in every other respect, your wishes will be consulted as far as possible, and nothing will be omitted, to make the situation as agreeable and as beneficial to you as circumstances can allow.

1815.
Æt. 30.

It is necessary to mention that the original arrangement made by Mr H. for giving up the tuition has been altered. He had agreed to resign it to us at Michaelmas. This agreement he has broken, and has induced the Master to suffer him to retain the pupils till Christmas, except the freshmen, whom he will give up to Brown and myself at Michaelmas. Though this alteration is naturally a subject of displeasure to us, on many accounts, yet it will make no difference in the department of which you are invited to accept: since Brown is decidedly of opinion that it is desirable for the person who gives the mathematical lectures to the freshmen, to begin with them in October.

I am aware that your health has not been strong, but I trust that it is now better, and at all events that there is no fatigue in the office alluded to which you need apprehend.

I shall leave College next Friday morning—but a letter directed to me here will follow me—so do not hurry in deciding, should you hesitate. But I am sanguine in hoping that we may have the benefit of your valuable assistance.

I am, dear Sedgwick, with great regard,
Ever your most faithful Servant,
J. H. MONK.

This letter must have been in every way gratifying to Sedgwick. The proposed office offered him a congenial occupation with less wear and tear than private tuition, and besides might lead eventually to a Tutorship. He did not, however, accept it without due consideration. The answer which he finally wrote—so modest, dictated by so nice a sense of honour, and so perfectly straightforward—must have confirmed Monk in the belief that he had made choice of a colleague who would do him credit.

DENT, *July* 28, 1815.
Dear Sir,

Our communication with the post town is so irregular that I have only had your letter a day or two, but should think myself unpardonable, if I any longer postponed my reply. I should indeed have written sooner if I had not been desirous of first speaking to some of my friends in this neighbourhood. They have all earnestly wished me to accept the mathematical lectureship.   The reluctance which I at first expressed to them did not arise, let me assure you, from any dislike to the appointment, more especially when the offer of it was made in terms so very flattering to my feelings ; but from an indescribable fear of not being able to discharge the duties of it properly.   You are aware that for the last three years my health has not allowed me to attend seriously to any mathematical subjects.   I am, however, now much better, and have besides most solemnly resolved within myself not to retain the office of lecturer unless I feel myself on trial quite equal to the duties of it.   If I acted otherwise I should show myself very little deserving the good opinion you have so handsomely expressed, and of which you and your colleague have given me so substantial a proof.   I hope you will both accept my thanks, and at the same time my congratulations on your appointment to the Tutorship.

It will of course be necessary for the Bishop[1] to confirm my appointment; but you or Mr Brown will know how to speak to him on the subject.   I hope you will have the kindness to write again when all is finally settled.

Since I left Cambridge I have been leading the most retired life possible.   But even in this corner of the world we are all overjoyed at the great events which have been passing on the Continent.   Every individual in the Island must have exulted at the exploits of our brave fellows.   By the way I had forgot that there may perhaps be one or two exceptions in the Combination Room of Trin. Coll.   I expect some friends when the moorgame season commences, in whose company I shall probably not spend a very sedentary life.   Early in September I propose to return to College.   In a few weeks after that time I shall hope to have the pleasure of meeting you.

Believe me, Dear Professor,

Your most faithful servant,

A. SEDGWICK.

[1] William Lort Mansel, D.D. Master of Trinity College, was Bishop of Bristol from 1808 to his death in 1820.

TRIN. COLL. *September* 23, 1815.

Dear Ainger,

I have been leading a very active, though perhaps not a very profitable, life since I received your last letter. Two friends from Nottinghamshire spent the latter half of August with me in Dent. We were constantly out on the moors, and killed a good many birds. About a fortnight after their departure I turned my face towards the South. I had several reasons for leaving Dent so soon. In the first place I have accepted the office of mathematical lecturer under our new Tutors, and I was desirous of having some time to prepare for my new duties. In the next place I was desirous of getting James up to College, where he will, I hope, adopt new habits. He has been doing nothing in the country this summer. Besides, I hoped to have some sport in this neighbourhood before the confinement of lectures.

I left our friends well, and as for myself I am in better health than I have enjoyed for three years before. Bland is returned from France. He does not open out freely, but he looks well. My sisters' school flourishes as well as they can expect. I attended almost every Sunday during the summer, and heard one of the classes. They are desirous of instituting some little rewards occasionally among the children, and have petitioned the assistance of some of the good folks in the parish. I have promised a guinea, which I mean to spend in books. Can you recommend me to any cheap shop? Does the National Society print any books which would answer the purpose? I intend, if they will admit such a heretic, to be a subscriber to the Society for Promoting Christian Knowledge. Could you propose me?

Yours ever,

A. SEDGWICK.

TRIN. COLL. *November* 29, 1815.

Dear Ainger,

I fear you will complain of me for not having written sooner. Indeed I have no excuse to offer. As Hudson still

continues in College, and the labours of the mathematical
lecturer are consequently divided between Brown and myself,
Euclid and the first part of Algebra have fallen to my share.
If you inquire what I have been doing, I can hardly tell you.
My health has, however, been much better than it was last
year. The run in the country last summer has quite set me
up. Father Bland and your other college friends are I believe
well ; though I am sorry to say that one or two cases of fever
have made their appearance in the University during this term.

It is now near 12 o'clock, and my fire is almost out. You
must therefore allow me to finish by assuring you that I
remain yours truly and affectionately

A. SEDGWICK.

In the summer of the following year (1816) Sedgwick
spent four months in France, Switzerland, part of Germany,
and Holland. Travelling was slow in those days, and,
though he was so long abroad, he did not see more than
would now be accomplished in less than half the time. Nor
has he left any detailed account of what he did see. This is
the more provoking, as he was a keen and accurate observer,
and was travelling at a time of special interest. He kept a
journal, it is true, but unfortunately it was only written up at
intervals—occasionally very long ones—and therefore is little
better than a record of places visited, people met, and discomforts endured in the inns and on the road. Page after
page, especially at the commencement, is filled with dis-
jointed notes, which make us suspect that, when he started,
he had still some idea of writing a book, and was jotting down
heads of paragraphs, and fragmentary details, to assist his
memory on a future occasion.

He embarked at Brighton on June 22nd, and reached
Dieppe on the following evening at 9 o'clock. There he
joined his friend Edward Valentine Blomfield, Fellow of
Emmanuel College[1], his pupil Lord Charles Murray, and

---

[1] A distinguished classical scholar (B.A. 1811), younger brother of C. J.
Blomfield. He died, of a fever caught abroad, 3 October, 1816, a few days after

another Englishman, and in their company journeyed by way 1816.
of Rouen to Paris, which was reached on the afternoon of Æt. 31.
June 27th. Here we will give a specimen of the diary :

"27. Early walk in Mantes. Church injured by the Revolution.
Churches in Normandy very beautiful. Down the valley of the
Seine to Meulan. Mons. Wastel the priest. Destruction of his
church and his sufferings. Dialect of the Normans. St Germain.
Royal Palace. Panorama from the top. Immense work on the
Seine at Marly. Enter Paris. Palais Royal. Dinner surrounded
by [disreputable characters]. A general description of the valley of
the Seine. Riches of the country—no appearance of depopulation
or misery."

A fortnight was spent in Paris ; but bad weather, a
rooted prejudice against the Roman Catholic religion, and a
cordial hatred of all the ways and works of the French,
prevented him from enjoying it as much as he did a few years
afterwards. Entries of this sort frequently occur: "A French-
man will never pretend to be ignorant; but he will rather lie
than make that confession. Insolence of the soldiers;" "Con-
temptible character of the French. No display of loyalty on
this occasion (a review of the National Guard by the King)";
and in one of his letters, though he admits that "the beautiful,
gay, and profligate city of Paris is a noble capital," he adds,
" but the people are so abominable and detestable that there
can be no peace for Europe if they are not chained down as
slaves, or exterminated as wild beasts." He took a master
in the French language, and worked hard at the usual sights,
among which the method of instructing the deaf and dumb
pursued by the Abbé Sicard—successor to the famous Abbé
de l'Epée—seems to have interested him more than Notre
Dame or the Louvre. He also visited the gaming-houses, then
so numerous, the cafés, and the principal theatres ; saw Talma
twice in *Manlius Capitolinus*[1] and Mlle. Mars as Elmire in

his return to Cambridge. An interesting memoir of him, by J. H. Monk, is in the
Museum Criticum, of which he had been one of the founders, i. 520.
[1] A tragedy in verse by Antoine de la Fosse, sieur d'Aubigny (born 1653,
died 1708) a contemporary and imitator of Racine. The play is poor, with a
single fine scene, in which Manlius by means of an intercepted letter discovers a

1816.
Æt. 31.

Tartuffe. This latter occasion was of unusual interest. The performance had been commanded by the Duc de Berri, on whose entrance, accompanied by his wife, there was a good deal of enthusiasm, repeated when the famous lines were spoken, which have been applied to so many different persons, and have glorified such opposite principles :

> Remettez vous, monsieur, d'une alarme si chaude,
> Nous vivons sous un prince ennemi de la fraude, etc.

About the middle of July, accompanied by William Hodge Mill[1], then a Junior Fellow of Trinity College, Sedgwick started for Switzerland. A weary journey of six days brought them to Lyons ; whence, after a brief rest, they proceeded to Geneva. The bad weather followed them, and, besides this drawback to their enjoyment, the place was "filled with a set of lounging impertinent English coxcombs, who appear to go abroad for no other purpose than to disgrace their country." There he met by appointment, John Haviland[2], Fellow of St John's College, with whom he made a tour through Switzerland, for the most part on foot.

"We started from Lausanne on the second of August, and walked to Vevay, a beautiful small town near the head of the lake. Next day we proceeded in a voiture to Martigny, through a valley infinitely more beautiful than anything my imagination had ever formed. Of the Alps I had formed a good general notion. One can conceive an outline varied in every possible form ; a man can imagine a mountain four times as high as any he has seen ; but of that exquisite perfection of scenery which arises from contrast and combination, no one can have any perfect notion who has not been in Switzerland. If I attempt to describe these delicious

conspiracy against his life, organised by his friend Servilius. He hands the letter to Servilius and bids him read it aloud. When he has finished Manlius exclaims "Qu'en dis-tu?" The great success of the revival is said to have been entirely due to the expression of Talma's face while Servilius was reading, and to the tone in which he uttered the above words.

[1] Regius Professor of Hebrew 1848—1854.

[2] Professor of Anatomy 1814—1817; Regius Professor of Medicine 1817—1851.

scenes, I should only use certain general terms which would
convey no distinct meaning.

"Next morning we started for Chamouni, with a guide and
three mules. After having ascended for some time we entered
the pine forests. These forests are constantly broken in upon
by small patches of cultivated ground. If the soil is capable
of producing anything, the Swiss are sure to find it out. You
observe on the very confines of perpetual snow small wooden
cottages, many of which are only inhabited during the summer.
They drive up a certain number of goats or cows sufficient to
consume the vegetation ; and when that is finished descend
again into the valleys. I was also much pleased with observing
the mode in which the Swiss have cultivated some mountains
which in any other country would have been quite unproduc-
tive. They have erected a number of strong walls parallel to
the horizon on the sides of their most rugged hills, which by
those means become divided almost from top to bottom into
a series of steps or platforms, the top of each wall being on
a level with the field immediately above it. Each of these
small slips of ground is cultivated with the utmost care ; and
the whole mountain-side presents the appearance of an immense sloping garden. With such habits of industry, and
such a country, the people can never be uninteresting. We
were much pleased with the honest simplicity and kindness of
this people, which was rendered still more agreeable by being
contrasted with the unmanly insolence of the French. I find
I am forgetting myself and running out into general observations. I must pull up, and proceed with my journey. After
having traversed the pines we reached the forests of larch
trees which in this country are always found near the extreme
limits of vegetation. Some of them were of an enormous size.
We could not help observing the effects of the winter storms
in these wild regions. Sometimes several acres of trees are
cleared away in one night. We remarked also a passage
formed through the forest by an avalanche of the preceding
winter, which had literally forced its way to the bottom of the

1816.   valley.   After having traversed this second valley we as-
Æt. 31. cended a second and higher ridge, and at length emerged
from the forest, and found ourselves in the regions of perpetual snow.   We soon gained the top of the Col de Balme,
and had before us perhaps the most glorious mountain
scenery in the world.   On the right were a ridge of lofty
mountains, whose pointed summits rose far above the limit
of perpetual snow, before us were the beautiful villages and
fields of Chamouni, and on the left were the pinnacles of
Mont Blanc, rising to a height of more than twelve thousand
feet above the level of the valley.   In our descent to the
village we passed three glaciers.   There is a very fine one
below the village, which we visited that day.

"Early next day we started for the sea of ice.   It takes
about two hours' good work to climb up to it.   You are,
however, well rewarded for your labour.   A few stunted
larches mark the limit of vegetation.   After you enter the
valley, everything is rude, barren, and desolate.   We descended on the ice, and were amusing ourselves with throwing
lumps of ice down the deepest crevices we could discover,
when the rain began to fall in torrents, and soon drove us
among the larches.   We afterwards descended by a steep
path along the side of the ice to the bottom of the valley.
The lowest parts of these enormous glaciers appear to me by
much the most interesting.   One cannot form any perfect
notion of the depth of the sea of ice, or of its general magnitude, but no one can see without astonishment huge blocks of
ice, some of them coming down into the even fields, piled
one upon another to the thickness of some hundred feet, and
extending for many leagues in the channeled sides of the
mountain.   In the lower part of the glaciers large masses are
continually rolling down the hill with a loud rumbling noise,
which adds much to the effect produced by such savage
scenery.   We returned next day, though not by the same
route, to Martigny.

"On the 7th, by the help of our mules, we ascended in

eleven weary hours to the Mont St Bernard. We were well
rewarded for our exertions. About half way up we met two
monks, one bearing a banner with a picture of the Virgin,
and another with a crucifix, heading about two hundred
people dressed in white. They consisted of a set of country
people who had gone up to the convent to kiss the image of
St Bernard, and to beg for his interest to get them some fine
weather. We travelled over snow for about three miles
before we reached the convent. The monks received us with
hospitality, and even with politeness. One of the monks
walked with us over a lake, at that time frozen four feet
thick, to the ruins of an ancient temple. We took a peep
into the north of Italy, returned, and dined, or rather supped,
in hall with our new friends. The Prior was fortunately
there—he does not commonly reside—a pleasant, well-
informed man as I should wish to meet. We retired early—
rose next morning at four, and were much astonished and not
a little pleased to find two honest monks up, with some warm
coffee and toast, to see us well off.

"I have a great deal more to say, but my paper is nearly
over, and I have already stolen an hour from my sleep. I
must therefore content myself with saying that we got safe
down; that we went up the Valais; crossed the Gemmi, and
were nearly frozen among the sleet and snow; that I had
nearly broken my neck, and that I did break my crupper in
endeavouring to follow a mad English sailor across a precipice; that we have seen the lakes of Thun and Brientz, the
glaciers of Grindelwald, and the town of Berne; that we
marched across the country from Berne to Lucerne by the
help of the sun and stars, inasmuch as we neither knew the
language of the people, nor the names of the towns we were
to pass through; that we reached Lucerne this morning[1];
that we are off for the Devil's Bridge tomorrow; and lastly
that I am

Yours ever, A. SEDGWICK.

[1] This letter, to the Rev. W. Ainger, is dated Lucerne, 17 August, 1816.

10—2

It is worthy of remark that the future geologist, though
filled with enthusiasm at the first sight of the Alps, says not a
word about their physical structure, nor does he appear to
have been more surprised than any ordinary tourist by the
novel spectacle of a glacier. After some further adventures
in Switzerland the travellers proceeded down the Rhine to
Cologne, whence Haviland returned to England (10 September) and Sedgwick went on alone to see something of Holland. Writing from Leyden (19 September), he says:

"The Dutch I have found a mighty comfortable, sober-
mannered, old-fashioned, people. In the towns you see great
signs of active industry, though there everything goes on in a
quiet orderly manner. It is however in his country-house
that you see the animal in all his glory. By the side of his
canals you see him enthroned amidst clipped hedges, sedge,
and duckweed; he is so grave and immoveable that at
first you might easily mistake him for a smoking automaton.
When you approach him you find his face the very picture of
internal comfort. I had a deal of conversation with one of
these comfortable-looking gentlemen in my way down the
canal to Amsterdam. From his appearance I should conjecture that he was first cousin to a burgomaster. He asked me
if I thought the Swiss villages as beautiful as the Dutch. I
answered that I thought the Swiss villages much more beautiful; and then proceeded to describe some of them. The
Dutchman puffed the tobacco once or twice with somewhat
more violence than before, and then observed that these things
were well enough to look at, but after all Holland was the
country to live in. The English are in great favour in this
country. I have met with the greatest civility in all the parts
of it I have seen. The inns are so excellent that I am more
than half a convert to the old citizen's opinion."

Sedgwick was always fond of art, and his diary shows that
he took considerable interest in the Dutch School, which
would of course be quite new to him. When he got to
Antwerp, he criticises Rubens and Vandyke with an acuteness

which shows a remarkable natural aptitude for grasping a painter's characteristics :

"One striking character of Rubens' pictures, is animation. He always chooses a moment when some great event is taking place, and represents it with vigour and truth. He groups well—but there is almost always a want of delicacy and variety in his female figures. He is a great master of colour, but often seems only to have painted for distant effect. I have, however, seen some pictures of his finished to the last degree, and which appear to me to equal anything I have ever seen in the richness, the disposition, and the harmony of the colours."

From Antwerp, though he had been unwell for some days, Sedgwick persisted in going to Brussels, to have a look at the field of Waterloo. A sharp attack of fever—due to a neglected cold, incessant exposure, and hard travelling—was coming on, and by the time he reached *La Belle Alliance* he was so ill that he could with difficulty hold up his head. Next day a Belgian physician prescribed herb-tea, which did more harm than good; and the result might have been serious had not an English physician been discovered, whose remedies, though severe, were efficacious. After nearly a week's confinement, the patient, sorely enfeebled, was allowed to travel, and proceeded by way of Calais to Dover, where he landed in safety, after a passage of only five hours and a half, on the 17th October.

TRIN. COLL. *March* 16, [1817].

Dear Ainger,

Some months have elapsed since we last parted, and I have still to reproach you for not having written to me. Pray what have you been doing? How do you get on with your new college[1]? How do you like your curacy, your living, &c. &c.?

My own history may be written in a very few words. Since we parted I have not been a single day out of College. During the Christmas vacation I was present at divers parties of whist, in which I did not join; I witnessed the scaling of

[1] The new Theological College of St Bees in Cumberland, of which Mr Ainger had just been made Principal.

1817.
Æt. 32.

many pies, of which I did not taste; and I saw huge bowls of punch emptied without venturing even to sip of them. Notwithstanding this system of mortification, I spent my time pleasantly enough, for my health was better than it has been for the four preceding years. During the greater part of this term I have been slightly indisposed, principally I believe from the fatigue of lecturing; I am beginning now to see land, for the Easter vacation commences before the expiration of next week. James will this week be very busy with the Fellowship examination. I am of course most anxious about his success. John is now here, but is not able to sit, as he was unfortunately elected to a Heblethwaite scholarship, which prevents his being a candidate for either of the Lupton Fellowships which are now vacant. Bland looks dismally; he has for some time been tormented with a jaundice; he is now, I hope, convalescent. Carr was up last vacation; he had only been married about a fortnight, and was apparently quite happy, and most anxious to be back to his wife. He came to take possession of a small College living[1] to which he had been presented a few weeks before. The old Knight[2], I think, died after you left us; Haviland has got the Regius Professorship. The Anatomical Professorship will be vacant next term. Clark and Woodhouse are again candidates. There is, I believe, no doubt whatever of Clark's success[3]. If you should be anywhere in this neighbourhood about the time of the election I hope you will come up and give him your vote.

Yours ever,

A. SEDGWICK.

[1] Hatfield Broad Oak in Hertfordshire.
[2] Sir Isaac Pennington, M.D., Fellow of St John's College. He was Professor of Chemistry 1773—93, and Regius Professor of Physic 1793 to his death, 3 February, 1817.
[3] William Clark, M.A., Fellow of Trinity College, and John Thomas Woodhouse, M.D., Fellow of Gonville and Caius College, had been candidates for the Professorship of Anatomy in 1814, when Haviland was elected. On this occasion Woodhouse retired, and Clark was elected without opposition.

Since Sedgwick obtained his Fellowship in 1810 the study 1817.
of Divinity had formed part of his programme of work, as Æt. 32.
often as he drew up that record of good deeds to come. He
was always going to begin; he intended to be ordained before
the summer was out; and the like. Now, however, it had
become impossible for that procrastinating spirit of his, which
furnished him with so many jokes at his own expense, to
frame any excuse for further delay. The stern voice of the
statutes under which Trinity College was then governed pro-
claimed that all the Fellows save two should be in Priest's
Orders within seven years from the full completion of the
degree of Master of Arts, under pain of forfeiting their
Fellowship. No time was therefore to be lost; and on July
20th, 1817, having obtained letters dimissory from the Lord
Bishop of Bristol, Master of Trinity College, he was ordained
deacon by Bishop Bathurst of Norwich. His companions in
a postchaise thither, were Charles Musgrave, and Mill, his
fellow-traveller for a portion of the previous summer. Mill,
who had already begun the oriental studies in which he afterwards obtained such distinction, beguiled the tedium of the
journey by translating, for the amusement of his companions,
a tale from an Arabic manuscript[1].

The greater part of the Long Vacation was spent in the
North, which he had not visited since 1815. There he began
to perform the duties of a clergyman, in the shape of writing
and preaching sermons at Dent. In the course of the summer
he found time for an excursion to the Lakes with Charles
Musgrave; paid a visit to Ainger at St Bees; and later in the
year visited Ambleside, where he provokingly just missed the
pleasure of being introduced to Wordsworth, with whom he
afterwards became so intimate[2]. Besides these occupations,
he got his usual shooting at Dent, for the last time, as events

---

[1] To Miss F. Hicks, 27 December, 1853. Sedgwick was admitted to Priest's
Orders on Sunday 15 February, 1818, at Quebec Chapel, London, by the Lord
Bishop of Salisbury.
[2] To Rev. W. Ainger, 6 November, 1817.

proved—but he never forgot the pleasure he had derived from
that sport. In 1866—forty-nine years afterwards—happening
to write to a friend on the first day of September—he said:
"In early life I used to count much upon this day, for I was
a keen sportsman till I became a professed Geologist. So
soon as I was seated in the Woodwardian Chair I gave away
my dogs and gun, and my hammer broke my trigger. My
sporting days ended with the autumnal season of 1817[1]." So
long as he was employed in this way we hear nothing about
his health; but as soon as he got back to Cambridge he
began as usual to feel ill again. The work was no doubt
severe. "I am as usual employed two hours every morning
in lecturing to the men of the first and second year, and every
other day we are engaged about two hours and a half more in
examining the men of the third year. We are besides employed at least three hours in the evening in looking over
their papers[2]." His relations had already urged him to
resign his lectureship, and rusticate for the rest of his days.
To those who knew him, idleness and Sedgwick is such a
strange conjunction, that it sounds wonderful that even
paternal solicitude should have suggested it. It is fortunate
that he turned a deaf ear to these admonitions; had he not
done so he might have missed the golden opportunity which
shortly presented itself.

Early in the Lent Term of 1818 it was whispered in
Cambridge that the Rev. John Hailstone, one of the Senior
Fellows of Trinity College, who had been Woodwardian
Professor of Geology since 1788, and must therefore have
reached the ripe age of fifty-eight, was proposing to take to
himself a wife—a step which would *ipso facto* render the
Professorship vacant by the provisions of the Founder's Will.
Sedgwick at once made up his mind to be a candidate, for
reasons which are best stated by himself in the following
letter:

[1] To Rev. J. Edleston, 1 September, 1866.
[2] To Rev. W. Ainger, 6 November, 1817.

TRIN. COLL., *March* 19, 1818.

Dear Ainger,

I sent a letter to St Bees about five months since which most probably never reached its address; I should otherwise most assuredly have had an answer from one of my most *punctual correspondents.* But change of place, and change of time, and change of circumstances, are enough to work stranger changes than even this, and may, after all, have broken in upon those punctual business-like talents for which my old friend was most deservedly in good repute. But enough of other people, let us talk about ourselves. If thou art a priest, so am I, and, if thou art a Professor, so I fain would be. I don't suppose you have so entirely forgot Cambridge as not to feel some interest in our proceedings.

We were very busy in the October term with the subjects of lectures: for, besides the ordinary course, we established additional examinations for the men who were going out. We have certainly reaped the fruits of our labours, for we turned out the Captain of the Tripos with eight other wranglers at his heels. Since that time we have got both the medals, the Pitt Scholarship, and the first on Bell's foundation. Notwithstanding this blaze of honours I am most heartily sick of my connexion with the Tuition, and only wish for an adequate motive for resigning all hopes in that quarter. Now such a motive will probably present itself; for it is generally expected in Cambridge that the Woodwardian Professorship will be vacant by the marriage of Hailstone. In case that event should take place I mean to offer myself as a candidate for the vacant appointment. It would be quite premature to commence a general canvass; I have therefore only written to my personal friends, requesting them to give publicity to my intentions, in a way too most likely to promote my interests. What do you think of the business? If I succeed I shall have a motive for *active* exertion in a way which will promote my intellectual improvement, and I hope make me a happy and useful member of society. I am not such a fool as to suppose that my present employment is useless; and my pecuniary prospects are certainly better than they would be if I were Woodwardian Professor. Still, as far as the improvement of the mind is considered, I am at this moment doing nothing. Nay I often very seriously think that I am doing worse than nothing; that I am gradually losing that little information I once had, and very sensibly approximating to that state of fatuity to which we must all come if we remain here long enough. If you were two hundred miles nearer you might perhaps serve me with a vote. As it is let me have your opinion of the matter in the first place, and your good wishes in the second. There will probably be several candidates. Evans of our college means to offer himself. Carrighan of St John's[1] has been written to. He is now at Rome, and is expected back in a month or two.

Yours ever

A. SEDGWICK.

His friend's answer was rather lukewarm. Ainger approved his purpose, but added : " I should be quite delighted with it if I did not find, on consulting the Cambridge Calendar, that the salary is only £100 a year: yea, indulge me, as a Benedict, in saying further, that I am sorry to find you must, if successful, resign your honours whenever you follow your predecessor's example! But, notwithstanding these drawbacks, you have my most hearty wishes for your success." The conclusion of the letter is significant, as showing the view then taken of Geology: " I really think the pursuit of mineralogy will suit you to a hair, as I take it for granted that it will sometimes lead you to pick up stones, as well as to range them in your lecture-room." Other intimate friends to whom he wrote at this early date were more enthusiastic. Armstrong and Duckworth began to canvass Members of the Senate in London, and were successful in obtaining numerous promises of support against all comers. Bickersteth for instance pledged

---

[1] Arthur Judd Carrighan, Fellow of St John's College, B.A. 1803.

himself at once: "I am quite sure," he said, "that Sedgwick would not propose himself if he did not judge himself to be the proper person; and, if that is his opinion, I have no doubt of the fact."

In the first instance the only candidate who appeared to have any serious intention of going to the poll was Robert Wilson Evans, Fellow and assistant-tutor of Trinity College. One of the two Musgraves had been thought probable; but there is no evidence that he had any such intention himself, and the same may be said of Carrighan of St John's College. Evans was a dangerous opponent. He was a man of high character, deservedly popular both in his own college and in the University. Before long a third candidate appeared, George Cornelius Gorham, Fellow of Queens' College, who afterwards became celebrated for his long doctrinal controversy with Bishop Philpotts of Exeter. He had been third wrangler in 1809, and could therefore show a better place in the Tripos than Sedgwick. Moreover he was reported to have "been studying Geology for a long time[1]"—an important point of which his friends did not fail to take full advantage. Sedgwick could make no such pretensions—nor indeed could Evans—but it was specially unfortunate that one of Sedgwick's two opponents belonged to his own college; for, as one of his most active supporters observed, it "destroyed that corporate spirit which induces men to inconvenience themselves to attain an object about which individually they care nothing[2]."

Professor Hailstone having sent in his resignation (1 May), Sedgwick issued a short circular, dated on the same day, addressed to those whom he thought likely to support him. After announcing the vacancy, he said:

"The kind assurances of support which I have received from many Members of the Senate, have induced me to declare myself a Candidate for the appointment. I am at the same time aware, that I have no right to found my expectations of success on support

---

[1] Pryme's *Recollections*, p. 135.   [2] From R. B. Armstrong, 8 May, 1818.

1818.
Æt. 33.

derived from the partiality of personal friends. Let me then assure you that no one can appreciate more highly than myself, the great responsibility attached to the office for which I am now soliciting. I venture, therefore, to ask for the honour of your Vote and Interest at the ensuing Election; pledging myself, in the event of my success, to use my best endeavours to discharge the important duties of the Professorship, and to carry into full effect the intentions of its Founder."

We have not seen the circular issued by either of Sedgwick's opponents, but both of them evidently took advantage of the vagueness of his pledges to state explicitly that they intended to deliver lectures—a move which created a diversion to their side—for Armstrong warns Sedgwick (9 May) that: "the promise to lecture given by your opponents is considered by their supporters as greatly in their favour; and perhaps with those that do not know you the maxim *dolus in generalibus* may do you some disservice." It therefore became necessary to correct the erroneous impression which had got abroad; and, the 21st day of May having been fixed by the Vice-Chancellor for the election, Sedgwick issued a second circular (14 May) in which he informed his supporters of this fact, and added:

"I have pledged myself, in the event of my success, to use my best endeavours to carry into full effect the intentions of the Founder of the Professorship. In making that pledge, I more especially wished to refer to a clause in the Will of Dr Woodward, by which it is provided that a Course of Lectures be annually read on some subjects connected with the Theory of the Earth. I am happy in having an opportunity of giving this additional explanation of my views respecting the important duties attached to the office for which I am now soliciting."

A postscript to this circular announces " that Mr Evans of Trinity College is no longer a candidate." He had retired from the contest after a comparison of votes, which Sedgwick, or some of his friends, had induced him to agree to. Similar proposals were made to Mr Gorham, but in vain. Sedgwick's friends therefore, though they felt that after Evans' resignation he was "out of all hazard", being compelled to continue the contest, determined to make his majority as large as possible; and, in addition to his resident supporters, arrangements were made for twenty London voters, most of whom were barristers, and could ill afford to lose even a single day, to travel to Cambridge in post-chaises or on horseback, record their votes, and return at night. The position of affairs, on the eve of the election, is graphically described by Gorham to his father:

1818.
Æt. 33.

QUEEN'S[1] COLLEGE, CAMBRIDGE,
17 May, 1818.

My dear Father,

Evans gave in on Thursday. I instantly got the Clare and Bene't resident Voters, but no others. St John's is against me. In fact, except a few stragglers, I have only Queen's, Catharine, Peterhouse, Clare, and Bene't. You shall have a note on Thursday night, though the event is not doubtful. I can only *reckon* on 50 votes, and Sedgwick has promises of 190. Nevertheless I will, *on principle*, carry on the contest, and go through the disagreeable business of the poll. Sedgwick is put up by a large College, merely as a *man of talent*, who *can* soon fit himself for his office. For myself, I feel a conviction that few persons in the University have followed up the Science more sedulously than I have. If, therefore, the Electors choose to dispose of Woodward's funds upon the shameful principle of *influence* against *qualification*, I will drive them to the necessity (which I know they wish to avoid) of recording their votes at a poll, which may be published if I like it—not that I intend to take that step. Some few (like our dear friend Farish[2]) were taken in by anticipation: but the greater number avow the precedent of Bishop Watson.

It has been clearly expected that I should give in; to obviate any such rumour my notice in the Cambridge paper was worded in the form in which you see it[3]. Clarke[4], I suspect, has given me his *name*, but not his *interest*. The Trinity men employed him on Friday to persuade me to agree not to call in out-voters. I rejected any such arrangement.

You may rely on the number of votes, 60 to 200, being nearly correct, even if I push my minority to the utmost. To say nothing of the prejudices against a small College, and a methodistical one— and my having little acquaintance in the University—I feel that I have been left to myself. While Sedgwick's printed letters were

---

[1] Mr Gorham always, on principle, wrote Queen's, not Queens', College.

[2] William Farish, Fellow of Magdalene College, then Jacksonian Professor. Charles Farish, Fellow of Queens' College, voted for Gorham.

[3] Cambridge Chronicle, 15 May, 1818. "We are authorised to state that the Rev. G. C. Gorham, Fellow of Queens' College, decidedly intends to continue the contest for the Woodwardian Professorship."

[4] Edward Daniel Clarke, Professor of Mineralogy 1808—1822.

1818.
Æt. 33.

underwritten by friends, only *one* of mine had that advantage—and, excepting Mr Holmes and Dr Ingle, I do not believe that *one* member of the Senate has canvassed for me.

Your affectionate Son,

G. C. Gorham.

The result showed the correctness of these anticipations. Sedgwick polled 186 votes to his opponent's 59, a conclusion on which Gorham made the following comment in the promised note to his Father : " In the result I feel perfect satisfaction, though certainly not in the circumstances connected with it. The plain fact is that Sedgwick had all the influence of his College, and that of St John's exercised their influence against me as being a Methodist. Disagreeable as the day has been, I am glad I drove the matter to a Poll[1]."

A summary of the votes taken gives the following results :

	S	G	Votes
Proxies	4	1	5
Peterhouse	4	7	11
Clare Hall	1	9	10
Pembroke Hall	5	0	5
Caius College	7	2	9
Trinity Hall	5	0	5
Bene't College	2	2	4
King's College	8	0	8
Queens' College	0	19	19
Catharine Hall	4	4	8
Jesus College	7	1	8
Christ's College	11	0	11
St John's College	32	4	36
Magdalene College . . .	7	2	9
Trinity College	67	3	70
Emmanuel College . . .	11	1	12
Sidney College	6	1	7
Downing College	2	0	2
Commorantes in villâ . . .	3	3	6
Total	186	59	245

It was natural that a disappointed candidate, smarting under a sense of undeserved wrong, should call Sedgwick's

---

[1] These letters—extracts from which are printed above—have been most kindly lent by the Rev. G. M. Gorham, Vicar of Masham, Yorks.

success "an instance of favouritism[1]." An examination of the
above summary, however, shows that he not only polled more
votes in his college than his opponent did in the whole
academic body, but that there was a general feeling throughout the University in his favour. The colleges of Pembroke,
Trinity Hall, King's, Christ's, and Downing voted "solid" for
him; he had a majority in those of Caius, Jesus, St John's,
Magdalene, Emmanuel, Sidney; those of Corpus Christi and
St Catharine were equally divided, contributing respectively
two and four to each side; while Gorham had a majority
only in his own college, in Peterhouse, and in Clare Hall.
Further, those who care to go through the names recorded in
the poll-book will find that Sedgwick's majority was not
merely strong in numbers; he had on his side most of those
who were distinguished in the University by their position or
their attainments.

Gorham is probably right in saying that his own claims
were never fairly considered. Nor is it unlikely that the
strong evangelical tone of Queens' College at that time,
taken in conjunction with Dr Milner's personal unpopularity
with Liberals, and with most of the Fellows of Trinity
College, the Mastership of which he had twice tried to obtain,
may have done him some disservice. But we may safely assert
that the election was virtually decided by Sedgwick's personal
character. In the next chapter it will be shown that the
successive Woodwardian Professors had done little or nothing
to justify their appointment. To this discreditable state of
things the University not unnaturally wished to put an end;
and Sedgwick, with his fiery energy, and reputation for
thoroughness in whatever he did, seemed to be the man most
likely to do this necessary work in a completely efficient
manner.

In attempting to form a just estimate of the qualifications

---

[1] These words occur in Gorham's note to his father, dated 21 May, 1818, from
which an extract has been already quoted.

of the two candidates, we must discard the ideas which we
now attach to the term geology, and recollect that at the
beginning of the present century it was regarded as little
better than a subordinate department of mineralogy, which,
from its practical usefulness, and the beauty of the substances
with which it dealt, had become popular at an early period.
It was considered to be the business of a geologist to investigate the mode in which the earth had originated, and the results
of these speculations may be seen recorded in various essays
called *Theories of the Earth*. When, therefore, we find Gorham
credited with "a long study of geology", and are told that
some of Sedgwick's personal friends, among whom was Mr
Pryme, thought his claims so strong that "it was only just to
vote for him", we are led to suspect that an acquaintance with
mineralogy must have caused this favourable opinion.   This
theory is supported by the fact that Dr Edward Daniel
Clarke, Professor of Mineralogy, voted for Gorham—though
in the above-quoted letter he makes light of his support—
and that the same person was also proxy for Sir Joseph
Banks, President of the Royal Society, whose action would
naturally be governed by the opinion of the official representative of mineralogy at Cambridge.   This, however, is mere
speculation, and may be erroneous.   On the other hand, we
have Gorham's own deliberate statement, in the letter quoted
above, that he had "sedulously" studied geology ; and we
have been informed that he had worked at the physical
structure of Scotland, and of parts of Yorkshire.   But, unfortunately for his own reputation, he had never published
any geological papers ; and, in the absence of the direct proof
of his acquirements which these would have given, the
worthlessness of his geological knowledge has been too
hastily assumed from Sedgwick's celebrated account of himself and his opponent, which is still remembered in the
University : "I had but one rival, Gorham of Queens', and he
had not the slightest chance against me, for I knew absolutely
nothing of geology, whereas he knew a good deal—but it was

all wrong[1]!" This remark, however, was not made seriously, and it would be unjust to Gorham's memory to quote it as a deliberate judgment, without making a large allowance for that departure from literal truth which is permitted to a brilliant antithesis. That he had a genuine love for natural science may be taken for granted, for he was a good practical botanist, and had formed a valuable collection of plants in the course of an extended tour in Switzerland in 1810 and 1811.

1818.
Æt. 33.

While the contest was proceeding Sedgwick is reported to have said : " Hitherto I have never turned a stone; henceforth I will leave no stone unturned," and his contemporary Mr Pryme amplifies the idea of thoroughness conveyed by this sentence into the following statement:

" The latter [Sedgwick] professed to know nothing of the subject, but pledged himself, if elected, to master it, and to resign the assistant tutorship in order that he might give the more complete attention to it[2]."

This passage contains several inaccuracies. If Sedgwick professed ignorance, he had the good sense to reserve such professions for private conversation with his intimate friends ; his public utterances contain no reference to it. Nor did he state, as we have seen, that he meant to master the science of geology. All he said was that he would deliver public lectures " on some subjects connected with the Theory of the Earth." Neither did he announce his intention of resigning the assistant tutorship if elected : we know that he had long been anxious to do so ; and, as a matter of fact, he did resign in the course of the following Long Vacation, and Mr Whewell was elected in his room. But he never pledged himself to such a course.

At the same time there is no evidence that he had ever troubled his head with any cosmical speculations. The word

---

[1] To this story Mr G. M. Gorham adds the following delightful anecdote: "Did not such logic warrant an ancient inhabitant of Dent, himself a stone-breaker, in his reply to my pilgrim-inquiry in 1874, 'Have you ever heard of a native here called Adam Sedgwick?'" "What ! d'ye mean the Perverser?"

[2] *Recollections*, p. 135.

S. I.   11

1818.
Æt. 33.

"strata" occurs in one of his letters from Switzerland, but, with that exception, there is no evidence that he had given things of the earth a moment's consideration. It has been recorded, on his own authority, that he collected fossils at Dent when he was a child; but, had he lived on the sea-coast, he would probably have picked up recent shells; and the former habit no more indicates a future geologist than the latter a future conchologist. The time that he could spare from mathematics he devoted to general literature. He had indeed, as we have seen, attended the lectures of William Farish, Jacksonian Professor of Natural and Experimental Philosophy, in 1815. But the object of those lectures was to exhibit "the application of Chemistry to the Arts and Manufactures of Britain," by means of a series of models of the machinery employed. The method was novel, and we have the authority of Professor Willis for stating that the illustrations were ingeniously contrived, and the lectures generally instructive; but, beyond the fact that Part I. of the Syllabus is headed *Metals and Minerals*, and that such subjects as *The structure of the Earth, Strata, Dislocation of the Strata*, appear as sub-headings, there is nothing in the whole course to suggest geology[1]. Professor Hailstone, Sedgwick's predecessor, did not lecture. An interleaved syllabus of Dr E. D. Clarke's lectures on mineralogy, enriched with copious notes, shews that he attended him for at least one course; but Professor Clarke was no geologist. Nor do any of his friends, when writing to him about his chances of success, refer to his special knowledge of the subject as a reason for supporting him. Mr Daniel Pettiward, for instance, says (May, 1818):

[1] A general view of the course is given in the *Camb. Univ. Calendar*, 1815, p. 38. Farish had been Professor of Chemistry, 1794—1813, and, on finding "the province of reading lectures on the principles of Chemistry already ably occupied by the Jacksonian Professor [F. J. H. Wollaston, Trin. Hall, 1792—1813] was therefore obliged to strike out a new line." When elected to the Jacksonian Chair he continued his former course, as may be seen by comparing the Calendar for 1815 with that for 1802, p. 24. See also : *A Plan of a Course of Lectures on Arts and Manufactures, more particularly such as relate to Chemistry*. By W. Farish, 8vo., 1821.

"It is my Intention to Enlist myself under your Banners, in great hopes that one of my Favourite pursuits, from your Activity of mind and the Genius you possess for General knowledge, may not be hid in a Napkin, but that the world may be better for the fruit of your Labours." 1818. Æt. 33.

This is very different language from what he would have used had his correspondent been thoroughly conversant with even the little geological knowledge of those days. Sedgwick's intimate friend Carr too, writing a letter of congratulation (26 May), says:

"I suppose you will be busily employed this summer in the pursuit of your new studies; for this purpose I venture to recommend the North as the most proper place, both as it abounds in those productions of nature which will now more particularly engage your attention, and as it will at the same time afford me the opportunity of seeing you."

Moreover Sedgwick himself, in his first letter to Ainger on the approaching vacancy, dwells on the "motive for active exertion," and the intellectual stimulus, which the Professorship would give him; but says not a word about his wish to cultivate a science which he had already begun. Nor should his impaired health, and the fatigue which mathematical teaching caused him, be left out of consideration. His eagerness to escape from an uncongenial occupation has been already mentioned more than once.

Precedents were not wanting at Cambridge for the election of a man of ability to a Professorship in a subject of which he knew nothing. Bishop Watson, to whom Gorham refers, was made Professor of Chemistry in 1764, and says of himself:

"At the time this honour was conferred upon me, I knew nothing at all of Chemistry, had never read a syllable on the subject, nor seen a single experiment in it; but I was tired with mathematics and natural philosophy, and the *vehementissima gloriæ cupido* stimulated me to try my strength in a new pursuit[1]."

Dr E. D. Clarke's knowledge of mineralogy was thoroughly unscientific, and in fact he was only saved from mistakes by

[1] *Anecdotes of the Life of Bishop Watson*, ed. 1817, p. 28.

11—2

1818.   the interposition of a friend, the Rev. John Holme, Fellow of
Æt. 33. Peterhouse, to whom he submitted his syllabus, and the
outline of his lectures[1]. Notwithstanding these drawbacks
both these gentlemen filled their lecture-rooms; Watson
advanced his subject scientifically; and Clarke was successful
in creating a general enthusiasm.

It is almost impossible, at a distance of just seventy years
from the period we are investigating, to obtain personal
recollections of Sedgwick at the time of his election. One
valuable testimony has, however, been placed in our hands by
the Rev. Leonard Blomefield, who, as the Rev. Leonard
Jenyns, established a high scientific reputation as a systematic naturalist. He attended Sedgwick's lectures "not more
than a year or two after his election to the Professorship," and
is under the impression "that he was not a mere learner
himself at the time; he seemed a master of the subject, and
his lectures were earnestly listened to, as well as earnestly
delivered." Mr Blomefield is further of opinion "that though
Sedgwick had not made Geology much of a study, nor learnt
its details to any great extent *practically* in the fields, before
he was admitted to the Professorship—he was fairly acquainted
with the subject in a general way, and took a great liking to
it, or he would not have offered himself for the chair[2]."

Our own opinion is, on the whole, for the reasons mentioned above, opposed to that of Mr Blomefield, and in
favour of the notion commonly accepted in Cambridge, that
Sedgwick got up his subject *after* his election. It will
be our business to trace, in subsequent chapters, the
gradual development of his geological knowledge; for the
present we will content ourselves with quoting a passage from
his latest work, dictated only a little more than three months
before his death, in which he sums up the purpose he set
before himself at the outset of his career, and which domi-
nated his long academic life:

[1] Gunning's *Reminiscences*, ed. 1855, ii. 195.
[2] From Rev. L. Blomefield, 20 August, 1887.

"There were three prominent hopes which possessed my 1818.
heart in the earliest years of my Professorship. First, that I Æt. 33.
might be enabled to bring together a Collection worthy of
the University, and illustrative of all the departments of the
Science it was my duty to study and to teach. Secondly,
that a Geological Museum might be built by the University,
amply capable of containing its future Collections; and
lastly, that I might bring together a Class of Students who
would listen to my teaching, support me by their sympathy,
and help me by the labour of their hands[1]."

[1] Preface to *A Catalogue of the Collection of Cambrian and Silurian Fossils
contained in the Geological Museum of the University of Cambridge*, by J. W.
Salter, 4to. Cambridge, 1873, p. xxxi. The preface, by Professor Sedgwick, is
dated 15 September, 1872.

CHAPTER V.

SKETCH OF THE LIFE AND WORKS OF DR JOHN WOODWARD. HIS
TESTAMENTARY PROVISIONS. ARRIVAL OF HIS CABINETS. A
ROOM CONSTRUCTED FOR THEIR RECEPTION. SEDGWICK'S
PREDECESSORS: CONYERS MIDDLETON; CHARLES MASON;
JOHN MICHELL; SAMUEL OGDEN; THOMAS GREEN; JOHN
HAILSTONE. ORDERS AND REGULATIONS SANCTIONED IN 1818.

WE have now reached a point in our narrative at which it
is desirable to sketch Dr Woodward's life, together with his
intentions in founding a Professorship, or, as he would have
called it, a Lectureship, in the University of Cambridge. As
a supplement to this, we shall briefly record what Sedgwick's
predecessors did, or, we might almost say, did not do, to carry
out the founder's instructions.

John Woodward was born in Derbyshire, 1 May, 1665.
His father is said to have been "a gentleman of a good family
in the county of Gloucester"; but, if such were the case, it is
strange that the son should have been apprenticed, on leaving
school at sixteen, to a linendraper in London. It is true that
the most original of Woodward's biographers, Dr Ward[1],
guards himself with an "as is said," while making this statement; but, on the other hand, it was certainly believed during

[1] *The Lives of the Professors of Gresham College*, by John Ward, Professor of
Rhetoric in Gresham College, and F.R.S. Fol. Lond. 1740, pp. 283—301. It
will be understood that the quotations in the following sketch are from this
work, unless other references are given.

Woodward's lifetime, and was used to his discredit by un-scrupulous opponents[1]. That he did go to London, while a mere boy, is certain; and while there had the good fortune to become acquainted with Dr Peter Barwick, physician to King Charles the Second, who received him into his house, and "took him under his tuition in his own family."

To this circumstance the general direction of Woodward's studies is obviously due; and it may be further conjectured that his interest in the University of Cambridge may have been inspired by Barwick. Barwick had been educated at St John's College, where his elder brother John, the sincere and courageous royalist, afterwards successively Dean of Durham and of St Paul's, was already Fellow. He proceeded Bachelor of Arts in 1643, Master of Arts in 1647, and Doctor of Medicine in 1655. As a London physician he had a large practice and a well-deserved reputation; while as a man of science he is known as the defender of Harvey's theory of the circulation of the blood. Barwick is also described as a man of sincere religious convictions, a strong churchman, and a daily attendant at service, either at St Paul's or at Westminster Abbey. From him therefore Woodward probably derived that religious tone of mind which led him to devote most of his scientific writings to the support of the Mosaic history of the Deluge.

Barwick has recorded in a testimonial dated 24 September, 1692, that he had then known Woodward "for above these eight years"; that he "had made a very great progress in learning" before he came to him; that he studied physic with him for nearly four years; and that subsequently he "prose-cuted his studies with so much industry and success that he hath made the greatest advance not only in physick, anatomy, botany, and other parts of natural philosophy; but likewise

---

[1] For instance, Dr Richard Mead, in his *Discourse on the Smallpox and Measles*, calls Dr Woodward " a man equally ill-bred, vain, and ill-natured, who, after being for some time apprentice to a linen-draper, took it into his head to make a collection of shells and fossils," etc. *Works*, ed. 1763, ii. 100.

in history, geography, mathematics, philology, and all other useful learning of any man I ever knew of his age." It would appear, therefore, that he became known to Barwick in 1684; and, as he was born in 1665, and remained at school till 1681, when he was sixteen, there remain only three years to be accounted for, during part of which—whatever may be the truth of the apprentice story—he is said to have pursued his studies "with great diligence and application." We are not, however, told what these studies were, nor is any hint given that he went through the ordinary course prescribed for candidates for a medical degree. By 1692, however, he had become sufficiently well known to obtain the Professorship of Physic in Gresham College, for which he was recommended not merely by Dr Barwick, in the testimonial already mentioned, but "by many gentlemen of figure in the learned faculties." It is much to be regretted that Dr Ward, who had seen these testimonials, should give no particulars of them, nor even record the names of those who wrote them. Had he been a little more explicit, we might have discovered the reasons which induced the electors to choose a young man of twenty-eight, who, so far as we know, had given no visible signs of fitness for so distinguished a position; and we might thus have learnt the nature and extent of Woodward's early studies.

In the following year (30 November, 1693) Woodward was elected Fellow of the Royal Society. Two years later, 4 February, 1695, he was made a Doctor of Medicine by Archbishop Tenison, and in the same year (28 June) the same degree was granted to him by the University of Cambridge; on which occasion he was admitted a member of Pembroke Hall—as Pembroke College was then termed[1]. Here again Barwick's influence may have disposed the

---

[1] He is so recorded in Dr Richardson's List of Degrees preserved in the Registry of the University. The Supplicat for his degree is: "Placeat vobis ut Johannes Woodward sit eisdem Gradu Honore et Dignitate apud nos Cantabrigienses quibus est per Literas Patentes Domini Archiepiscopi Cantuariensis"; endorsed, "Ad. Lect. et Conc. 28 Junii 1695. Non subscripsit." The records of Pembroke College do not mention Woodward.

University to admit his friend and pupil; but on this point we can only form a probable conjecture, and as to Woodward's reasons for selecting Pembroke, we are completely in the dark. He was admitted a candidate of the College of Physicians 25 June, 1698; and Fellow 22 March, 1702—3. He held the office of Censor there in 1703 and 1714; and in January 1710—11 delivered the Gulstonian Lectures *On the Bile and its Uses*[1].

It would be beside our present purpose to investigate Woodward's claims to distinction as a physician, or to do more than allude to his quarrel with Dr, afterwards Sir Hans, Sloane, in 1710, which led to his expulsion from the Council of the Royal Society[2]; or to his controversy with Dr Freind and Dr Mead on the new treatment of the smallpox suggested in his treatise, *The State of Physick and of Diseases*, published in 1718[3]. For the same reason we will be silent about his antiquarian pursuits, and "poor Dr Woodward's shield[4]," by which "he ingaged the attention of the learned for a considerable time." Those who wish to enjoy a hearty laugh at his expense should turn to the third chapter of the *Memoirs of Martinus Scriblerus*, where, under the transparent disguise of Dr Cornelius Scriblerus, the misfortunes of the learned owner

[1] *Roll of the Royal College of Physicians of London;* by William Munk, M.D. 8vo. Lond. 1861, ii. 3.

[2] Weld's *History of the Royal Society*, 8vo. Lond. 1848, i. 337; Brewster's *Life of Sir Isaac Newton*, Chapter XXI.

[3] This was the occasion of the attack made on Woodward by Mead. According to the account which Woodward sent to *The Weekly Journal* for 20 June, 1719 (printed by Nichols, *Lit. Anecd.* vi. 641), Mead followed him to the gate of Gresham College, and there made a pass at him with his sword from behind. Woodward drew, and was defending himself, when his foot slipped, and he lay at the mercy of his adversary, who bade him ask his life. The encounter was presently terminated by the interference of other persons. "Had he been to have given me any of his physic," said Woodward, "I would, rather than take it, have asked my life of him; but for his sword it was very harmless."

[4] Lord Castledurrow to Dean Swift, 4 December, 1736. The shield, of iron, 14 in. in diameter, is a cinquecento Italian work. It was bought, at the sale of Woodward's Collections (see below, p. 186, *note*), by Colonel King. After his death in 1767, aged 84, it was sold for £40 (*Nollekens and his Times*, by J. T. Smith, i. 39). It is now in the British Museum, Department of General Antiquities.

of the shield, reputed to have once been wielded by Camillus, are chronicled with infinite humour[1]. His geological speculations, on the other hand, deserve careful examination, for, though they are in many parts absurd, and warped throughout by the necessity for making the observed facts fit a preconceived theory of a universal deluge, "he appears to have had some very correct notions as to the general structure of the globe, and the proper method of pursuing the investigation of it[2]."

The circumstances which led him to these studies have been recorded by himself in the *Preface* to the *Catalogue of the English Fossils* in his own collection.

It may not be improper or unseasonable, before I proceed to the brief Account I am going to give of the Bodies in the following Catalogue, to take notice that I began my Observations and Collections in Gloucestershire; whither I was invited by Sir Ralph Dutton, along with his Lady's Father Dr Barwick, under whose tuition I then was, very happily, he being a Man of great Sagacity, Learning, and an Encourager of all ingenuous Studies. Here I had very generously allow'd me all Conveniencies and Assistances for the furthering of Comparative Anatomy, in which I took great pains; and had all the several sorts of Brutes, of Birds, of Fishes, that this noble and plentiful Country afforded, readily brought to me for Dissection. I had here likewise opportunity of carrying on my Botanic Studies, of which, being then young, I was very fond. Not that I confin'd myself so much to this part of Natural History as not to be ready, forward, and desirous to look into any other; and the Country about Sherborne, where Sir Ralph Dutton's Seat was, and the neighbouring parts of Gloucestershire, to which I made frequent Excursions, abounding with Stone, and there being Quarries of this laid open almost everywhere, I began to visit these, in order to inform myself of the nature, the situation, and the condition of the Stone. In making these Observations, I soon found there was incorporated with the Sand of most of the Stone thereabouts, great plenty and variety of Sea-shells, with other marine Productions. I took notice of the like, lying loose in the Fields, on the plough'd Lands and on the Hills, even to the very top of the highest thereabouts....This was a Speculation new to me; and what I judg'd of so great moment, that I resolv'd to pursue it through the other remoter parts of the Kingdom; which I afterwards did, made Observations upon all sorts of Fossils, collected such as I thought remarkable, and sent them up

[1] Pope's *Works*, ed. Roscoe, v. 160.
[2] *Edinburgh Review*, xxix. 316.   The article is by Dr W. H. Fitton.

to London. Some others were afterwards given me by such curious and intelligent Persons, as being appriz'd of the usefulness of these Studies, turn'd their Thoughts to such Searches[1].

The results of these observations are recorded in *An Essay toward a Natural History of the Earth and Terrestrial Bodies, especially Minerals; as also of the Sea, Rivers, and Springs, with an Account of the Universal Deluge, and of the Effects that it had upon the Earth*, first published in 1695, as the forerunner of a larger work, which, however, was never written. The author tells us that in order to inform himself of the present condition of the earth, he travelled through the greatest part of England, enquiring "for intelligence of all Places where the Entrails of the Earth were laid open, either by Nature (if I may so say), or by Art, and humane Industry. And wheresoever I had notice of any considerable natural Spelunca or Grotto, any digging for Wells of Water, or for Earths, Clays, Marle, Sand, Gravel, Chalk, Cole, Stone, Marble, Ores of Metals, or the like, I forthwith had recourse thereunto; and taking a just account of every observable Circumstance of the Earth, Stone, Metal, or other Matter, from the Surface quite down to the bottom of the Pit, I entered it carefully into a Journal, which I carry'd along with me for that purpose." The English tour being finished, Woodward wished to extend his travels beyond sea; but was prevented by "the Commotions which had then so unhappily invaded Europe," and had to content himself with the observations of others, for whose use, he says, "I drew up a List of *Quæries* upon this Subject, which I dispatch'd into all parts of the World, far and near, wherever either I myself, or any of my Acquaintance, had any Friend resident to transmit those *Quæries* unto[2]."

---

[1] *An Attempt Towards a Natural History of the Fossils of England; in a Catalogue of the English Fossils in the Collection of J. Woodward, M.D.* 2 vols. 8vo. Lond. 1729. Vol. I. Part II. p. I.

[2] *Brief instructions for making observations in all parts of the world; as also for collecting, preserving, and sending over natural things: Being an attempt to settle an universal correspondence for the advancement of knowledge, both natural and civil:* 4to. Lond. 1696. An abridgement, called: *Brief Directions for making observations and collections, and for composing a travelling Register of all*

These observations led him to the following conclusions: that in all parts of the world, "the stone and other terrestrial Matter" was "distinguished into *Strata*, or Layers; that those *Strata* were divided by parallel Fissures; that there were enclosed in the Stone, and all the other denser kinds of terrestrial Matter, great numbers of Shells, and other Productions of the Sea;" that the "Shells, and other marine Bodies, found at Land, were originally generated and formed at Sea;" and "that they are the real spoils of once living Animals." The arguments by which he establishes this truth—in those days a startling novelty—show that he had carefully examined both recent and fossil forms.

Unfortunately for Woodward's reputation, he felt obliged to account for the presence of these bodies where he found them, and therefore promulgated the geological romance with which his name is associated. He imagined the centre of the earth to be a spherical cavity—the Great Deep of Genesis—filled with water; that when the Flood took place these waters burst forth; that by their agency "the whole Terrestrial Globe was taken all to pieces and dissolved;" that "Stone, and all other solid Minerals, lost their solidity, and that the sever'd Particles thereof, together with those of the Earth, Chalk, and the rest, as also Shells, and all other Animal and Vegetable Bodies, were taken up into, and sustained in, the Water; that at length all these subsided again promiscuously, and without any other order than that of the different specifick Gravity of the several Bodies in this confused Mass, those which had the greatest degree of Gravity sinking down first, and so settling lowest; then those Bodies which had a lesser degree of Gravity fell next, and settled so as to make a *Stratum* upon the former; and so on, in their several turns, to the lightest of all, which subsiding last, settled at the Surface, and covered all the rest; that this very various Miscellany of

sorts of Fossils, was printed after Woodward's death in: *Fossils of all kinds, digested into a Method suitable to their mutual Relation and Affinity*, 8vo. Lond. 1728.

Bodies being determined to subsidence in this Order meerly by their different specifick Gravities, all those which had the same degree of Gravity subsided at the same time, fell into, and composed, the same *Stratum;* so that those Shells, and other Bodies, that were of the same specifick Gravity with Sand, sunk down together with it, and so became inclosed in the *Strata* of Stone which that Sand formed or constituted : those which were lighter, and of but the same specifick Gravity with Chalk (in such places of the Mass where any Chalk was), fell to the bottom at the same time that the Chalky Particles did, and so were entombed in the *Strata* of Chalk ; and in like manner all the rest[1]: " in proof of which he maintains that the shells usually found in sandstone are heavier than those found in chalk. Further, he scornfully rejects the notion that there have been " Changes and Alterations in the Terraqueous Globe" since the Deluge, except such as are due to the agency of man.

While engaged in the researches which led to this theory, he amassed a vast collection of specimens, all of which he terms fossils—though the collection is partly petrological, partly zoological. It is accompanied by an elaborate catalogue, in which the specimens are carefully described, and their localities noted. Here again Woodward was far in advance of his age ; and, had not his mind been predisposed to theory, he might have anticipated, by a century, the discoveries of William Smith. Instead of this, as Sedgwick pointed out, " he formed a magnificent collection of organic remains, and he separated from the rest a series of fossils of the Hampshire coast, and was aware that many of the species were the same as those of the London Clay ; but this fact, and many others of like kind, were with him but sterile truths ; and, being led astray by his theory, he knew nothing either of the real structure of the earth, or of any law regulating the distribution of organic forms[2]."

[1] *An Essay,* etc. p. 29.
[2] *Address to the Geological Society,* 18 February, 1831, on announcing the first award of the Wollaston prize.

The *Essay* achieved great popularity, passing through at
least four editions in England during Woodward's life, besides
being translated into Latin at Zurich, by Dr J. J. Scheuchzer[1].
After Woodward's death it was translated into French (1735)
and published simultaneously at Paris and Amsterdam; and
subsequently into Italian (1739).

It was not to be expected that views so novel should
escape attack; and accordingly we find them controverted,
"partly by occasional remarks in other writings, partly by
pamphlets written directly against" the *Essay*. Among the
latter is a tract by Dr John Arbuthnot, published in 1697.
He fully admits "that though Dr Woodward's Hypothesis
seems to be liable to many just exceptions, the whole is not
to be exploded;" but good-humouredly hits him in his
weakest point when he adds: "I cannot forbear to wish that
People were more diligent in observing, and more cautious in
System-making. First, the World is malicious, and when
they write for an Opinion it spoils the Credit of their Observations: They have then taken their Party, and may be
suspected for partial Witnesses. In the next Place, Mankind,
in these Matters, is naturally too rash, and apt to put more
in the Conclusion than there is in the Premises; yea, some
there are so fond of an Opinion, that they will take Pleasure to
cheat themselves, and would bring every Thing to fit their
Hypothesis[2]." The illustrious Ray, though he could not
make up his mind as to the real nature of the "formed
stones," as they were termed, rejected the notion that their
position had been regulated by their specific gravity, as "not
generally true," pointing out "that they are often mingled,
heavy with light, in the same Bed or *Stratum*." At the same
time he shows, with much acuteness of observation, that

---

[1] His translation is called: *Specimen Geographicæ Physicæ Quo agitur de Terra
et Corporibus Terrestribus Speciatim Mineralibus* [etc.]. 8vo. Tiguri, 1704.

[2] *An Examination of Dr Woodward's History of the Deluge*, in *Miscellaneous
Works of the late Dr Arbuthnot*, Lond. 1770, ii. 230. The same collection
contains (i. 166), *An Account of the Sickness and Death of Dr W-dw-rd*, a
satire on the Doctor's medical theories.

many parts of the earth had been changed in comparatively recent times, and that Woodward was by no means accurate in maintaining that it was now in the state in which the Deluge had left it[1].

These authors, and others whom we need not enumerate, treated Woodward seriously; but he met with not a little ridicule. Gay presented him on the stage, in a farce called *Three Hours after Marriage,* as Dr Fossile—"the man who has the Raree-show of Oyster-shells and Pebble-stones"—; and the wrath of the doctors against his medical theories manifested itself in a plentiful crop of scurrilous pamphlets. We have neither space nor inclination to recount the history of "Don Bilioso de l'Estomac[2]," as Woodward is called in one of these; but in *Tauronomachia: or a description of a Bloody and Terrible Fight between two Champions, Taurus and Onos, at Gresham College*[3], his theory of the earth is hit off so happily that we cannot forbear transcribing a few lines. After introducing us to *Onos,* "a fam'd Empirick of the Town," who knew how everything was created, and who

> "of Atoms what, can tell
> *Echinites* made, and *Cackle-Shell*[4],"

the satire gives a humorous description of the Abyss, the Deluge, and the struggle of the various substances to get to their proper places as the waters subsided.

> Each thought himself as good as other,
> And with confounded Stir and Pother,
> Strove to accelerate his Pace,
> And shove some other out of 's Place.
> But cross-grain'd Levity combin'd
> With Fate to make some lag behind;

---

[1] *Three Physico-Theological Discourses,* by John Ray, ed. 1713, pp. 165-167, 206-295.

[2] *The Life and Adventures of Don Bilioso de l'Estomac. Translated from the Original Spanish into French; done from the French into English. With a Letter to the College of Physicians,* 8vo. Lond. 1719.

[3] *Tauronomachia* etc. Lond. 1719. Small folio, pp. 6.

[4] A note on this word in another line says, "By a figure of speech peculiar to *Onos.*" It was evidently intended to ridicule his affected pronunciation, which other writers allude to.

For some, tho' immensely large and huge,
Were *Naturally Centrifuge ;*
Whilst others, Atoms, yet their Weight
Inclin'd them to be *Centripete.*
Oh! had you heard what dreadful Moans
Were made by Marle, and Coals and Stones,
And Seeds of Trees; that had not Power
To sink themselves two Inches lower;
How Chalk and Soil did curse and swear,
That they must lye in open Air;
You'd been amaz'd, to find this Worldly
Frame in so d—d a Hurly-burly.
Thus I've observed, *pro re natâ,*
A Kitchin-Wench of Bread lay *Strata,*
Eggs, Suet, and Plums in plenteous store;
But, in a Moment of an Hour,
Milk in a *Deluge* vast comes flowing,
And *dissipates* all she'd been doing :
But, when the Streams began t' asswage,
And quiet grow, and free from Rage;
Then to my Sorrow have I spy'd
Whole Troops of Plums with speed *subside.*

Woodward took no notice of any adverse criticism until 1714. Two years before Dr Elias Camerarius, Doctor of Medicine in the University of Tübingen, had published a volume of essays on various points of physic and medicine[1], in some of which he had disparaged Woodward's theories, but without acrimony or severity. There seems to be no special reason why Woodward should have broken his silence on this occasion in particular; but perhaps he was glad of an opportunity of addressing the learned world, and especially the learned world of Germany, where his own works had made many converts. Accordingly he published, with a dedication to Thomas, Earl of Pembroke, a short Latin essay, in which he examines the points challenged by Camerarius, and defends his whole theory, but, it must be admitted, without any great novelty of argument[2]. The essay concludes

[1] Eliæ Camerarii *Dissertationes Taurinenses Epistolicæ, Physico-Medicæ.* 8vo. Tubingæ, 1712.
[2] Johannis Woodwardi *Naturalis Historia Telluris Illustrata et Aucta. Una cum Ejusdem Defensione; Præsertim contra Nuperas Objectiones D. El. Camerarii Med. Prof. Tubingensis.* 8vo. Lond. 1714.

with a *Classification of Fossils*, prefaced by an epistle to Sir Isaac Newton, at whose suggestion, says the writer, the work had been undertaken.

One other scientific work of Woodward's must be briefly noticed. In June, 1699, he laid before the Royal Society *Some Thoughts and Experiments concerning Vegetation*[1]. This remarkable paper shows that the author should be ranked as a founder of experimental plant physiology, for he was one of the first to employ the method of water-culture, and to make refined experiments for the investigation of plant-life.

Woodward's object was to controvert a theory then prevalent, that water, and not mineral matter, was "the only *Principle* or Ingredient of all natural things;" and that there was "a direct transmutation of water into plants and other bodies." The supporters of this view had investigated, experimentally, the growth of various selected plants, and had shown in the first place that "mint and other plants prosper and thrive greatly in water"; and secondly, that, if a plant be placed in a given weight of earth and allowed to grow for a considerable time, at the end of the experiment the earth will be found to have experienced no loss of weight, thus proving that all the nutriment must have been obtained from the water. Woodward proceeded as follows. A number of glass phials of fairly equal size were filled with water, and weighed. Each was then covered with a piece of parchment, pierced with a hole just large enough to admit the stalk of a plant. Sprigs of spear-mint (*Mentha viridis*) and other plants, having been carefully weighed, were inserted through the holes in the parchment, and the phials set in a window, under the same conditions of air, light, and sun. As the water evaporated, the phials were replenished, account being taken of the weight of the water added. He placed in his phials: (1) pure water; (2) water containing soluble matter in varying proportions; (3) water

---

[1] *Philosophical Transactions*, 1699, Vol. xxi. pp. 193—227.

artificially mixed with earth. At the conclusion of the experiment the plants were weighed a second time, and their growth calculated in proportion to the weight of the water used. The most important results, as stated in Woodward's own words, were the following:

1. "The Plant is more or less nourished and augmented in proportion as the Water in which it stands contains a greater or smaller quantity of proper terrestrial Matter in it."

2. "The much greatest part of the Fluid Mass that is drawn off and conveyed into the Plants, does not settle or abide there: but passes through the Pores of them, and exhales up into the Atmo sphere."

The first of the above conclusions was a sufficient answer to those who supported a contrary theory; the second shows that Woodward had discovered what is now called *Transpiration*, which has so important a bearing on plant-life. In the light of this result he proceeds to discuss the effect of vegetation on climate; and concludes that "so continual an emission and detachment of water in so great plenty from the parts of plants affords us a manifest reason why countries that abound with trees and the larger vegetables, are very obnoxious to damps, great humidity of the air, and more frequent rains than others that are open and free;" and that this evaporation is dependant on temperature, for "much less quantity of water was exhaled in the colder months."

Woodward was not popular with his contemporaries. Thoresby, the well-known antiquary of the last century, calls him "very ingenious, yet not the best-tempered;" and, in another place, "that ill-natured piece of formality[1]." Nor did foreigners judge him more favourably than his own countrymen. Dr Christian Heinrich Erndl, or Erndtel, who visited England in 1706, says:

The said Doctor owns an inestimable treasure of minerals and petrified shells, partly collected by himself in Britain, partly obtained with much diligence from all corners of Europe, as specimens of the rarer minerals and petrifactions. He has likewise a very choice library of books on medicine and philosophy. It is to be regretted

[1] Nichols, *Illustrations*, i. 800, 806.

that this celebrated man should be very ignorant of Latin, which he speaks with difficulty; and it is wonderful how chary and churlish he is in showing his cabinet of curiosities. If you do get a peep at it, mind you do not touch the smallest object with so much as the tip of your finger. Nor may you look into a single volume, unless he holds it in his own hands[1].

His eccentricity and vanity are amusingly described in Uffenbach's account of a visit paid to him in 1710; but allowance must be made for the writer's evident vexation.

30 *October.* In the morning called on Dr Woodward for the fifth time, and at last found him at home; but were shown into an ante-chamber. When we had stood there a good quarter of an hour, he first sent his boy to ask our names: after another quarter of an hour the boy came back, saying, 'His master was still in bed, as he had sat up somewhat late the night before; it might be half-an-hour before he got up, if we could wait so long.' We left our interpreter and servant behind, with orders to summon us, when it was convenient to the man, and meanwhile drank a cup of coffee in the next coffee-house. When one of them came for us, we set off at once, but must again wait some half-hour in the ante chamber. At last his boy called us, and led us through two rooms to the precious Mr Doctor. He stood stiffly up in his silk dressing-gown, and with an affected air and screwed-up eyes, asked who we were, and where we came from. But when we begged for a sight of his cabinet, he excused himself, saying that in half-an-hour he had to attend a consultation, which he could not possibly put off, and prayed us to come again the next afternoon at three.

When we were about to take leave, he begged us to stay awhile, and called to his lad, 'make haste,' intending, as we supposed, to offer us chocolate, according to his custom. For, as we had been assured, he presents it to all strangers, and that with such ridiculous fuss and ceremony, that one can scarce refrain from laughing. For till the chocolate comes he keeps urging the boy with every variety of expression; a shouting to which, much to our disgust, we were forced to listen some half-hour. But this time we had not the honour to drink a cup with him; for though the boy brought a silver can and a cloth, it was only for shaving; and we were to be favoured with the privilege of looking on. We had heard already of more than four foreigners, who had received the same treatment. But we excused

[1] C. H. E. D. *De Itinere suo Anglicano et Batavo Annis MDCCVI et MDCCVII facto relatio ad amicum D. G. K. A. C.* Amsterdam, 1710, p. 41. A second edition, published 1711, omits the passage "It is to be regretted—own hands." An English translation of the first edition appeared in 1711, entitled : *The Relation of a Journey into England and Holland in the Years 1706 and 1707.* By a Saxon Physician. 8vo. Lond. 1711. This translation is badly done, and has not been exactly followed in the above passage.

ourselves, and said we would not detain him, and got away, though he several times begged us to stay.

31 *October.* In the afternoon we drove again to Dr Woodward, and at last attained our end, to see his things. Yet he kept us waiting, as his way is, again a good half-hour in his ante-chamber; and then complained that we were not quite punctual, and had not come half-an-hour before. This is said to be the uncivil compliment which this affected, learned charlatan, pays to all strangers that come to him.

He showed us first all kind of precious stones found here and there in England; then some minerals, and then petrefactions, his strong point. Not only was the quantity amazing, but the specimens were select and fine. Amongst others he showed us shells filled and partly overgrown with stone of all kinds, even the hardest flint. Specially curious was the collection in which he showed us the whole growth of the *conchylia* from first to last. He had also many stones containing fossil plants of all kinds; shells covered with metals and ores, and partly also filled with them; amongst the rest very many fine ammonites. He had a cabinet filled with ancient urns and vases. In another were great fossil snails and ammonites. In another he had a good number of MSS., chiefly relating to the Natural History of England, which, as he professed, were mostly of his own writing. Among these books was a volume, in which he had had all his *conchylia* tolerably well drawn. Again, a fine *herbarium vivum anglicanum* of his collection, in which the plants were quite fresh and well-preserved. Dr Woodward showed us all his things with such an affected air, and such screwing-up of the eyes, that one cannot help laughing; though he suffers you to laugh as little as to speak, requiring every one to listen to him as an oracle, approve and extol all. You must listen to his opinion *de diluvio et generatione antediluviana et lapidum postdiluviana,* till you are sick of it. He repeats whole pages of his works, accompanying them with running panegyrics. The maddest thing of all is, that he has many mirrors hanging in every room, in which he constantly contemplates himself. In all he does he behaves like a woman and a conceited fool[1].

Woodward died of a decline, in his apartments at Gresham College, 25 April, 1728, in the sixty-third year of his age. His Will records a wish that he may be buried "in the Abbey Church of Westminster, with as little Pomp and Expences as may well be." On May-day following this wish was carried out. His grave is close to that of Newton,

---

[1] Zacharias Conrad von Uffenbach, *Merkwürdige Reisen.* Dritter Theil. pp. 228, 235. The translation is by the Rev. J. E. B. Mayor, M.A. St John's Coll. A few lines have been omitted in transcription.

JOHN WOODWARD, M.D. F.R.S.

Born 1665; *died* 1728.

From a contemporary oil-painting in the Woodwardian Museum, Cambridge.

on the north side of the entrance to the choir. No inscription marks his resting-place, but near the west end of the nave, on the same side of the church, an elaborate monument of white marble, erected to his memory by the pious care of his friend Colonel King, bears the following inscription :

M.S.

JOHANNIS WOODWARD,

Medici Celeberrimi,
Philosophi Nobilissimi,
Cujus
Ingenium et Doctrinam
Scripta per Terrarum ferè orbem
Pervulgata ;
Liberalitatem verò et Patriæ Caritatem
Academia Cantabrigiensis,
Munificentiâ Ejus aucta,
Opibus ornata,
In Perpetuum declarabit.
Natus Kal. Maij A.D. MDCLXV.
Obiit VII. *Kal. Maij* MDCCXXVIII.

RICHARDUS KING
Tribunus Militum, Fabrûmque Præfectus,
Amico optimè de se merito
D. S. P.

Our portrait of Woodward is taken from an oil-painting in the Woodwardian Museum. Its history is unknown, but, from the style, it is evidently a contemporary picture, and is believed to have been sent to Cambridge at the same time as the collection, or shortly afterwards.

Woodward's will is dated 1 October, 1727. He names as his executors the Honourable Dixie Windsor[1]; Mr Hugh Bethell, of Swinton in Yorkshire ; Mr Richard Graham ; and

---

[1] Of Trinity College; B.A. 1694, M.A. 1698, and M.P. for the University, 1705—1727.

Colonel Richard King, of the Office of Ordnance in the Tower of London. They are directed to convert into money his personal estate and effects, including his library, and his antiquarian collections; to purchase land of the yearly value of one hundred and fifty pounds, and to convey the same to the University of Cambridge. Out of this yearly income £100 is to be paid, in four quarterly instalments, to a Lecturer, to be chosen in the first instance by the executors, and after their decease by "the Lord Archbishop of the Province in which the said University is, who, it is to be presumed, besides his favouring of learning and all useful knowledge, will think himself under obligation to have special regard to this University"; the Lord Bishop of the Diocese in which the said University is; the President of the College of Physicians; the President of the Royal Society; the two Representatives of the University in Parliament; and the whole Senate. The six persons first-named, together with the Chancellor of the University, are to have the privilege of voting by proxy.

The Lecturer is to be a bachelor; "and in case of the marriage of any of the said Lecturers afterwards, his election shall be thereby immediately made void, lest the care of a wife and children should take the Lecturer too much from study, and the care of the Lecture." This condition was evidently borrowed by Woodward from the statutes of Gresham College. In choosing the Lecturer, a layman is to be preferred to a divine, "not out of any disrespect to the clergy, for whom I have ever had a particular regard, but because there is in this kingdom better provision, and a much greater number of preferments, for the clergy than for men of learning among the laity"; he is to be "further subject to such rules, orders, and directions, not interfering with those hereinafter particularly specified and set forth, as the electors, or a majority of them, shall from time to time think fit to make"; he is not to hold "any preferment, office, or post, whatever, that shall any ways so employ and take up his time as to interfere with his duty herein set forth, and in particular that shall require his attendance out of the University"; if he accept such, his post is to become vacant; he is not to be absent from Cambridge for more than "two months in the year, and those to be in the long vacation in the summer"; he is there to "read at least four Lectures every year, at such times, and in such place of the said University, as the majority of the said electors shall appoint, on some one or other of the subjects treated of in my *Natural History of the Earth,* my *Defence of it against Dr Camerarius,* my *Discourse of Vegetation,* or my *State of Physick,* at his discretion, but in such language, viz., English or Latin, as shall be appointed from time to time by the Chancellor, Vice-Chancellor, Provosts, and Masters of the several Colleges and Halls belonging to the said University; the said Lectures, or at least one of them, at the Lecturer's own free choice and election, to be published in print every year."

In the next place he bequeaths to the University his collection of English fossils, with the two cabinets containing them, and their catalogues, copies of which are to be "reposited in the publick Library of the said University, for greater security that the said Fossils be preserved with great care and faithfulness." The executors are to "cause and procure the same to be lodged and reposited in such proper room or apartment as shall be allotted by the said University"; the Lecturer is to "have the care and custody of all the said Fossils and the catalogues of them"; he is to "live and reside in or near the said apartment so to be allotted for repositing the said Fossils"; he is to "be actually ready and attending in the room where they are reposited, from the hour of nine of the clock in the morning to eleven, and again from the hour of two in the afternoon till four, three days in every week (except during the two months in the long vacation, wherein he is allowed to be absent) to show the said Fossils, gratis, to all such curious and intelligent persons as shall desire a view of them for their information and instruction"; and he is to "be always present when they are shown, and take care that none be mutilated or lost." For the sake of additional security the Chancellor, Vice-Chancellor, and Heads of Colleges are to appoint "two discreet and careful persons" before the admission of every Lecturer, and also once every year, "who shall inspect and examine the said collections of fossils, and compare them with the catalogues." These inspectors are to "give under their hands a report of their examination," and to receive "for their care and trouble," £5 a piece out of the testator's estate. Besides these precautions, the Lecturer, before his admission, is to give such security for the safe-keeping of the fossils as the electors shall think proper; and, further, he is to receive £10 in each year, "to be laid out and employed by him, from time to time, in making observations and experiments, keeping correspondence with learned men on the subjects directed to be treated of in the Lectures, and in procuring additions to the Collections of Fossils...he rendering annually to such of the electors as shall be in the University an account in writing of the ways in which the said sum hath been disbursed and employed[1]."

Notwithstanding these minute directions and limitations, Woodward clearly intended his benefaction to be modified from time to time; for he directs that a further sum of £10 "be appropriated for a dinner, on the first day of May, or, if this fall on a Sunday, then on the second day of May," for the Lecturer, Inspectors, Chancellor, Vice-Chancellor, and Heads of Colleges, "to the end that they may then confer and consider of the methods to improve the design and use of the said donation by me hereby made. And I greatly wish that these things that are of so much use and importance, and which I have with great diligence and expense collected, may by this settlement, the care of the electors, and the diligence of the Lecturer, be made serviceable to the setting

---

[1] The addition of this £10 virtually raised the annual salary to £110, an amount often mentioned as though it had been specified by Woodward.

forth the wisdom of God in the works of nature, to the advancement of useful knowledge, and to the profit and benefit of the publick."

Lastly, the balance left in hand after the discharge of the sums above specified, is to be spent by the University in "the payment of taxes, or any other necessary contingencies"; and any further surplus "in such manner as the said University shall think fit; but in hopes, that for the honour of the University, and the benefit that will thence accrue to the publick, if the design of this donation be rightly carried on, that the said University will be pleased to dispose of the said residue in making experiments and observations, in correspondence, in natural collections, books, or other things that may serve to the promoting the good ends of this donation."

It is evident that Woodward's primary object in this foundation was the permanent commemoration of himself and his researches without limitation of subject. Geology was not, in his eyes, more important than Medicine or Botany, provided his collections—the monument of his industry and sagacity—could be preserved, extended, and displayed to the public. This point having been secured, he did not bind his lecturer to their illustration. He might be a Botanist or a Physician, provided he took the Woodwardian utterances on those subjects as his text-book, and provided always that he was willing to act as an honest and efficient curator of the Woodwardian cabinets.

Woodward's estimate of the value of his works is worth quotation. After directing his executors to treat his unfinished writings as they shall think fit[1], he proceeds :

But for such others of my writings as I have at any time in my life caused to be published, the property and copyright of all which is in myself, and also all such others of my writings as my executors may hereafter appoint to be printed, I say of all those and these I do give and devise one moiety of the said property and copyright, and the benefit and profit thence arising, to the said University ; and the

---

[1] Notwithstanding this it was found that the MSS. had been placed in a box by Woodward's order, with a request, bearing a date anterior to that of his Will, that they might be destroyed, which was accordingly done.

other moiety to the said Lecturer and his successors from time to time, upon this special trust and confidence, that the said University and the said Lecturer and their successors do take care that all my said works from time to time be printed as soon as the former edition of the same or of any part is sold off or become scarce, and that they and he do not by any contracts to be made for the republishing thereof so enhance the price as to prejudice the sale and divulgation of any of the said copies to be reprinted.

The executors lost no time in carrying out Woodward's instructions. He died, as we have seen, in April, 1728, and by the following September the two cabinets containing the English fossils had been sent to Cambridge[1]. The University seems to have been fully aware of the importance of the new foundation, for at the beginning of the following year (26 February, 1728—29), the Senate agreed to purchase, for a sum not exceeding one thousand pounds, two other cabinets containing "foreign fossils," and "additional English fossils" respectively. These cabinets had at first been included by the testator among the effects which his executors were directed to sell, but by a subsequent clause they were empowered to make any arrangements they thought proper respecting them, or even to give them away. They decided, however, to dispose of them by public auction, and the Grace recommending their purchase dwells at some length upon the importance of preventing a separation of collections so valuable, and collected at so great an expense.

There is no evidence to show where the cabinets were bestowed on their arrival at Cambridge; and it was not until 1734, during the Vice-Chancellorship of Dr Roger Long, Master of Pembroke College, that a definite place was devised

---

[1] Grace of the Senate, 17 September, 1728. "May it please you that the acquittance now read to you be given to the executors of the late Dr Woodward, sealed with your common seal." This Grace can only refer to the collections, for the estate was not bought until 1731; and in the University Accounts for the year ending 3 November, 1728, we find : " Paid Colonel King the carpenters bill and other charges in packing Dr Woodward's boxes and two cabinets £11. 15s. 0d." His Library and Antiquities of various kinds were sold by auction. See *A Catalogue of the Library, Antiquities, etc. of the Late Learned Dr Woodward*, [etc.] 8vo. pp. 287. The sale began 11 November, 1728, and occupied 33 days, 28 of which were devoted to the Library.

for their reception. In a letter dated 13 April, 1734, the executors express to Dr Long their "thanks for the room which you have been so good as to appoint for the better standing of the Cabinets; and hope you will be pleased to order the fitting it up for that purpose." Dr Long was renowned for his mechanical contrivances, and it is probable that he himself suggested the ingenious scheme which was completed in 1736. By dividing off from the north end of the Arts School a space about fifteen feet in length, a room was contrived, now the Novel-Room of the Library, of convenient size, and fairly well lighted. The comfort of the Lecturer, who was supposed to spend twelve hours of each week in it, was provided for by a fireplace, curtains to the windows, and other luxuries. The whole work was superintended by Mr James Burrough, of Gonville and Caius College, the popular amateur architect of the day[1]. So long as the geological collections belonging to the University were contained in Woodward's four, or five, cabinets, this room was probably not ill-adapted for its purpose; but even then it was impossible for the Lecturer to "live and reside in or near the said apartment" as the Will directed. As time went on, and new acquisitions had to be displayed, it was found to be wholly inadequate, and we shall have to notice, as we proceed, several abortive attempts to provide a proper Museum.

The acquisition of an estate of the exact annual value specified in the Will proved a somewhat difficult matter, and was not effected until 1731, when a property near Beccles in Suffolk was conveyed to the University. The annual value was slightly in excess of £150, and the proportional difference in the purchase-money was made up partly by the generosity of Colonel King, Woodward's residuary legatee, partly by a loan from the University.

This matter having been settled, the four executors drew up a formal document under their hands and seals, dated

[1] University Accounts and Vouchers for 1735 and 1736.

30 July, 1731, by which they appointed Conyers Middleton, D.D., formerly Fellow of Trinity College, to be the first Lecturer. Middleton was a good scholar, and a distinguished man of letters. He wrote an English style of which it has been said that "for elegance, purity, and ease, it yields to none in the whole compass of English literature[1]." His love of music had gained for him the epithet of "musical Conyers"; which, as he himself played on the violin, was contemptuously changed by Dr Bentley to "fiddling Conyers[2]." He was a well-bred, courteous man of the world; and, having married a lady of good fortune[3], his house had become, to some extent, the centre of Cambridge society. His pamphlets against Bentley, and the subsequent degradation of his opponent on a question which he had been foremost in raising, had made him a prominent person in the University, and the office of *Protobibliothecarius*, or Principal Keeper of the University Library, had been created for him by the Senate in 1721 as a mark of gratitude to a man whom they regarded as their champion. But he had no knowledge of any department of science, and he probably owed his appointment either to his general distinction as a scholar and a gentleman, or to a personal acquaintance with Woodward, on which he insists in more than one passage of his inaugural lecture. This composition, an elegant piece of Latin, was printed in 1732[4]. As might be expected, it refers to science only in language borrowed from Woodward's own writings, without expansion or criticism. It may be described as a string of well-turned compliments to Woodward, to the executors, and to the University. Woodward had brought science out of the depths of the earth, as Orpheus brought Eurydice; Woodward might claim a place by Newton's side.

[1] Monk's *Life of Bentley*, ii. 67.   [2] Ibid. ii. 38.

[3] Mrs Middleton died 19 February, 1730.

[4] It is entitled: *Oratio de novo Physiologiæ Explicandæ Munere, ex celeberrimi Woodwardi testamento instituto, habita Cantabrigiæ in Scholis Publicis a Conyers Middleton, S.T.P. Academiæ Cantabrigiensis Protobibliothecario et Lectore ibidem Woodwardiano:* 4to. Lond. 1732.

Newton had explained the nature of light by study of the sun; Woodward had made light shine out of darkness—a conceit which may have been suggested by Dr Bentley's well-known lines:

> Who Nature's Treasures wou'd explore,
> Her Mysteries and Arcana know,
> Must high, as lofty *Newton*, soar,
> Must stoop, as searching *Woodward*, low[1].

From the compliments to the executors, though hardly less rhetorical than those to Woodward, the interesting information may be extracted that the University was principally indebted to Mr Richard Graham for Woodward's benefaction. He had been educated at Cambridge, though he did not proceed to a degree, and actuated by love for his old University he had constantly urged his friend to entrust his collections to Cambridge, as a place of note where they would be seen and valued.

Middleton held the office for rather less than three years. He resigned, 7 April, 1734, either from a sense of his own unfitness, or because he was meditating the second marriage which he shortly afterwards contracted.

The executors next appointed the Rev. Charles Mason, M.A., Fellow of Trinity College, apparently at the suggestion of the then Vice-Chancellor, Dr Roger Long. Some passages from their letter to him, dated 13 April, 1734, will be found interesting:

" Mr Vice-Chancellor,

Dr Middleton having transmitted to us his Resignation of Dr Woodward's Professorship, in Form, dated the 7th of this month, we proceeded forthwith to the Nomination of a fit Person to succeed him: and have unanimously made choice of Mr Mason: being confirm'd in the good Opinion we all have of his Abilities and

---

[1] Johnson (*Life of Cowley*, Works, ed. 1787, ii. 43) speaks of these lines as "the only English verses which he [Bentley] is known to have written." For Johnson's admiration of them, see Boswell's *Life*, ed. 1823, iii. 468. The whole poem is printed in Monk's *Bentley*, ii. 174, from Dodsley's *Collection*, ed. 1765, vi. 189. We have followed an earlier text as given in *The Grove; or, a Collection of original Poems, Translations, etc.*, 8vo. Lond. 1721. In the 4th line Dodsley reads 'delving' instead of 'searching.'

sufficiency for that Post, by the Character you have been pleas'd to give him.

There is nothing, Sir, we have more at heart, than the firm Establishment of this Professorship. And therefore we make it our Request to you, that as the Professor is to receive his Salary by the hands of the Vice-Chancellor for the time being, no part of the said Salary may be paid to him till he shall produce a Certificat, sign'd by two Masters of Arts, that he has duly read the Lectures, according to the Institution; and at the expiration of every year shall present to the Vice-Chancellor one of them Printed, giving him (at the same time) a particular Account, in writing, how the £10 annually allow'd for Correspondence, Experiments, etc., has been expended for the Year past; as is directed by the Founder's Will."

Two contemporary accounts of Mason have been preserved. The first is by the Rev. William Cole, his "particular friend":

"He is looked upon as rather unhewn, rough, and unsociable.... He is a very ingenious Man, an excellent Mechanic, and no bad Geographer: witness a most accurate Map of Cambridgeshire, which he has made from a personal Visitation of almost every Spot in the County. He has also large Collections for an History of the same County[1]."

The next is by Mr Richard Cumberland:

"A man of curious knowledge in the philosophy of mechanics, and a deep mathematician; he was a true modern Diogenes in manners and apparel, coarse and slovenly to excess in both; the witty made a butt of him, but the scientific caressed him; he could ornament a subject at the same time that he disgusted and disgraced Society. I remember when he came one day to dinner in the College hall, dirty as a blacksmith from his forge, upon his being questioned on his appearance, he replied—that he had been *turning.* 'Then I wish,' said the other, 'when you was about it, friend Charles, you had *turned* your shirt[2].' "

Mason was Woodwardian Lecturer for twenty-eight years. During that period he printed a single Latin lecture[3] (in 1734). Like Middleton, he devotes the greater part of it to praise of Woodward and his executors; and then, after commending

---

[1] MSS. Cole xxxiii. 156 (Add. MSS. Mus. Brit. 5834). A full account of Mason is given in the *Architectural History of the University of Cambridge*, ii. 674—677.

[2] *Memoirs of Richard Cumberland*, 4to. Lond. 1806, p. 106.

[3] *Oratio de Physiologiæ Explicandæ Munere, ex celeberrimi Woodwardi Testamento Instituto. Habita Cantabrigiæ in Scholis Publicis a Carolo Mason, M.A., Coll. S. S. Trin. Soc. et Lectore ibidem Woodwardiano.* 4to. Cant. MDCCXXIV. (*sic*).

the clause in the above letter to the Vice-Chancellor which makes the Lecturer's stipend depend on his reading lectures, he ends with a promise to devote his best energies to the work. Notwithstanding this engagement, we believe that all he did was to make a considerable private collection of fossils, which was sold by auction after his death. It is noteworthy that he was Vicar of Barrington in Cambridgeshire from 1742 to 1747, and Rector of Orwell in the same county from 1747 to his death. As no man can be in two places at once, and as Woodward had expressly forbidden his Lecturer to hold any preferment, office, or post, which might interfere with his duty as set forth in the Will, it is evident that Mason must have neglected either his parish in favour of his lectureship, or his lectureship in favour of his parish. In 1762, when he was between sixty and seventy years of age, "he quitted Senior-Fellowship, Professorship, and Liberty, for a Lady of small Fortune, but of great Accomplishments[1]," and took up his abode at Orwell, where he died, 18 December, 1770.

On Mason's resignation, Colonel King, Woodward's last surviving executor[2], appointed the Rev. John Michell, B.D. Fellow of Queens' College, a man of talent, who had already distinguished himself by his scientific writings. In 1750, while still a Bachelor of Arts, he had published a *Treatise of Artificial Magnets*[3]; and in 1760 he read to the Royal Society, of which he became subsequently Fellow, *Conjectures concerning the Cause, and Observations upon the Phenomena, of Earthquakes*[4], in which 'he advanced many original and philosophical views respecting the propagation of subterranean movements, and the caverns and fissures wherein steam might be generated[5].' At the outset of this paper he describes

---

[1] MSS. Cole, *ut supra.*

[2] *Cambridge Chronicle*, 10 December, 1762. Mason's marriage had taken place 5 November.

[3] A tract of 81 pages, 8vo. Cambridge, 1750.

[4] *Philosophical Transactions*, 1760, pp. 566—634. It was reprinted in full in Tilloch's *Philosophical Magazine*, 1818, Vol. lii.

[5] Lyell: *Principles of Geology*, Ed. 1867, i. 61; ii. 150, 152.

the general appearance and structure of stratified countries with such remarkable accuracy that so far as principle is concerned, a foremost place may be claimed for him among the founders of modern geology. That he was acquainted with the details also of the beds in certain parts of England is proved by a memorandum in his handwriting discovered in 1810 among the papers of Mr Smeaton, then in the possession of Sir Joseph Banks. In this memorandum several of the principal beds are enumerated, from the chalk down to the coal; and, in two instances, detached portions, several miles distant from each other, are associated under the same name[1]. On the other hand "he was ignorant of the importance of organic remains, and did not use them as a means of identifying strata[2]." Michell vacated the Lectureship by marriage, in September, 1764[3], having held it for rather less than two years.

We have not been able to discover that he ever delivered lectures, much less published any. Cole says of him:

"He is a little short Man, of a black Complexion, and fat, but having no Acquaintance with him, can say little of him. I think he had the Care of St Botolph's Church while he continued Fellow of Queens' College, where he was esteemed a very ingenious Man, and an excellent Philosopher[4]."

After his marriage he held more than one piece of preferment, and does not appear to have revisited Cambridge. Nor did he continue the geological studies which he had commenced with so much promise. His subsequent communications to the Royal Society are on astronomical subjects. He died 29 April 1793, at Thornhill, near Dewsbury, in Yorkshire, of which place he had been rector since 1767.

[1] This estimate of Michell's geological attainments is derived, in the main, from *Notes on the History of English Geology*, by W. H. Fitton, M.D. in the *Philosophical Magazine*, 1832, i. 268. The memorandum alluded to was first published in Tilloch's *Philosophical Magazine*, 1810, xxxvi. 102.

[2] Sedgwick's *Address to the Geological Society of London*, 18 February, 1831, on announcing the first award of the Wollaston prize, pp. 4, 5, and *note*.

[3] *Cambridge Chronicle*, 8 September, 1764.

[4] MSS. Cole, xxxiii. 156 (Add. MSS. Mus. Brit. 5834).

The Lectureship being vacant for a fourth time, Colonel King, now a very old man, appointed the Rev. Samuel Ogden, D.D., Fellow and President of St John's College. Dr Ogden had been master of the grammar-school at Halifax from 1743 to 1753, when he returned to Cambridge, where he resided until his death, 22 March, 1778. He held the livings of Stansfield in Suffolk and of Lawford in Essex, and was vicar of St Sepulchre's, Cambridge, from March, 1759, to May, 1777. For his personal appearance we will again quote Cole:

"Dr Ogden is a bald, swarthy, black Man: of a most extraordinary Turn of Humour, great Vivacity, odd, whimsical, and like no one else: a great Epicure, and very parsimonious: a very ingenious Preacher, and on that account his Church of St Sepulcre at Cambridge is usually so thronged as to be difficult to get a Place[1]."

This summary of Dr Ogden's peculiarities may be supplemented by the following note by Gilbert Wakefield:

"I heard Dr Ogden preach most of those discourses, which were afterwards made public. His manner, and person, and character of composition, were exactly suited to each other. He exhibited a large, black, scowling figure; a lowering visage, embrowned by the horrors of a sable periwig. His voice was growling and morose, and his sentences desultory, tart, and snappish. His sermons are interspersed with remarks, eminently brilliant and acute, but too epigrammatic in their close....He was a good *scholar*, a liberal-minded *Christian*, and an honest *man*.

His uncivilized appearance and bluntness of demeanour were the great obstacles to his elevation in the Church. He kept a public Act for his Doctor's degree at the Installation of the Chancellor, the late Duke of Newcastle, in 1749, with distinguished applause. The Duke was willing to have brought our divine up to Court, to prefer him; but found, as he expressed it, that the Doctor was not a *producible* man[2]."

Ogden had a turn for writing verse, and his name appears in three of those volumes which, in the seventeenth and eighteenth centuries, the University used to address to

[1] MSS. Cole, *ut supra* p. 137. The account is dated 19 June 1770.
[2] *Memoirs of the Life of Gilbert Wakefield*, 8vo. Lond. 1792, p. 95. Gunning's *Reminiscences*, ed. 1855, i. pp. 215—219, contain several amusing stories of Dr Ogden. See also *Sermons*, by Samuel Ogden; with *Life*, by S. Hallifax. 8vo. Camb. 1814: and Whitaker's *Loidis ad Elmete*, p. 387.

the sovereign on important occasions. In 1760 he mourned
the death of George the Second in Latin elegiacs; in 1761
he hailed the marriage of George the Third in English
stanzas; and, in the following year, the birth of George
Prince of Wales in Arabic. These curious changes of
language were satirised in the following lines[1]:

When Ogden his prosaic verse
  In Latin numbers dressed,
The Roman language proved too weak
  To stand the critic's test.

To English rhyme he next essayed
  To shew he 'd some pretence;
But ah! rhyme only would not do,
  They still expected sense.

Enraged the Doctor swore he 'd place
  On critics no reliance,
So wrapt his thoughts in Arabic,
  And bid 'em all defiance.

The extraordinary caprice of choosing an eccentric divine
to fill a scientific Lectureship probably gave rise to the story
that Ogden had obtained it by a pecuniary gratification to
Colonel King, or to one of his female relatives[2]. It is almost
needless to add that the Woodwardian Lectureship was a
sinecure during the fourteen years that it was held by Dr
Ogden. For two or three years before his death he was
" much broken with Gout and other Complaints[3]".

The next Lecturer was the Rev. Thomas Green, M.Λ. of
Trinity College, elected by the Senate 7 May, 1778[4]. Our
information about him is limited to the solitary fact that he
was Librarian to Trinity College from 1763 to his death. He
" added some valuable organic remains to the Woodwardian

---

[1] They were written, according to Cole (MSS. xxxiii. 157. Add. MSS. Mus.
Brit. 5834) by R. Pepper Arden, of Trinity College, B.A. 1766, afterwards
Fellow, created Baron Alvanley, 1801.

[2] MSS. Cole, *ut supra*. Nichols, *Literary Anecdotes*, ix. 612. The sum
paid is fixed by the former authority at £150, by the latter at £105.

[3] MSS. Cole, *ut supra*, p. 157.

[4] *Cambridge Chronicle*, 9 May, 1778.

cabinets",[1] and at his death, which took place 7 June, 1788[2], when he was still a comparatively young man, he bequeathed some books for the use of the Lecturer.

On this occasion two candidates came forward: John Hailstone, M.A., Fellow of Trinity College, and the Rev. Thomas Newton, M.A., Fellow of Jesus College. The election took place on Saturday, 7 June, 1788, when the Senate selected Hailstone by one hundred and twenty-seven votes to forty-three. Hailstone was then in his twenty-eighth year—having been admitted[3] in 1778, at the age of eighteen—and he had been second wrangler in 1782. The University had, therefore, good reason to expect that a young man, who had taken a distinguished degree, would apply himself with energy to the work of his office. Nor did he wholly disappoint their hopes.

He proceeded, by permission of the University, to study the progress which *Mineralogy*, as he terms his science, had made in Germany, where he attended one or two courses of lectures by Professor Werner. In 1792 he published: *A Plan of a Course of Lectures on Mineralogy, to which is prefixed an Essay on the different kinds of Mineral Collections, translated from the German of Professor Werner.* In the preface he apologises for the defects in "the Geognostical part" of the syllabus ; "Geognosy, or the knowledge of the Earth's internal structure", being "a Science yet in its infancy, when ingenious Men may with much more ease fabricate systems than confute them". The syllabus shows that he proposed to lecture on minerals and rocks—both of which he calls fossils, just as Woodward did—and on their systematic arrangement. Under the head Geognosy, "the Strata of the Earth", divided into *Primary, Secundary, Alluvial, Volcanic,* are treated according to "their relative

[1] *University Calendar,* 1820. The article, by the style, is evidently written by Sedgwick.

[2] *Cambridge Chronicle,* 14 June, 1788. Green had proceeded B.A. 1760, M.A. 1763. He was admitted Librarian of Trinity College 12 September, 1763.

[3] He matriculated as a pensioner of Catharine Hall 17 December, 1778, and removed to Trinity College in the following year.

13—2

Antiquity and order of Stratification"; and "Petrifactions" are considered separately, without any reference to the beds in which they occur, unless we except the general statement that "for the most part they appear to have been generated, lived and died, in the beds wherein they are found"; and that "the Fossil Bones of Siberia and the Ohio" are noticed under the heading "Alluvial Strata".

Notwithstanding this elaborate syllabus, we believe that Hailstone never lectured. In fact its publication was probably not intended to serve any other purpose than "to excite the attention of the University to a Branch of Knowledge, which, although honoured with an establishment in this place for a considerable number of years, has hitherto been suffered to languish in unmerited obscurity".

To the Museum, on the other hand, Hailstone paid considerable attention. He held that mineralogy was neglected at Cambridge through "a want of opportunity to consult and examine the different productions of Nature"; an obstacle which the "institution of a public Museum under proper custody and regulations" would remove. "The Woodwardian collection, which was made near a century ago", was, he thought, "ill-calculated to promote the study of mineralogy in its present state of improvement"; and, he might have added, the restrictions imposed upon its inspection by Woodward himself, must always have prevented its being generally studied. Accordingly he procured in Germany a typical series of rocks and minerals, and on his return home founded a separate collection, assisted by "the munificence of various friends of the University", among whom he gratefully commemorates the Duke of Grafton, and Mr John Hawkins of Trinity College, both of whom contributed specimens. The accounts of the Woodwardian Estate show that during his tenure of office nearly one hundred and fifty pounds were spent on fossils, and nearly seventy pounds on books and cabinets. When Sedgwick became Professor he found that this collection was "composed of many rare and beautiful simple minerals, and of specimens illustrative of the physical structure both of the British isles, and of some portions of the continent[1]." Nor did Hailstone omit teaching altogether. We are told that though

"No *systematic* Lectures are delivered, but the Professor constantly attends to demonstrate and explain the subjects of this Branch of Natural History to such curious persons, whether residents or strangers, as are engaged in the study of them. Much of his time is of course devoted to this part of his duty, as applications to this effect are numerous and frequent[2]."

On Hailstone's resignation there was evidently a feeling in the University that Woodward's bequest had not produced the results which might have been anticipated, and, as it was provided in his Will that new regulations might be framed from time to time by the electors, advantage was taken of the vacancy to appoint a Syndicate (8 May, 1818), "to consider what rules and orders should be framed for the development of Doctor Woodward's intentions". The Syndics lost no time in carrying out these instructions, for their report, which they call *Statement and Resolutions*, is dated 19 May following.

I.   It appears, that the clear annual income of the Woodwardian Estates is about £430, of which the sum of £108. 6s. 5d.[3] is paid to the Lecturer for his own use, and about fifty pounds are applied to other purposes, in conformity with the Will of Dr Woodward.

II.   That there is an accumulation of about £1200, which has been invested in the public Funds.

III.   That the Room, in which the Fossils and Minerals are at present kept being too confined to exhibit them to advantage, or to receive many more with convenience, it is desirable that a larger should be built with a contiguous room for the accommodation of the Lecturer.

IV.   It is proposed, that to effect this object as soon as possible, the surplus annual income shall be added to the above accumulation, with the exception of such sums as it may be judged proper to apply

---

[1] *Cambridge Calendar*, 1820.   Compare also what Sedgwick says, *Commission Report*, 1852, *Evidence*, p. 116.

[2] *Cambridge Calendar*, 1803.   This passage is repeated annually until 1820, when the whole account is replaced by a different article, evidently written, as mentioned above, by Sedgwick.

[3] The £8. 6s. 5d. is the surplus of the original rental (£150), left after paying the other charges.

to the purchase of Fossils and Books, and to other necessary purposes.

V.   That to entitle the Woodwardian Lecturer to the receipt of his annual stipend, it shall be certified to the Vice-Chancellor, that Lectures have been given.

VI.   It is agreed, that the knowledge of Fossil organized bodies, and of the Constitution of the Earth's Strata having been very much extended since the time of Dr Woodward, it would conduce to the diffusion of science, and to the credit of the University, as it would certainly be in perfect conformity with the Will of Dr Woodward, that a Course of Lectures should be read upon these subjects ; and if, after a new room has been built, the Professor, in addition to the lectures and duties prescribed by the Founder, should give such a course, it is proposed that his stipend be increased by one hundred pounds a year, and that all Members of the University have free admission.

This " good and stringent " report, as Sedgwick terms it[1], was passed by the Senate without opposition two days afterwards.   The recommendation that the Lecturer's stipend should depend on the delivery of lectures was merely a revival of what Woodward's executors had pleaded for so far back as 1734.   The affirmation of such a condition in 1818 is therefore only a proof that the University had determined to insist upon the performance of so special a part of his duties.   The curious proviso that his stipend should not be increased until a proper building had been erected was soon rescinded, as will be related in the next chapter.

[1] *Evidence, ut supra*, p. 116.

CHAPTER VI.

(1818—1822.)

Excursion to Derbyshire and Staffordshire (1818). First
course of lectures. Visit to Isle of Wight with
Henslow. Foundation of Cambridge Philosophical
Society. Visit to Suffolk coast. Commencement fes-
tivities. Geological tour in Devon and Cornwall.
Henslow's work in the Isle of Man (1819). Geo-
logical tour in Somerset and Dorset. Acquaintance
with Rev. W. D. Conybeare. Death of his mother (1820).
Visit to the Isle of Wight. Geological tour in
Yorkshire and Durham (1821). Controversy respecting
Professorship of Mineralogy (1822—1824).

WHATEVER may have been the amount of geological know-   1818.
ledge which Sedgwick possessed when elected to the Wood-   Æt. 33.
wardian chair, it must have been derived from study, and
not from experience. But he was the last man in the world
to take facts at second-hand, and therefore, so soon as the
Easter Term was over, he set out to use his eyes in the
field. This excursion must not be confounded with his subsequent systematic explorations. In 1818, his object was to
learn, not to instruct others; and it is not impossible that he
returned, as Darwin fancied he might himself return from
his first geological expedition, "very little wiser, and a good
deal more puzzled[1]," than when he started. We believe that

[1] *Life of Charles Darwin*, i. 189.

1818.   this first attempt at field-work is not alluded to in any of
Æt. 33.   Sedgwick's scientific papers ; and that, in fact, the only record
of it is contained in the following letter, the tone of which,
it may be remarked, is very different from the depression to
which his correspondents had lately been accustomed.

TRIN. COLL. *October* 23, 1818.

Dear Ainger,

My excursion for this summer is ended. I have
been about twenty four hours in Cambridge, during the greater
part of which time I have been employed in packing and
unpacking, till every table and chair in my room is nearly
filled with the spoils of my labours this summer. I have once
or twice thought of sending you some account of my operations ; but I have always been too lazy or too busy to take up
the pen for any such purpose.

I did not leave Cambridge before the 30th of July. The
weather had been so dreadfully hot the early part of that
month that I hardly ventured from under my own roof. After
spending a day in the neighbourhood of Mount Sorrel, I
advanced to Matlock, the immediate place of my destination.
In that neighbourhood I remained about five weeks. My
mornings were spent in professional pursuits ; that is, in
following the strata of the different rocks, collecting specimens,
and diving into the mines. The last operation was often
attended with no little fatigue, for the *rake veins*, (i.e. vertical
fissures, filled with spar and lead ore), are sometimes excavated
to an enormous depth. What the miners call *climbing shafts*
are formed in these veins, by which you descend to the
works ; not in buckets as in the coal-mines in your neighbourhood, but on cross-bars of wood (called *stemples*) which are
placed, like two perpendicular ladders, on opposite sides of the
pit. Between these you descend in a straddling position. I
let myself down in this way to the bottom of several of the
most remarkable mines in the county. In one or two of them
the works were nearly 1000 feet below the surface of the earth.

Matlock is one of the most beautiful spots under heaven, and was sometimes during my stay filled with very gay company. After being there a day or two I was advanced by the right of seniority to the chair, and in right of the same office was Master of the Ceremonies at the balls which took place every other evening. You see therefore that my employments have been not a little diversified. On the whole, I believe I got through my most arduous duties better than could have been expected.

After leaving Matlock I travelled on foot with a knapsack to the copper-mines in Staffordshire, by far the most wonderful excavations of the kind I have ever seen. I afterwards made Buxton my head-quarters for a fortnight, and finally found my way to Dent by the way of Macclesfield, Chester, Liverpool, and Lancaster. I remained one day at Northwich to visit the famous salt-mines in that neighbourhood. But I have no time to describe them.

My Father is becoming very thin; but his health is good, and his spirits do not indicate any of the infirmities of old age. I have not time for a word more, as I am off to a supper party. Give my best respects to Mrs Ainger, and believe me very busy, and very truly yours,

A. SEDGWICK.

The reference to Sedgwick's social duties at Matlock in the above letter is thoroughly characteristic. He always contrived to combine a large amount of amusement with business. 'That lively gentleman Mr Sedgwick,' as he was called by a stranger who met him in a stage-coach, had a happy knack of making himself agreeable to everybody with whom he happened to be brought into contact, and his geological tours gave him a wide and varied experience of mankind. With all sorts and conditions of men, quarrymen, miners, fishermen, smugglers, shepherds, artisans, grooms, inn-keepers, clergy of all denominations, squires, noblemen—he was equally communicative, and soon became equally popular. He could make the most silent talk, and could extract information and amusement out of materials that seemed at first sight destitute of either quality. It may be questioned whether his adventures would have been as diverting had they happened to anybody else ; he had a happy knack of meeting with strange experiences and untoward incidents ; and his return to Cambridge, after a summer's excursion, was eagerly looked forward to by his friends, for the sake of the budget of fresh stories with which he was certain to regale them.

The weeks spent in Derbyshire may be credited with at least one good result ; they convinced Sedgwick that in selecting geology as the work of his life he had made a wise choice. In a letter to one of his nieces, dated 12 August, 1854, thirty-six years afterwards, he gives some interesting and amusing particulars of this then distant period of his life : " When I was a young man, this was always a joyful day, the opening day of the grousing season ; and for a week before the 12th of August I could hardly sleep for thinking of the coming sport. Why we should take such delight in killing God's creatures is more than I can tell you ; but so it was, and so it is, and so it will be. I think it proves that by nature man is a savage carnivorous creature. Is he not ? The last time I ever fired a gun at a heath-cock was August 1817, thirty-seven long years since ! The year following I threw down the gun, and took to the hammer, and I enjoyed my new sport so much, that in 1818 (my first year of professional geology) I heard the sportsman's gun on the heaths of Derbyshire without a thought of regret or envy. And that year I was a dancing-man, and I fell three-quarters in love ; but, as you know, did not put my head through love's noose. But alas! times are sadly changed with me. I am now a gouty old man in my seventieth year[1]." The bright particular star of the Matlock assemblies, whose charms had nearly deprived Cambridge of her new geological Professor, married a goldsmith in Glasgow, and when Sedgwick was there in 1848,

[1] To Miss Fanny Hicks, 12 August, 1854.

he called at her husband's shop, in the hope of renewing his acquaintance. But, as ill-luck would have it, the lady was away at the sea-side, and her former partner had to content himself with an interview with her son[1].

Soon after his return to Cambridge Sedgwick delivered the first of those annual courses of lectures which became so celebrated, and were never interrupted—except for very brief intervals—until 1872, when he was compelled, by failing health, and the advance of old age, to appoint a deputy. It is provoking that no contemporary reference to this first course should occur either in his letters, or in any other source of information to which we have had access. In later years Sedgwick was fond of enumerating the courses of lectures he had given ; and he used to describe the one on which he was engaged according to its place in the series. " I am delivering my 40th course," or "my 47th course," and so forth. Had he been a man of scrupulous accuracy, it would be easy, by merely counting backwards, to discover in what year he began to lecture ; but, unfortunately, we frequently find two different courses denoted by the same numeral. On three occasions, however, he tells his correspondents that he began to lecture in 1819, and in two of these letters he states explicitly that the course was delivered in the spring. In 1851 he says: " The load of sixty-six years tells upon me, and I am not so fresh as I was when I gave my first course of geological lectures in the spring of 1819[2];" in 1859 he describes himself as " a toothless Professor who has been lecturing every year since he began his first course in the spring of 1819[3]"; and in 1861 he says : " I am trying to wind up my last course. It may well be my *last*; for I began to lecture in 1819[4]." These direct statements can hardly be erroneous, and we may safely assume that his first course was delivered

---

[1] To the same, 21 August, 1848.
[2] To Miss F. Hicks, 12 December, 1851.
[3] To Miss Malcolm, 29 November, 1859.
[4] To Rev. B. P. Brodie, 29 November, 1861.

1819.   either in the Lent or Easter Term of 1819, probably in the
Æt. 34. former.

The Easter vacation of 1819 was spent in the Isle of
Wight[1]. Sedgwick was accompanied by Mr J. S. Henslow
of St John's College, who became, in after years, Professor
first of Mineralogy and then of Botany, and who deserves
grateful recognition as one of the founders of the modern
school of Natural Science at Cambridge. As a boy he
had achieved considerable distinction in zoology, and while
still an undergraduate had found leisure to learn as much
of chemistry and mineralogy as was then possible at Cambridge. He had proceeded to the degree of Bachelor of Arts
in the previous January, and had therefore leisure to learn
something of the cognate science of geology. Mr Henslow's
brother-in-law and biographer, Mr Leonard Jenyns, speaks
of him as Sedgwick's pupil. In a certain sense this is undoubtedly true. Sedgwick was his senior by ten years—and
therefore superior to him in experience, and general knowledge both of science and of letters. As a practical geologist,
however, he could have known little more than Henslow,
while in the special subjects which the latter had already
studied with success Sedgwick had much to learn. The
expedition was successful in more ways than one. The
characters of the two men were very similar: they differed
for a time in politics, for Henslow began life as a conservative;
but in religion, love of truth, and hatred of wrong, they were
in exact agreement; and their intercourse—begun almost by
an accident—ripened into a warm friendship which was terminated only by Henslow's death. As regards the special
science they went out to study, Henslow learnt enough to
work out the geology of another part of England by himself
in the course of the following summer—a subject to which
we shall return presently—and Sedgwick brought home his
usual practical result in the shape of "a very large collec-

[1] Our principal authority for this tour is a *Memoir of the Rev. John Stevens Henslow*, by the Rev. L. Jenyns, 8vo. Lond. 1862, pp. 13—20.

tion" of geological specimens, intended for the Woodwardian 1819.
Museum, but for which at that time it was impossible to find Æt. 34.
room[1].

This short excursion deserves an honourable place in
the annals of Cambridge for another reason. In the course
of it the two friends discussed the want of some place to
which those interested in Natural Science might resort, with
the certainty of meeting men of the same or kindred tastes
with themselves—and where they might learn what was going
forward abroad. In these days of cooperation, when there is
almost a plethora of societies and associations more or less
learned for the promotion of every sort of object, it is difficult
to realise that barely seventy years ago there was a complete
dearth of such bodies; and that at Cambridge, which is now
taking the lead in Natural Science, there were only two lecture-
rooms for the scientific Professors—the one appropriated to
chemistry, the other to anatomy—no class-rooms, no museums,
no collections, except the Woodwardian, and the mineralogical
series then the private property of Dr E. D. Clarke.

The project thus started was eagerly prosecuted, after their
return to Cambridge, by the two energetic men who had
originated it. At first they proposed to establish a Corresponding Society, and with this idea they not only consulted
the residents likely to favour such a scheme, but solicited by
letter the cooperation of men of science at a distance. At the
beginning of the Michaelmas Term they laid their views
before Dr E. D. Clarke, who gave them such cordial support
that Sedgwick always spoke of him as one of the founders of
the Society which was presently established. At his suggestion the following notice was issued:

CAMBRIDGE, 30*th Oct.*, 1819.
The resident Members of the University, who have taken their
first degree, are hereby invited to assemble at the Lecture Room

---

[1] *Report of the Inspectors*, May 7, 1819. It will be understood that these and
other similar documents are preserved in the Registry of the University, unless
otherwise stated.

1819.   under the Public Library, at Twelve o'clock, on Tuesday, Nov. 2, for
Æt. 34.   the purpose of instituting a Society, as a point of concourse, for
         scientific communications.

This notice was signed by thirty-three persons, among whom were the Heads of the following colleges: Clare, Gonville and Caius, Queens', Christ's, Magdalene, and Trinity ; Professors Haviland, Monk, Cumming, Sedgwick, and Lee ; and ten tutors, or assistant-tutors, of colleges. Among the latter occur the names of Peacock and Whewell.

The proceedings at the meeting were not reported, but we learn from a second notice, issued on the day following, that the second Resolution : "That a Society be instituted as a point of concourse for scientific communication," was proposed by Sedgwick ; and that he was appointed a member of a committee to frame " such regulations as shall appear to them to be proper for the proposed institution." We can easily imagine, from the speeches made by him in after years, when he had to commend to audiences either lukewarm or hostile some scheme in which he was interested, the fire and energy with which he addressed the members of the Senate assembled on that November afternoon ; and we are not surprised to learn that the resolution entrusted to him was passed unanimously.

The Committee lost no time in discharging the duties assigned to them. Their first draft of the rules, endorsed,

" Report of the Committee appointed to form the Regulations of a Society, to be instituted in this University, for Philosophical Communication, to be read at the first meeting of the Society, on Monday, November 15, at one o'clock, in the Lecture Room under the Public Library,"

is dated 8 November, and, at the meeting therein announced, Sedgwick moved its adoption. This motion having been carried, those present voted themselves a Society, to be called, *The Cambridge Philosophical Society*, and the officers and Council were appointed. Professor Farish was the first President, Professors Sedgwick and Lee the first Secretaries. The

first formal meeting of the Society was held in the Museum 1819.
in the Botanic Garden, 13 December, 1819, when, by request Æt. 34.
of the Council, Dr E. D. Clarke read an address, explaining
the objects of the Society.

The first rule had originally run as follows :

"That this Society be instituted for the purpose of promoting
Scientific Enquiries, and of facilitating the communication of facts
connected with the advancement of Philosophy."

At this meeting the words "and Natural History" were
appended to this sentence. The addition is important, because
it determined, for many years, in fact so long as Henslow
resided in Cambridge, the direction of the labours of the
Society. We may justly credit Sedgwick with disarming
opposition, and launching the Society so successfully, that
before the end of the year 1820 it could boast of 171 members ;
but it was Henslow's patient devotion to zoology which
enabled it to form an excellent Museum, long the only
zoological Museum in the University, and the legitimate
parent of that large family of Museums which have grown up,
and are still growing up, in the old Botanic Garden[1].

Sedgwick always spoke with great delight of the share
which he had taken in the founding of this Society. The
annual dinner was one of his red-letter days; and no matter
how ill he might be, or imagine himself to be, he made a
point of attending it, and of making a speech after dinner.
This speech was one of the events of the academical year ;
and was, with most of the members, the principal, if not the
sole, reason for attending the dinner. It is no easy matter to
describe a speech—especially when it depends, as those
delivered by Sedgwick did, on the personality of the speaker;
but these particular postprandial orations have found a graphic
chronicler in the Lord Bishop of Carlisle :

[1] This account is derived from the following authorities: a complete set of the
early notices, etc., of the Society, preserved by Prof. Sedgwick: Henslow's *Life*,
ut supra, pp. 17—19: Otter's *Life of Dr E. D. Clarke*, ed. 1824, p. 649: *The
Cambridge Chronicle*, 5 November, 1819: *The Cambridge Portfolio*, pp. 121—129;
Macmillan's Magazine, April, 1880, p. 478.

1819.
Æt. 34.

"His speeches were the most remarkable things of the kind I have ever heard ; they sometimes began with a wild exuberance that nearly touched upon the region of nonsense, and then, apparently without effort, they rose to the solemn and almost. to the sublime ; the combination, without incongruity, of lofty morality with almost boyish fun was quite wonderful, and almost Shakespearean. It must have been on getting up at one of these dinners, that he explained the nervousness often felt on standing up to speak by maintaining that the vital spirits were very much in the nature of a fluid ; as long as you were sitting all was right, but the moment you stood up they left your head and went down into your boots. He used to tell us that the first conception of the Society was that of an organisation for the study of natural history ; and he somewhat regretted that the overwhelming mathematical bias of Cambridge had, to a great extent, changed the original design, and that our Memoirs were so exclusively mathematical as they then were. He was, however, proud of Cambridge mathematics, and I remember to have heard him express his satisfaction thus : 'I rejoice in the progress of mathematical science ; I measure it in this way ; I am a stationary kind of being with regard to mathematics ; the progress of the science may be measured by the small amount of that which I am able to understand ; and I give you my word of honour that I have not been able to understand a single paper that has been read before this Society during the last twenty years[1]'."

At the present day, when the study of Natural Science has been so long accepted in Cambridge, it is amusing to find Sedgwick recording the alarm which the establishment of this very harmless Society seems to have aroused.

To J. F. W. Herschel, Esq.[2]

TRIN. COLL. *February* 26, 1820.

Dear Sir,

I ought before this to have conveyed to you the thanks of our Society for your communication[3]. It will be

[1] *Macmillan's Magazine, ut supra*, p. 479.
[2] John Frederick William Herschel, Fellow of St John's College, B.A. 1813. Sedgwick informs him, 14 November, 1820: "The first meeting of our Philosophical Society took place yesterday evening. We elected several new members, and among the rest the Rev. J. Wood, D.D., Master of St John's. This was more than we expected, and certainly more than Dr Wood intended last year. It seems as if we had risen in his good opinion."
[3] *On certain remarkable instances of deviation from Newton's scale in the tints developed by Crystals with one axis of Double Refraction, on exposure to Polarized Light*, read 1 May 1820, and printed Trans. Camb. Phil. Soc. i. 21.

read at our next meeting. Now that we are launched I have little fear: we shall, I doubt not, go on and prosper. Among the senior members of the University some laugh at us; others shrug up their shoulders and think our whole proceedings subversive of good discipline; a much larger number look on us, as they do on every other external object, with philosophic indifference; and a small number are among our warm friends. We may count on the zeal of our members for a sufficient number of communications; we may also venture to found some hopes on an active spirit infused by a new system. When you visit Cambridge, I shall hope to have the pleasure of seeing you. Peacock presents his kindest regards.

1819.
Æt. 34.

Believe me, Dear Sir,
Yours most faithfully,
A. SEDGWICK.

At the beginning of the Long Vacation—in June, 1819—Sedgwick went into Suffolk to study the geological structure of the coast. Before he could set to work, however, he met with a serious accident, the nature of which has not been recorded, but the following extract from a letter, written in 1866 to the Reverend Osmond Fisher, then Vicar of Elmstead near Colchester, shows that it must have been severe. It is, we believe, the only record of the occurrence.

CAMBRIDGE. Good Friday Morning.
[30 March, 1866.]

"I do hope within the next two or three months to see the Chillesford beds, and to have a look at the beds near Orford, Aldborough, etc., where we have the lowest (so-called) Coralline Crag. I had just touched these when I nearly lost my life, and, instead of working at Aldborough, I was lying on a sofa for nearly three weeks, and only left the room to move away in a post-chaise; and since then (i.e. *June, 1819*!) I have never seen these lower Crag beds. So bravo for a merry meeting, some time hence, at Chillesford, etc."

S. I.   14

Sedgwick was detained at Cambridge until the end of July—partly by the aforesaid accident—partly by his social duties at the Commencement, which was attended by the Chancellor, the Duke of Gloucester, accompanied by his Duchess, and his sister the Princess Sophia Matilda. Sedgwick was appointed one of the "managers," as they were styled, of the public breakfast given by the University to these distinguished persons. It was held in Nevile's Court at Trinity College, and Sedgwick presided at a table in the north cloister[1]. When all was over he started for his summer's work.

This year, 1819, may be regarded as the commencement of Sedgwick's geological career, not only as an academic teacher, but as an original investigator. In the spring, as we have seen, he began to lecture ; and in the summer he undertook the first of those tours, to which, either in England or on the Continent, he usually devoted several months of each year during the most active period of his life. These journeys were undertaken on a regular system, for the investigation of some definite group of rocks, and the results—the most important of which, as we shall hope to show, have been confirmed, rather than shaken, by subsequent research—were duly recorded in a series of papers. In the present chapter and the next we propose to consider the period from 1819 to 1827, during which he explored the west of England in the first instance, and next Yorkshire, Durham, and the Lake District.

The tour of 1819 is described, in part, in the following letter:

TAVISTOCK, August 14, 1819.

Dear Ainger,

It is now nearly a month since I left Cambridge on a mining expedition to the West of England. I have not yet

---

[1] Cooper's *Annals*, iv. 524; *Grace of the Senate*, 29 April, 1819; *Report of Syndics*, 3 July, 1819.

reached Cornwall, the great object of attraction, though I have now for about a week been hovering on its confines. Before I proceed any farther, let me request you to write to me, at Penzance, by return of post. I long to know how you all are, and what has been the fruit of your last year's labours. I shall probably be in the neighbourhood of Penzance in about a fortnight, and shall remain there some time. If your letter should come after I have left that place, I may probably miss it, though I shall order the Postmaster to forward it to some town in the north of Cornwall.

Half-past ten o'clock. I am just returned from a long and fatiguing expedition to some mines on the banks of the Tamar. A cup of tea has so far refreshed me that I think I may have it in my power to finish my sheet before my eyelids come together. I started on my western course the week after our Commencement festivities. The vicinity of Bristol detained me four days. I saw it in the company of Dr Gilby to great advantage. He has paid great attention to geology, and has published two papers on the structure of that neighbourhood[1]. I was therefore enabled by his assistance to observe everything best worth seeing. I afterwards rambled on foot all over the Mendip and Quantock hills, and examined almost all the cliffs on the north-west coast of Somersetshire. They afford fine specimens of the contorsions exhibited by that rock to which geologists have given the name of greywacké. What a delightfully sounding word! It must needs make you in love with my subject. The country I have been just describing wants some of the grander features, but in beauty, luxuriance, and variety, yields to none. The rugged cliffs which rise perpendicularly on both sides of the Bristol Channel are in many places exquisitely contrasted with the fine lawns and rich foliage which go sweeping down to the very edge of

---

[1] W. H. Gilby, M.D. published *A Geological Description of the Neighbourhood of Bristol* in Tilloch's *Philosophical Magazine*, Vol. xliv. (1814); and *On the Magnesian Limestone and Red Marl or Sandstone, of the Neighbourhood of Bristol*, Trans. Geol. Soc. Lond. 1817, iv. pp. 210—215; besides other papers.

1819.   the water.   As for the people of Somersetshire, they seem a
Æt. 34. mighty stupid good sort of people, who have not wit enough
to cheat a stranger.   The men get drunk with cider, and the
women make clotted cream.   I remained a week with an old
friend and relation of my father[1], and then proceeded to
Plymouth by the way of Exeter and Ivy Bridge.

Devonshire has rather disappointed me.   I had heard too
much of it, and perhaps have not seen the finest part of the
county.   The country about Plymouth is, however, in its way
of unrivalled beauty.   You of course have often heard of
the breakwater.   When completed it will form a mound of
solid masonry a mile long, in many places seventy feet high,
and more than five hundred feet wide.   The blocks of which
it is composed weigh from three to nine tons each, and are all
procured in the quarries of limestone which fortunately
abound immediately on the Plymouth shores.   More than six
hundred day-labourers, and upwards of forty transport-vessels,
are employed in this enormous work.   I was greatly delighted
in observing the operations of the workmen in the quarries.
They commence operations by clearing away till they have a
kind of terrace.   They then proceed to form a hole in the rock
in the usual manner, but not of the usual dimensions, for they
penetrate to the depth of five or six feet, and run down eight
or nine pounds of gunpowder into one opening.   One of these
blasts, when well-placed, will sometimes bring down as much
as forty tons of limestone.   It is a sight well worth seeing to
observe these masses rolling down into the lower part of the
works from the height of 150 feet.   The number of explosions
adds greatly to the effect, for the fires are communicated by a
signal through the whole line of the works.   These masses are
elevated by cranes into low waggons which are conveyed by
machinery into the vessels moored close by.   They are in due
time conveyed away to the breakwater, and discharged upon
it by appropriate machinery.   You may imagine that no

---

[1] The Rev. James Sedgwick, of Curry Rivell near Taunton, a descendant of
the Sedgwick of Bankland.   See above, p. 35.

time is lost when I tell you that in this way a mass of stone 1819.
is daily deposited on the work equal to the burden of two Æt. 34.
of our largest Indiamen.

I have no time to tell you how an alarm of fire threw the
whole house in confusion—how I found my way to Tavistock
—how I mean to get back again—how I purpose to proceed
to Cornwall—what I mean to see there &c. &c. So no more
at present.

A. SEDGWICK.

At Penzance Sedgwick fell in with the Rev. John Josias
Conybeare—of whom and his brother we shall have occasion
to speak at length in connexion with the tour of 1820—and
impressed him very favourably, to judge from the following
passage, which occurs in a letter from Conybeare to Buckland,
written shortly afterwards :

' In the line of Geology the best thing I have done is to contract
a sort of *liaison* with the new Woodwardian Professor of Cambridge,
Mr Sedgwick, whom I met here. He had in his company another
Fellow of Trinity College, Mr Gilby, cousin to Dr Gilby of Clifton.
Mr Sedgwick appears a remarkably clever, active man, and had *done*
all he had gone over in a very accurate and masterly manner. Having
been for some time head Mathematical Tutor of Trinity, he brings
to the study of Geology all the *subsidia* that a thorough knowledge
of mathematics and natural philosophy can give him[1].'

The geological results of this tour were communicated to
the Cambridge Philosophical Society in two papers in 1820
and 1821, and from them we learn that it concluded with a
thorough exploration of a great part of Cornwall, especially
the district adjoining the Lizard. Unfortunately no hint is
given of the reasons which induced Sedgwick to select this
part of England for exploration, but it may be conjectured
that his friendship with the Reverend W. R. Gilby, who was
his companion throughout, may have had something to do
with it. Mr Gilby had relations at Bristol, and therefore
probably knew the West of England, or part of it, himself;

[1] The letter is dated Penzance, 25 September, 1819.

1820.
Æt. 35.

and, besides this, he would be able to urge upon Sedgwick the advantages to be obtained from his cousin's local knowledge.

While Sedgwick was thus employed in Cornwall, Henslow was doing similar work in the Isle of Man. His visit to the island happened to coincide with the first discovery there of the great pre-historic elk—commonly called the Irish Elk (*Megaceros hibernicus*). One nearly perfect skeleton was sorted out from the mass of bones obtained, and passed into the hands of a local blacksmith. Sedgwick was eager to obtain it for his own museum ; and Henslow, no doubt at his suggestion, returned to the island in the following March, in the hope of being able to buy it. After a stormy passage of 30 hours from Liverpool he landed in Ramsay Bay, and on the following day—having borrowed a horse of one man, and a saddle of another—he rode to Bishop's Court. The ingenious possessor of the elk, though ignorant of anatomy, had put the bones together by comparing them with the mounted skeleton of a horse, and had placed his prize in a caravan for exhibition. After examining it Henslow wrote :

" You know I am not much given to the marvellous, but I really think I never saw a more magnificent sight of the kind in my life, and doubt if the Petersburg Mammoth[1] would surpass it. The only parts missing are half of one hoof, and the end bones of the tail; the rest is in the highest preservation. I could not have conceived it would have cut half so good a figure, and the fellow has really put it together with very great ingenuity[2]."

The negotiation to buy the skeleton failed, and it was ultimately acquired for the Museum at Edinburgh ; but the attempt to secure it for Cambridge forms an interesting episode in the early history of the Woodwardian Museum and

[1] The nearly complete skeleton of a Siberian Mammoth, acquired by an Englishman named Michael Adams in 1807, and mounted in the Museum of St Petersburg. It is described (with a complete history of the discovery) in the *Mémoires de l'Académie Impériale des Sciences de St Pétersbourg*, for the year 1812, v. 406. The paper, published only four years before Henslow wrote, had greatly interested all men of science.

[2] Henslow to Sedgwick, 31 March, 1820. The letter is dated 1819, but that it

shows that Sedgwick had already begun to pay attention to Palæontology.

In the following summer (1820), Sedgwick resumed work in the west of England.

To Professor Monk, Trin. Coll. Cambridge.

MAIDENHEAD BRIDGE, *Monday Evening*,

[17 *July*, 1820].

Dear Professor,

About the time that we parted I mentioned to you a subject on which I am now requesting your good offices. I am so much in want of money that without some assistance I shall be obliged to abridge my summer's labours for want of funds to carry me through them. By the Will of Dr Woodward, of blessed memory, I am entitled to a quarterly payment of my salary[1]. The Vice-Chancellor knows the fact as well as myself; and will, I have no doubt, order twenty-five or thirty pounds to be placed to my credit at Mortlock's Bank. That sum, with what I now have, will enable me to pay my way among all the Oolite Beds in the south-west of this island. I hoped before this time to have been fairly at work, but a violent diarrhœa has retarded my progress. I am half disposed to rejoice in my misfortunes as they have induced me to spend a day (and God only knows whether I shall ever spend another day), with our dear friend Mill[2]. He rejoices in his appointment, because it holds out to him a rational expectation of being enabled to employ his great talents in promoting the highest interests of his fellow-creatures. He is now dining with his Rector; I excused myself on the score of bodily infirmity. Tomorrow I hope to be in travelling condition, in which case I shall endeavour to reach Devizes.

should be 1820 is evident from Henslow's *Supplementary Observations to Dr Berger's Account of the Isle of Man*; Trans. Geol. Soc. Lond., 1821. Vol. v., p. 502, *note*.

[1] See above, p. 182.

[2] The Rev. W. H. Mill, Fellow of Trinity College, was at that time curate of Taplow. His rector was the Rev. Edward Neale.

1820.   There I shall unpack old Thor, and commence a furious
Æt. 35.   assault on all the solid materials I may meet with on the
surface of the earth.

You will greatly oblige me by mentioning the subject I
began to write about to the Vice-Chancellor.

Believe me, Dear Professor,

Yours ever,

A. SEDGWICK.

P.S. You fire your thunderbolts[1]; other men must be
content with squibs and crackers.   I have several times
thought of getting up a geological article for the *Quarterly*.
Some interesting foreign works (French and Italian) have
appeared on the subject, which might be made the foundation
of a dissertation.   My health is never to be depended on; I
have no facility of composition; what is more I am much
engaged; still I think that I might be able to bring together
some remarks not entirely undeserving of insertion in that
Journal.   Perhaps you would have the goodness some time or
other to mention this to the editor.   *Vale.*

CHARLTON MACKEREL, September 5, 1820.

Dear Ainger,

I am now halting with my old friend Sharpe[2], who
has been married for several months, and is bearing him-

---

[1] This expression refers to the controversy, in 1818—19, between Professor
Monk and Sir J. E. Smith, President of the Linnean Society. The latter had
been authorised by the Rev. T. Martyn (who had been Professor of Botany since
1761, but had not lectured since 1796, nor resided since 1798) to deliver a course
of lectures on Botany. The Vice-Chancellor, Dr Webb, Master of Clare Hall,
had given his consent, and the lectures had been announced to begin on Monday,
6 April; when on Saturday, 4 April, a remonstrance was forwarded to the Vice-
Chancellor, signed by eighteen Tutors of Colleges, who objected to their "Pupils
attending the Public Lectures of any Person who is neither a Member of the
University, nor a Member of the Church of England." After this Sir J. E. Smith
declined to lecture, and wrote an indignant pamphlet, which Monk answered. In
July, 1818, an article in favour of the course pursued by Monk and his fellow-tutors
appeared in the *Quarterly Review*, Vol. xix. pp. 434—446. Smith wrote a
pamphlet in answer to it, and Monk a counter-pamphlet. Here the controversy
ended.

[2] William Sharpe, Trin. Coll., B.A. 1807.

self with all meekness after his exaltation to the Benedictine 1820.
honours. Perhaps your brother will have informed you that I Æt. 35.
started on an expedition to the West of England in the second
week of July. My first object was to examine all the strata
from the foot of the Wiltshire chalk-downs to their termination
in the ravines near Bath. This part of my summer's labours
employed me about three weeks. I was, however, interrupted
by a severe cold, which yielded immediately to the effects of
the hot bath. Though I started alone, this part of my expedition has not been quite solitary. I was joined near Bath by
an Oxford gentleman with whom I formed an acquaintance
on the road. I also experienced the greatest possible kindness from Mr Conybeare, an Oxford Professor and a stone-
eater. After leaving Bath I went to the house of Mr William
Conybeare, brother of the aforesaid Professor (perhaps you
will like these men better when you know that they are grand-
sons of Bishop Conybeare)[1], who accompanied me in my
expeditions for three weeks, during which time we examined
the most interesting portions of the country to the north of
Bristol. If I were to give a minute account of our labours I
should be obliged to use language which would hardly be
understood, and, if understood, would not, I fear, be very
amusing. I must therefore leave the lower regions of the
earth and talk of its surface. The whole face of the country
north of Bristol, and up to the banks of the Severn, is most
lovely. The hills are not very lofty, but they are beautifully
broken, and the woodland scenery is among the richest in
the world. Part of the country you are, I believe, acquainted
with.

About the time that I purposed leaving Mr Conybeare's
house I was attacked by my old enemy; after being raked
fore and aft for eight and forty hours, I really thought that
the whole vessel was going to the bottom ; the internal hurri-
cane, however, suddenly abated; a state of calm for three days
succeeded, and after that time I was able to spread my sails,

---

[1] John Conybeare, D.D., Bishop of Bristol 1750—1755.

and steer with a fair wind for the Mendips. Mr Conybeare accompanied me in that excursion. We spent five days in examining a most interesting mountain ridge. We have here on a small scale examples of every variety of stratification, and of almost every species of secondary rock. The great mass of rock is the same limestone we have about Sedbergh, and in the district of the caves. The strata of the Mendips are, however, much more highly inclined, being in some instances nearly vertical, and are surmounted by many newer strata which ride upon their edges in an horizontal position. This singular conformation is exhibited in many of the deep glens which traverse the strata near the eastern end of the chain.

Mr Conybeare and I parted on the top of the Mendips on Saturday last. I descended to Wells. The Cathedral is small, but very perfect ; and the west front is ornamented by a great many good examples of ancient gothic sculpture, which the fanatical zeal of our blessed reformers has fortunately spared. All revolutions are accompanied with violence, which is an evil great enough in some instances to counterbalance the good of change. It was, however, better that our protestant ancestors should break the heads of stone images than of men and women, after the manner of our neighbours on the other side of the water.

Sharpe is coming down to breakfast, so that I must cut short my narration. My next great move will be to the coast near Sidmouth, from which place I intend to face about, and trudge my way back by the coast as far as Portsmouth. Thence to Cambridge. Best respects to Mrs A. and the young ones.

Yours ever,

A. SEDGWICK.

TUXFORD, NOTTS, *November* 10, 1820.

Dear Ainger,

I wrote to you during the labours of the Long Vaca-

tion, but from what place I do not at this moment recollect. 1820.
I presume that I gave you some account of what I had been Æt. 35.
about, and what I intended to do. How uncertain are all
our expectations! I have not yet reached the University.

After an examination of some parts of the country near
Taunton, which I visited last year, I crossed the Black-
downs to Sidmouth, and there commenced a laborious
examination of the coast. The cliffs are on a most magnificent
scale, abound in organic remains, and are of great geological
interest. Almost the whole coast of Dorsetshire presents a
succession of rugged precipices of varied forms, arising from
the peculiar disposition of the strata. Weymouth detained
me three weeks. The geological map of that district is so
erroneous that I resolved to rectify it as far as my time would
allow, and I succeeded almost to the extent of my wishes.
From thence I found my way to the Isle of Wight by
Southampton, and there heard, from my brother James,
such an alarming account of my mother's health that I immediately recrossed the Channel, and hastened down to the
North with all the expedition which I could command.

I should have been very thankful had it pleased God to
have allowed me to arrive in time to receive my poor mother's
last blessing; but that melancholy satisfaction was denied me.
She died early on Sunday the 15th, after an illness of a week.
I rejoice to say that she possessed her self-possession almost
to the moment of her death, and expressed her entire resignation to the will of God. We have all reason to thank Him for
His great mercy in so long sparing our dearest friends. This
is the first affliction with which our family has been visited.
My father and sisters are quite well. My father looks much
better now than he did last spring when I was in the North.
He is much afflicted at losing the companion of his old age,
but on the whole is less weighed down than I could have ventured to hope. It may still, I hope, please God to spare him
to our family for some years.

I am now resting on my way to Cambridge with my old

1820.   friend Mr Mason. I hope to be in Cambridge tomorrow, and
Æt. 35. shall find abundant employment for the term in my professional duties. My spirits are not good, but I hope to rally
sufficiently to give a few public lectures before the Christmas
vacation. My sister Ann was married about two months
since to Mr Westall jun. I have no doubt you heard of the
engagement when you were in the north. God bless you and
yours.

Yours ever,

A. SEDGWICK.

The touching allusion in the first of the above letters to
the Reverend William Hodge Mill, Fellow of Trinity College,
then about to leave England as Principal of the newly founded
Bishop's College at Calcutta, is a proof of the breadth of
Sedgwick's sympathies, and of the strength of his affections.
No two men could have differed more widely—Mill was a
High Churchman—Sedgwick, if he belonged to any party
in the Church, an Evangelical. Yet they exchanged letters
frequently during Mill's absence in India, and after his return,
when he resided in Cambridge as Regius Professor of Hebrew,
notwithstanding the very decided attitude taken up by both
in the religious controversies of that day, their friendship did
not suffer any interruption.

Sedgwick's meeting with the brothers Conybeare, briefly
mentioned in the second letter, is specially noteworthy. With
John Josias Conybeare, Professor first of Anglo-Saxon, and
next of Poetry at Oxford, then rector of Batheaston, close
to Bath, he had become acquainted while exploring Cornwall the year before, as mentioned above[1]; but his brother,
William Daniel Conybeare, afterwards Dean of Llandaff, who
then held a lectureship in the church of Brislington near Bristol,
seems to have been unknown to him until this occasion[2].

---

[1] See above, p. 213.

[2] In a letter written to the Rev. John Charles Conybeare, second son of the
Dean of Llandaff, dated 15 October, 1858, Sedgwick refers to his first acquaintance
with the family: " Tell Mrs Conybeare that I love all ladies who pass under the

The introduction led to important results. Not only was the foundation laid of a lifelong friendship, continued to the younger members of the Conybeare family long after their parents had been laid in the grave, but Sedgwick obtained so large a measure of geological instruction that he used to speak of Mr W. D. Conybeare as his master in the science. The two brothers were both fond of geology, and while still at Oxford, in conjunction with Buckland, had established a sort of club for the study of it. Thus educated in the methods of the science, and, we may be certain, well-informed respecting the geology of the district in which they resided, they would both be eminently qualified to serve as guides to one who until he met with them had had little except his natural talent to help him. The extent of his obligations, as realised by himself at the time, is forcibly expressed in the last letter which he wrote to Conybeare from Weymouth, detailing his discoveries and difficulties. "My Long Vacation is now ended, and I go home with the conviction of having completely accomplished the great objects of my summer's tour. I may add, with great truth, that I consider the acquaintance I have formed with you among the most fortunate and agreeable circumstances of my vacation. If I had not been under your tuition for three weeks, I should, I fear, never have been able to disentangle the difficulties of this neighbourhood."

By a strange coincidence this letter—throughout which Sedgwick expresses himself in a tone of elation at his well-merited success—was written on the very day of his mother's death. How deeply he felt this sudden calamity—'the first deep domestic pang I ever endured'—is shown by the letter in which he informed Ainger—who must have known her well— of his irreparable loss; but the depth of his sorrow may be

name of Conybeare; and tell little Mary that I mean to love her as dearly as I loved another Mary [the Dean's only daughter] whom I first knew at Brislington in 1820, when little Johnny was on the knees. It was the year I first became acquainted with your Father. I knew his brother John before, having been acquainted with him in Cornwall. But what ancient memories these are now!"

estimated more truly still from the casual references to his mother which are scattered through his later correspondence. Thirty years afterwards, when one of his friends was anxious about her own mother, he wrote: "The word Mother has a charm in its sound; and there was a blank in the face of nature, and a void in my heart, when I ceased to have one;" and again, to the same; "It was my first great domestic sorrow, and deeply did I feel it. I pity the man who has no remembrance of a mother's love. The memory of my dear mother and my dear old father throw a heavenly light over all the passages of my early life[1]."

During the months which elapsed between Sedgwick's return from Dent in 1820, and the commencement of the Long Vacation of 1821, he was fully occupied with what he termed 'professional pursuits.' Not to mention lecturing—digesting the information acquired in the course of the previous summer—and writing at least one paper for the Philosophical Society—he had plenty to do in his Museum. His report (dated 1 May, 1821) speaks of three months spent in arranging the contents of seven large cases of "specimens from all the strata of Somersetshire, Gloucestershire, and Dorsetshire which appear between the old red sandstone and the chalk;" "the arrangement of a series of specimens from all the English strata, commencing with the granite of Cornwall, and ending with the alluvial deposits of Suffolk, which the Professor has employed three vacations in collecting;" and he gratefully acknowledges "the assistance of Mr Henslow in arranging the simple minerals[2]."

It is pleasant to be able to record that Sedgwick's delivery of a course of lectures—in excess of the minimum of four prescribed by Dr Woodward—did not pass unrewarded. It

---

[1] To Miss Malcolm, 7 December, 1856; 21 December, 1859.

[2] It must be remembered that the Woodwardian Audit, at which the Professor made his annual report, was held on the first day of May. As Sedgwick usually left Cambridge soon after the commencement of the Long Vacation, most of the work chronicled in these reports must have been done between his return in October and May of the year following.

will be recollected that on the day of his election the Senate had accepted a supplementary scheme for the government of the Professorship. In this document three points were mainly insisted upon : (1) that a Museum, with an apartment for the Professor thereto adjoining, ought to be built without delay; (2) that the payment of the Professor's stipend should be contingent on his delivering lectures ; (3) that if he gave additional lectures, he ought to receive an additional stipend, but only after the erection of the aforesaid buildings. Two years had elapsed, but nothing had been done ; nor was it probable that any building-scheme would be accepted for some time. Sedgwick was naturally unwilling to see his promised increase of stipend adjourned *sine die*, and therefore addressed a letter, dated 22 June, 1820, to the Vice-Chancellor and Heads of Colleges. This letter contains some interesting biographical details, and therefore we may be excused for quoting a considerable portion of it. After giving an analysis of that portion of Dr Woodward's Will which dealt with the annual income of his estate, he proceeds,

"In consequence of the great change in the value of money since 1727, the Woodwardian estate now rents for £430, but none of the specific payments have received any augmentation whatever. The present Lecturer does not therefore stand in the situation intended by the Founder, the stipend he receives being virtually not much more than one-third of the sum intended.

In the same Will it is directed that the Woodwardian Cabinets be 'reposited in a proper Room or Apartment allotted by the University,' and, 'that the Lecturer reside in or near the said Apartment.' When therefore the University accepted the Cabinets, they contracted an obligation to find an Apartment, suitable at the same time to the Collection and to the accommodation of the Professor, out of their own funds. It certainly does not appear to have been the intention of Dr Woodward that any part of the rents of his estate should be held back as a building-fund. As however there is at present an accumulation of fourteen or fifteen hundred pounds out of the Woodwardian estate, the said accumulation might, with the addition of such sums as the University shall think fit, be employed in erecting a 'proper apartment' for the reception of the Geological Cabinets, in fitting up a Lecture-Room, and, if thought expedient, in building rooms for the Lecturer, contiguous to the said Apartment.

1820 to   It is further obvious, from the said Will, that Dr Woodward
1821.   intended his Lecturer to perform important duties[1]....In the year
Æt. 35— 1727, a salary of £110 was a sufficient remuneration to a member
36.   of the Senate for performing the conditions in question. At present
the same sum is not sufficient.

The Woodwardian Lecturer wishes finally to observe that, since
his appointment, he has endeavoured to comply with the severest
clauses of the Founder's Will.

1. He resigned, on his appointment to the said Lectureship,
offices and employments in College of the yearly value of £200.

2. He has read 22 public Lectures to the University and
gratuitously, and will engage to give at least that number annually.

3. He is preparing to print the substance of two of the said
Lectures.

4. He has been always ready to exhibit the Museum, and
during term has spent more hours in it than are specified in the
Founder's Will.

5. He has, at a personal expense of between three and four
hundred pounds, made a Geological Survey of several parts of
England, by which he has been enabled to deposit large, and, he
believes, important additions to the specimens in the University
Cabinets; for an account of which he refers to the report drawn up
under the sanction of the Inspectors and presented to the Auditors
on the first of May[2]."

The persons addressed took six months to consider this
letter, but early in the following year (24 January, 1821), a
Grace passed the Senate, by which, after several well-turned
compliments to Sedgwick's energy and capacity, it was
provided that £100 should be paid to him for the extra
lectures delivered in the past year; and £100 in each future
year, on the condition that he delivered fifteen lectures, at
the least, in each year, in addition to the four stipulated for
by Dr Woodward.

The next three letters describe the employment of the
Long Vacation of 1821. For once, it may be remarked, we
hear nothing of his health—perhaps he was too busy to
pay attention to it.

---

[1] In the omitted passage Sedgwick enumerates the principal duties imposed by
Dr Woodward, as related in the previous chapter.

[2] The only copy of this letter which we have seen is in the rich collection of
University Papers formed by the late Dr Webb, Master of Clare Hall 1815—56,
now in the University Library.

TRIN. COLL., *April* 13, 1821.

Dear Ainger,

Your letter was left on my table on Tuesday last. When I saw the direction I believe I should have blushed if my complexion would have allowed it. For my conscience told me that I had been your debtor three or four months, both for a letter and for a small book of goodly admonition. Let me thank you for one or both of them now, if it be not too late. I have myself turned author since we met last. I will send you a copy of my paper in the *Cambridge Transactions* when I have an opportunity. Not that I wish you to read it. I will therefore give it to Mrs A. to tie up sugar-plumbs for my god-daughter. But hold—I am quite out of all order. I must first congratulate you and Mrs A. on your new accession of domestic honours, and then express my own joy in the hopes of establishing this spiritual relationship with the young stranger. Both my god-fathers are old bachelors, and my god-mother (God be with her), is as arrant an old maid as ever whispered scandal round a tea-table. My own destinies were therefore fixed at the font, and I already feel myself fast sinking in the mire of celibacy. I hope I shall bring no evil on my charge. But do contrive to have some one joined with me who is either married or given in marriage, otherwise we cannot answer for the consequences. *Experto crede.*

Geology has not, I hope, dried up all the social affections; it has, however, left me very little time for the exercise of them. I am too much engaged to be down this short vacation. The last week in May and the first week in June I shall be engaged with our examination in Hall. A stupid companion you will find me if here at that busy time; still let us meet if possible. I will promise to do my best during the intervals. Immediately after that troublesome business is over I start for the Isle of Wight[1], to relieve my brother who

---

[1] Whewell, writing to his sister, 18 June, 1821, says: "I am going for a short time to the Isle of Wight. I expect to join there Professor Sedgwick, a very intimate friend of mine." *Life*, p. 64.

is going down to the North. In the months of July, August,
and September I must make a regular geological tour, but my
actual destination is not quite fixed. You may suppose that
my hands are full when I tell you that besides the care of our
college examination I have to give public lectures four times
a week during the next term, and that I am now preparing a
syllabus of my course for the press. Present my kindest
remembrances to Mrs Ainger, and believe me,

Yours ever,

A. SEDGWICK.

WHITBY, *September* 2, 1821.

My dear Ainger,

For the life of me I cannot tell when or where it was
that I last wrote to you. It must surely have been since my
return from the Isle of Wight. Let us now therefore take it
for granted—that I did return from the Isle of Wight—that I
did sojourn about three weeks in the University of Cambridge
—that I then took a tour through Coventry, Kenilworth,
Warwick, Birmingham, and Lichfield, where I saw churches,
ruined castles, hardware manufactories, and Cathedrals—that
I attended an auction of old bones, of which I bought enough
to fill fat Lambert's coffin—that I then found my way thro'
Derby, Nottingham, and Lincoln, to the residence of my
brother John[1]—that after three days I left the lodgings of my
said brother, and was conveyed in a steam-packet down the
waters of the Humber as far as Hull—that I left Hull on the
top of a coach—that I was set down by the said coach at the
door of my old friend Gilby, who is now married and performing the duties of a parson and a magistrate in a small village
about five miles from Bridlington. All this I give in the way
of summary, because: 1st, it is no easy matter to give it in
any other form ; 2dly, I do verily believe that I have given
you the greater part of it before.

[1] At that time curate of Stowe in Lincolnshire.

With Gilby I remained three or four days, during which 1821.
time I made an excursion almost as far as Spurn Head. Æt. 36.
Holderness is well cultivated and well inhabited, but as dull as
the fens of your native county. Its physical structure did not
supply me with a single new fact. The sea is making terrible
encroachments on the whole district. In one place the cliff
cuts through an ancient burial-ground, and the upper face of
the precipice is literally studded with human bones. Gilby
accompanied me on foot almost as far as Scarborough. North
of Bridlington the character of the country is completely
changed. We crossed the great chalk range, which at its
northern termination is nearly ten miles broad. As we went
along the top of the cliff we frequently had on our right hand
a naked precipice of chalk more than three hundred feet high.
The chalk cliffs of the Isle of Wight must, I think, yield to
these in grandeur. But in the wolds of Yorkshire we entirely
want that combination of woodlands which makes the scenery
in the Isle so peculiarly beautiful. North of the chalk range
the cliffs are of less elevation, but more varied in form, and
perhaps more beautiful than those of Flambro' head. The
bay of Scarborough viewed from the south is not inferior to
any part of the coast which I have seen. It is bounded to the
north by a fine mass of perpendicular rock, which is crowned
with the ruins of the Castle. So far I have got on without
taxing your patience with any account of my peculiar craft,
and I hope to finish without doing so. I am, however, partly
disqualified from enjoying the picturesque beauties of this
country, as I have nearly lost the use of one eye in my combats
with the rocks. A splinter struck it with such violence that it
has for the last three or four days been of very little use
to me[1].

I am sitting in the travellers' room, and they are beginning

---

[1] Sedgwick's eye never recovered from this accident. Writing to Lady
Augusta Stanley, 23 July, 1865, he says : "My old eyes—I ought to say my old
eye, for one of my original pair has struck work ever since 1821, when I offended
it by a splinter of rock which flew from my hammer in Robin Hood's Bay—
work badly by candle-light."

to be so noisy that I really hardly know what I write. Let me then tell you at once that I have paced my way to this place along a most rugged coast, and that I hope to proceed with my tour on Tuesday next. Whitby is a dirty, stinking, town in a very picturesque situation, but I have no time to describe it. Pray write to me Post Office, Sunderland. Give my best respects to Mrs Ainger.

Yours ever,

A. SEDGWICK.

P.S. *Monday morning.* I concluded last night rather bluntly. Wine and tobacco seemed to excite some of my companions above all measure. Among the rest an old, fat, deep-mouthed Scotsman became so enthusiastic in his admiration of the native genius of his country, that he roared, ranted, or sung, whole pages of Burns. The entrance to the harbour of Whitby is by a narrow opening in the cliff. It winds up two or three miles into the country flanked on both sides by very steep hills. The town is disposed on both sides of this estuary just at its entrance into the sea. St Hilda's domain was on the south side of the estuary. The ruins of the abbey are still very imposing, and in a very beautiful style of architecture. The choir and transept are nearly perfect, and in the Early English order; the nave is more ruinous, yet has some exquisite arches of the more modern Gothic. About two-thirds of the great tower is standing[1]. The old Scotsman is again come into the room coughing, and breathing hard, so I must conclude. I hope to emerge from this district to Stockton in about a week. I shall then go up the coast of Durham. The greater part of the expedition must be performed on foot. *Vale.*

DENT, October 16, 1821.

Dear Ainger,

My memory does not improve, for I am as much abroad as I was before in regard to the place from which I directed my last letter. It was I suppose from some place or

---

[1] These remains of the great tower fell at 1 p.m. 25 June, 1830.

other on the Yorkshire coast. My sisters and Miss Davoren[1] are so noisy that my poor confused brain is not much aided in its recollection. I will not give you any details respecting the cliffs of alum-shale, and the mode of extracting the salt, lest I should bore you with a second edition of what I told you in my last. These cliffs in some places rise to the elevation of 600 feet, and then take a sweep round into the interior, forming a magnificent natural terrace overlooking the fine flat district of Cleveland and part of Durham. From Stockton I made a two days' excursion up the Tees, and called upon our old friend Wallace[2]. He looks charmingly for a man of fifty, and is as much alive as ever. Old time, that unmerciful scratcher of faces, has, however, worn a few lines in his face since you were of his household. The coast of Durham, in a picturesque point of view, is very inferior to the north-east cliff of Yorkshire. The rocks prevailing in that district are composed of a magnesian limestone which performs more freaks in its mode of aggregation than any mineral substance I have yet examined. The mouth of the Tyne has an interest peculiar to itself. The river finds its way into the sea through a chasm in a rock which on the Northumberland side is of great elevation, and crowned by a very picturesque ruin of an old abbey. The eternal bustle on the river, which a little above Tynemouth is wider than the Thames, reminds one of the scene below London Bridge. I found my way up this river to Newcastle in a steam-boat; and in getting out of it had my ribs nearly staved in by a fall in the hatchway. Old Robert Foster now lives there, and is very hoary. I returned by Durham, and spent a day or two with Carr, who has now three children, and will soon (D. V.) have a fourth. From

[1] Miss Jane Davoren, niece to Mrs Brownrigg, a lady who had resided for several years at Broadfield in Kirthwaite, married the Rev. John Sedgwick in April, 1822.

[2] Dr Wallace, a physician who lived at Sedbergh when Sedgwick was a boy, and was a great friend of his. Ainger boarded in his house. A few years before the date of this letter Dr Wallace removed to the neighbourhood of Barnard Castle.

there I went to Darlington, to take possession of a horse
which the Doctor had bought for me; on this beast I went up
the higher part of Teesdale. It is perhaps more beautiful
than any valley in the North of England. I had before
examined the country on the borders of the river near Rokeby,
which has been so well described by Scott. In the highest
part of the valley, which is ornamented with foliage, the whole
river is precipitated over a fine mass of columnar basalt, and
forms a fall not inferior to any in the Lake district. The
columnar basalt is also found on both sides of the valley for
some miles above the High Force, and is arranged in magnificent clusters of pillars. I was driven from this interesting
district by the most incessant rain with which I was ever
persecuted, and found my way to Dent by the way of Brough
and Kirkby Stephen. When I reached Sedbergh I was
soaked with wet like a piece of blotting-paper.

My friends here are all well, but complain of the shortness
of my visit. I arrived last Friday morning and intend to
start for Cambridge the day after tomorrow. Under the
circumstances of my visit you will see that it was impossible
for me to reach St Bees. I shall be happy to assist in the
procession when you keep your Act[1]. The ladies, who are as
noisy as ever, now desire their kind regards. For the last
quarter of an hour I have not been able to see a word that I
am writing, and Miss D. is just beginning to rattle the keys of
the piano, so I must needs conclude. My best regards to Mrs
Ainger and my love to my god-daughter. Yours ever

A. Sedgwick.

Mr Robert Foster, whom Sedgwick met at Newcastle, was
a friend of his boyhood at Sedbergh, and he has sketched his
portrait with singular vividness among his reminiscences of
those early days.

" The next person who rises before my mind's eye is Mr Foster
of Hebblethwaite Hall, a beautiful property a little more than two

---

[1] Mr Ainger proceeded to the degree of B.D. in 1822.

miles above Sedbergh. He was of the Society of Friends, and sometimes when he drove over to visit the brotherhood in Kirthwaite, or
at other times when tempted by the bright weather to make a short
cut over the hills to the old vicarage of Dent, he would halt a few
hours in friendly intercourse with my father. I remember his
presence well, when I was but a little boy : his dark complexion
which had been made darker by a tropical sun; his small and regular
features; his dark and bushy eyebrows ; his earnest and grave look,
which at first sight gave me an impression of sternness. But all
that feeling went off when he began to speak; for his voice was
pleasant, and his discourse at once earnest and genial. Even in my
childhood I felt joy whenever he came to the vicarage ; and I used
to creep behind his chair that I might hear him talk. He wore a
broad-brimmed hat, and a grave outer garb of a quaker cut ; but I
never thought that he looked quite like a quaker. He had not the
soft, bland, expression of a good old quaker statesman ; and he had
a confirmed habit of slovenliness, which was utterly unlike the
precise and perfect neatness of all other men of his grade in the
Society of Friends.

While at Sedbergh School he soon outstripped all the boys of his
class in making his way through the standard authors in Greek and
Latin; and he outstripped them quite as much in audacious deeds of
eccentric waggery. His mind became inflamed by dreams of foreign
lands, and thoughts of enterprise ; and while in such moods, spite of
the beautiful scenery of his native home, his yearnings were little
satisfied by the thoughts of settling down into the placid life of a
leading quaker statesman. So he one day packed up bag and
baggage, and walked off to seek his fortune ; and a few days after
wards (I think at Liverpool) entered himself in a foreign-bound vessel
as a common sailor. He set to work in his new life with all the
energy of his ardent will ; and the master of the vessel, who was
a man of good sense and humanity, marked the boy's style and
manner, took him to his cabin, and drew from him his secret. 'You
have done wrong in leaving your parents,' said the captain ; ' but
spite of that I like your spirit, and I give you the choice of two
things : If you have the heart to go on with this profession you must
leave this ship, and be rated as a midshipman in a man-of-war, and I
have a friend in the royal navy to whom I will send you, and you
will be put, as a young gentleman, in a right position. If this do
not suit you, I have no choice left but to put you under arrest, and
send you back to your father.'

There could be no doubt which alternative the boy would choose.
He *was* rated as a midshipman in a man-of-war ; and by an
enthusiastic devotion to all the duties and studies of his profession,
he gradually became an accomplished sailor ; and during one of the
early wars of the reign of George III. performed in the West Indies
such acts of well-timed and daring courage, that he obtained the
commission of lieutenant much sooner, I believe, than would be
compatible with the rules of modern service.

Once or twice during the intervals of active service he came down to Hebblethwaite Hall; and it is said that he appeared at Briggflatts Meeting-house, with his laced cocked-hat on his head, and a cutlass by his side; perhaps to the suppressed admiration of the younger Sisterhood, but certainly to the horror of the venerable and peaceful Fathers of the Society. Every effort was made to win him back to a peaceful life. He loved his friends, and he loved the Dales; but he resolved to continue in that profession in which he had already won some glory.

At another interval in the service he again came down to Sedbergh, and mingled once more with the tried friends of his early youth: and then it was that he proved in his own person—what he had read of in the poets of antiquity—that love is in conflict mightier than fire and sword. He was smitten by one of the youthful Sisterhood, as by a fire from a masked battery, and brought to the ground, never again to rise in his former strength. His courage was gone, for no heart was left in him. His dearest friends seized the opportunity; and by every entreaty of duty, by the power of youthful passion, and by the prospect of realising new dreams of happiness in the immediate possession of the family estate and the lady of his first love—by the might of all these motives acting together, he was conquered, and struck his flag for ever. His visions of future glory vanished like the colours upon an air-bubble, and he collapsed into the condition of a country gentleman, much honoured in the Dale, and of a leader in that Society in which fate had first placed him.

These events happened long before I was counted among the inhabitants of the Dales; and after the lapse of many years, while I boarded with a kind quaker family, we often saw Mr Foster; and greatly rejoiced when we were invited to spend a half-holiday at Hebblethwaite Hall. He loved the society of boys who had risen to the upper classes of the school; and he had resumed his studies of the classics, and become a very accomplished Latin scholar. Sometimes he half alarmed us, when he took down some ancient classic, and began to discuss a point of criticism. We thought we had enough of such matters when before our schoolmaster. But our fears were of short duration; for he was soon carried on by his love of the author; and then, in a way peculiar to himself, he would roll out a noble translation of some favourite passage. It might be from one of the orations of Cicero, or some pregnant and pithy chapter out of the works of Tacitus; or it might be some burst of indignant scorn and mockery out of one of the old Roman Satirists. These were days of delight to the schoolboys who had the honour of being admitted to such genial and healthy visits.

Sometimes, but rarely, he and my father had discussions at the vicarage on subjects of religious ordinances; but I think I may say with full assurance that no word of bitterness ever escaped from the tongue of one or the other. They agreed in many of the great essentials of Christian truth: and they agreed that the end of all religious ordinances was to bring the heart—the fountain-head of all true religious emotion—into conformity, both in thought and outward act, with the revealed will of God.

The last time I saw Mr Robert Foster was at Newcastle, I believe in the year 1821, while I was upon a geological tour. The load of years had then been resting upon him : but his heart had not become cold; for the old man received me with the warmest welcome ; and then he walked with me (no longer with his firm step of former years), and shewed me some of the neighbouring establishments on the river Tyne. He seemed to be again in his own element ; and all the persons connected with the shipping interests of the river treated him with marked respect and confidence. After a while he said, ' We will go and rest ourselves at the study of one of my friends. You will like to know him, for he is a man of genius, and a great humourist.' It was Bewick, the well-informed naturalist, and celebrated engraver upon wood ; and we had a long and delightful interview with that great artist[1]."

The geological work of the next four months—as well as the results of the tour of 1821,—are well described in Sedgwick's report to the Woodwardian auditors dated 1 May, 1822.

"The Professor spent the month of June and the early part of July in examining the structure of the Isle of Wight, the coast of Hampshire, and part of Oxfordshire. The spoils obtained during this excursion were conveyed to the University in four large cases.

"He afterwards went to Lichfield for the purpose of attending an auction of fossils, and was fortunate in obtaining some very valuable specimens at what he considered a reasonable price. They were conveyed to the University in one very large case.

"He then employed between two and three months in a geological survey of the coasts of Yorkshire, Durham, and Northumberland. During this excursion (the greatest part of which was necessarily made on foot) he collected many illustrative specimens, which were sent off from Whitby, Sunderland, and Newcastle, in four large packing-cases. He also succeeded, through the assistance of a clergyman at Whitby, in purchasing some valuable spoils of the *Ichthyosaurus*, which have been conveyed to Cambridge in three large cases. On his return from the coast of Northumberland in the month of October, he examined the basaltic and mining districts of High Teesdale, and collected many interesting specimens.

"Of the preceding collections a part is at present in the progress of arrangement. The remaining specimens must be returned to the packing-cases until some more accommodation is found for their reception : as there is at present hardly a single drawer in the Museum which is unoccupied.

---

[1] *Supplement to the Memorial, ut supra*, pp. 54—59. The passage has been slightly compressed in transcription.

1822.
Æt. 37.

"The Museum has also received a very valuable accession in a collection presented by Mr Henslow, which consists of nearly 1000 specimens carefully selected during a geological survey of the Isle of Anglesea, and illustrated by a memoir and sections which will be published in the next number of the Cambridge Transactions[1]. Mr Henslow has undertaken the arrangement of this collection, which occupies twenty-four drawers.

"The Woodwardian professor begs finally to add that he has this year read twenty-eight public lectures to the University, and that he is endeavouring to comply with a clause in the Founder's Will in preparing for the Press the substance of two of the lectures[2]."

Most men in the position which Sedgwick now held, with an annual course of lectures to deliver, the value of which had received a substantial acknowledgment from the University— a Museum to maintain—and the almost boundless field of geology before him—a *terra incognita* of which he had just commenced the exploration—would have devoted themselves to their new duties with a singleness of purpose which would have excluded most other interests. But this was what Sedgwick never could bring himself to do. He had no intellectual self-control; he could never shut his eyes and ears to what was going on around him; and we shall continually find his geological work laid aside for long intervals, because he had allowed himself to be carried away by something foreign to what ought to have been the real purpose of his life —something which others less occupied than himself would have done as well, or better, than he did. At one time he appears as a member of a Syndicate appointed to provide temporary accommodation for Viscount Fitzwilliam's collections, and to consult and report to the Senate on the erection of a permanent Museum, pieces of business which led those who took them in hand into long and tedious negotiations; at another we find his name on *The University Branch Committee for promoting a subscription in Aid of the Greeks*[3],

---

[1] Mr Henslow's *Geological Description of Anglesea* was read to the Cambridge Philosophical Society 26 November, 1821, and is published in their *Transactions*, Vol. I., pp. 359—452.

[2] These lectures were never published.

[3] *Address of the Committee*, 20 Nov., 1823. The Rev. G. A. Browne, Fellow of Trinity College, was specially active in this matter, and it was probably through

whose 'holy cause,' as it is termed, was no doubt specially
dear to so true an Abolitionist as Sedgwick, and occupied a
proportionate space in his time and thoughts. The consequences of these distant excursions may be easily imagined.
Geological memoranda which ought to have been arranged
when the subject was fresh in his mind were laid aside;
specimens remained for months—sometimes for years—un-
determined, or even not unpacked; promised papers were
not finished—perhaps not begun. These remarks, which
apply to his whole life, have been suggested by what took
place in 1822. Hardly had his report been laid before the
Woodwardian auditors when he felt it his duty to plunge
into a University controversy, which, between discussions with
friends, and pamphlets against opponents, must have occupied
a considerable portion of his time for nearly two years. The
matter—especially Sedgwick's share in it—is of sufficient importance to demand a brief notice in this place.

On the death of Dr Edward Daniel Clarke, Professor of
Mineralogy, 9 March, 1822, there was some difference of
opinion as to the expediency of continuing a Professorship of
Mineralogy in the University. The title of Professor, it should
be remembered, had been conferred on Dr Clarke in 1808,
but no Professorship had ever been formally established.
Meanwhile Sedgwick's intimate friend, Mr Henslow, had
announced his intention of becoming a candidate, should the
Professorship be continued; and it soon became apparent
that, if he had a chance of coming forward, he need have no
fear of the result. At last, 15 May, a Grace passed the
Senate which may be thus translated :

"Whereas by the death of Edward D. Clarke, late Professor of
Mineralogy, that office is now vacant: may it please you that another

his influence that Sedgwick, together with his friends Pryme, Romilly, Whewell,
and Lodge were all induced to join the Committee. In 1824 Pryme invited two
of the Greek deputies, who had come to England to negotiate a loan, to stay with
him at Cambridge for the Commencement, and Sedgwick met them at his house.
Recollections, p. 143.

1822.
Æt. 37.

Professor be elected by you to discharge the duties of the said office[1]."

As this Grace was copied, *verbatim*, the name only being altered, from a Grace passed 23 January, 1732, for continuing the Professorship of Botany (which, like that of Mineralogy, had become vacant by the death of the first Professor), Members of the Senate assumed that the election of a Professor to succeed Dr Clarke would be conducted in the same manner as the election of the Professor of Botany had been, namely, by open poll. It was with great surprise, therefore, that they learnt, a few days after the Grace had passed, that the Heads of Houses,—at the instigation, as it was believed, of Dr Webb, Master of Clare Hall, and Dr Chafy, Master of Sidney Sussex College[2]—intended to conduct the election in the mode observed at the election of Vice-Chancellor—in other words, to nominate two persons, one of whom the Senate would obviously be constrained to elect. Alarmed at what was certainly an innovation, and apprehensive that, if it were acquiesced in, a similar claim would be made respecting all Professorships founded by the University, a large number of members of the Senate met at the Red Lion Inn, and 'organised a very pretty rebellion'[3]. The sense of the University was evidently with them, for seventy-four signatures were immediately affixed to a dignified *Representation* to the Vice-Chancellor and Heads of Colleges (24 May), praying them to abandon a position which, as the memorialists clearly shewed, was contrary to all precedent. To this *Representation*, though those who signed it were among the most distinguished men in the University, and represented, in numbers alone, at least three-fourths of the resident academic body[3] — no reply was vouchsafed, but,

[1] As the subsequent controversy turned, in great measure, on the wording of this Grace, it shall be cited in the original Latin: 'Cum per mortem Edwardi D. Clarke nuper Professoris Mineralogiæ munus istud iam vacans existit; Placeat vobis ut alius ad idem munus exequendum a vobis eligatur.'
[2] Whewell's *Life*, p. 76.
[3] Professor Christian, from whose *Explanation of the Law of Elections in the University of Cambridge*, 8vo. Camb. 1822, many of the facts here recited have

three days afterwards (27 May), a notice was issued by the 1822.
Vice-Chancellor, Dr French, Master of Jesus College, to the Æt. 37.
effect that at the congregation on the ensuing day, a Grace
would be offered to rescind the previous Grace—or, in other
words, to discontinue the Professorship. It is strange that
the Heads, who are said to have been by no means unanimous in favour of the claim to nomination, should have taken
so high-handed a course when the Senate was so deeply
irritated. As might have been expected, the Grace was rejected by forty-three votes to seven. Thereupon the Heads
at once nominated Mr Henslow and Mr Lunn, both of St
John's College, and the Vice Chancellor announced that the
election would take place on the day following, at two o'clock.
In the interval the members of the Senate who had drawn up
the *Representation* met again to consider their position. It
was determined to select a candidate of their own; and,
in the event of their votes for him being rejected, to take legal
measures for the vindication of their rights.

This conflict reveals a state of things so different from that
to which we are now accustomed to in the University, that
a few words of explanation are necessary. By the statutes of
the 12th year of the reign of Queen Elizabeth it was provided,
that when certain offices were vacant, the Heads of Colleges
should nominate two persons, one of whom was subsequently
elected by the Senate. Among the officers so nominated was
the Vice-Chancellor. By statute the Heads might nominate
whom they pleased, but in practice they invariably chose one
of their own body, who was only too glad, on all occasions of
difficulty, to shelter himself behind the other members of his
own order. Gradually, therefore, the Heads had acquired the
position of assessors to the Vice-Chancellor, and were, practically, the rulers of the University. Whether they executed
these high functions moderately, or tyrannically, need not

been borrowed, has the following passage (p. 25): "The whole number of a
different opinion could not probably be 25, for the whole number who voted for
and against the petition against the late Catholic Bill were 66 to 33."

here be discussed; all that need be pointed out is that they were regarded with jealousy by the Senate, as an oligarchy anxious to retain privileges which depended on custom rather than on statute, and eager to embrace every opportunity of extending them.

Through the whole course of the dispute Sedgwick was a prominent member of the opposition. *Que diable allait-il faire dans cette galère?* is naturally the first question of every one interested in his more important pursuits. The answer is not far to seek. In the first place, before the controversy began he had had an interview with Dr French, on Henslow's behalf, which seemed friendly enough at the time, but which, as events proved, brought him into the very front of the controversy. Secondly, as has been already stated more than once, Sedgwick had a horror of wrong, or even the semblance of wrong, and he had succeeded in persuading himself that the Heads were making an unjustifiable attempt to deprive the Senate of one of its privileges. It is impossible to approve the whole of his subsequent conduct—especially his personal controversy with Dr French; but in the initial stages of the affair he showed an independence of spirit combined with a courteous demeanour towards those who differed from him, which cannot be too highly praised, and which is the more striking when contrasted with the uncompromising defiance of his opponents.

Every effort was made by the opposition to avoid a conflict. After their last meeting, a deputation waited on the Vice-Chancellor, urging him to assemble the Heads, and devise some plan of conciliation. He replied that it was now too late to get them together before the election, but that on the following day he would consult those who might be present in the Senate House. Six Heads only came to the Congregation, and after a long discussion, the Vice-Chancellor decided to proceed with the election, with the proviso, it is said, that if no member of the Senate voted for either of the candidates nominated, the election should be adjourned to that day fortnight. The Senior Proctor having read out the names of the persons nominated by the Heads, Sedgwick and Mr Carrighan of St John's College handed in a written protest against the form of the election about to take place. A single vote was then recorded for Henslow, and a considerable number for a third person. These were disallowed. At the close of the election the Vice-Chancellor declared Henslow duly elected, and admitted him with the usual formalities.

The Heads, as well as their opponents, must have known that the matter could not end in this unsatisfactory fashion. Two days afterwards (30 May) a committee of their opponents met, and decided that any legal measures which they might resort to against the Heads should be conducted 'in the spirit of the utmost amity and courtesy'. Sedgwick does not state explicitly that this proposal emanated from himself, but the tone of his *Letter* to the *Cambridge Chronicle*, and some expressions in his first pamphlet against Dr French, warrant us in assuming that such was the case. At a subsequent meeting, held 15 June, Sedgwick, Mr Carrighan, and Mr Lodge were deputed to wait on the Vice-Chancellor, in order to propose that the question should be tried by a joint application to the Court of King's Bench. The answer, however, was not conciliatory, and therefore the Court was moved by the committee alone, who obtained a Rule (21 June) calling on the Vice-Chancellor to show cause why a Mandamus should not issue for the admission of their nominee to the Professorship. In these proceedings Sedgwick took a prominent part, with some unwillingness, as would appear from what he himself says:

"In the commencement of the legal proceedings, I expected to be called upon to make an affidavit on the intention of the party which proposed the Grace (of May 15, 1822), and on the construction put upon it, at the time, by the Senate. But the affidavit connected with the general merits was prepared for another Member of the Senate. In consequence of an unlooked-for delay in his arrival in London, it was re-modelled and sworn by myself. Thus, by mere accident, I was placed in the position of plaintiff; and in all the future proceedings I have watched the progress of the cause with

1823.
Æt. 38.
deep interest, but with no feelings of ill-will towards those who were opposed to the Senate[1]."

The hearing of the case was not concluded until the end of April, 1823; and, for reasons into which it is needless to enter, judgment was neither given, nor applied for. At the moment, however, when the University seemed on the point of hearing the last of a painful controversy, a personal conflict broke out between Sedgwick and Dr French, which, though it extended through 1823 and into 1824, had better be disposed of in this place. In May 1823, shortly after the conclusion of the case in the Court of King's Bench, Sedgwick unfortunately thought proper to address a long letter to the *Cambridge Chronicle*, with the view, as he said, of supplying some omissions in his affidavit. In this letter, which had better have been never written, he gave his own version of his conversation with Dr French at Jesus College Lodge; urging particularly that no hint of doubt as to the mode of election had then been dropped, that the Senate had accepted the Grace of 15 May, 1822, under the belief that it would be followed by an open poll, and, this being the case, that Dr French ought not to have joined those 'who endeavoured to force upon the University a construction of his Grace which was at variance with his own meaning when he proposed it, and with the understanding of the Senate when they accepted and ratified it'. In reply to this letter Dr French issued (18 June) *An Address to the Senate*. This curious composition is written in the third person, as though the author was in so exalted a position that he could not even use the same pronouns as the rest of the University—an assumption of dignity which becomes ridiculous when it is remembered that he was Sedgwick's junior by three years. At the same time the pamphlet is not deficient in ability. But, clever as the author certainly showed himself in handling a bad case, he made one very damaging admission :

---

[1] *A Reply to an Address to the Senate, published by the Master of Jesus College*, 8vo. Camb. 1823, p. 78.

" Dr French neither intended that his Grace should give the right of Election *more burgensium*, nor did he intend the contrary. Aware that a difference of opinion existed as to the proper mode of Election in such cases as this of the Professorship of Mineralogy, he intended simply to ascertain, without prejudice to the claims of any party, whether the Senate were desirous of continuing the office. When a Grace to this effect had passed, he determined, under these circumstances, not to proceed further without the sanction of the Heads. And, accordingly, as soon as there was a majority of the Heads of Colleges in the University, Dr French, as Vice-Chancellor, called a meeting, for the express purpose of asking their deliberate judgment upon the proper method of proceeding[1]."

1823.
Æt. 38.

This paragraph granted all that Sedgwick was contending for, namely, that he had been allowed to carry away the impression that the Grace to continue the Professorship of Mineralogy, being in the same form as that used on a similar occasion for Botany, would be followed, as that was, by an open poll. Had Dr French been more cautious, or more candid, he would have taken good care to avoid a mis-understanding on so important a question; and, above all, he would not have selected a form of words for his Grace with which the very meaning he did not wish to convey would infallibly be associated.

Sedgwick's feelings on reading Dr French's pamphlet will be best understood from the following letter, addressed to Dr Monk, then Dean of Peterborough, one of those who had signed the *Representation.*

TRIN. COLL., *October* 23, 1823.

Dear Mr Dean,

I am just returned to the University after an absence of more than five months. You will, I doubt not, have seen Dr French's reply to the letter which I addressed to the Senate. I am about to commence my reply to it this morning. My opponent has come out with a bold tone, and has taken a lofty flight. Unless I am most egregiously mistaken I can easily bring him down from his perch—not by swaggering invectives and solemn asseverations, but by a plain unvarnished

---

[1] *An Address to the Senate*, p. 10.

1823.
Æt. 38.

tale which he will have reason to remember to the last day of his life. I now formally accuse him of mental reservation towards myself, and of unfair and disingenuous dealing towards Henslow and the Senate. I am, however, resolved not to let the strength of my phrase go "beyond the staple of my argument", and to conduct the controversy with proper forbearance. On the abstract merits of the question in litigation I shall not be able to speak at length, but I shall endeavour to notice them some way or other. Have you any information to communicate on the subject? Would you have the goodness to give me a synoptical view of the proofs by which the rank of Professors is established, and their distinction from the *Lectores* of the 40th Chapter of Elizabeth's statutes is made out[1]? I am now very anxious to get forward with my pamphlet. You will therefore greatly add to the obligation by sending your information as soon as you can possibly make it convenient. Did I not know your zeal in a good cause I should not have ventured to trouble you.

Present my best remembrances to Mrs Monk, and believe me, Dear Mr Dean,

Yours ever,

A. SEDGWICK.

Sedgwick was as good as his word. Before the end of term he had produced a pamphlet of eighty-six closely-printed octavo pages, which may still be safely recommended to the perusal of those who care for University history. It is a straightforward, dignified, composition, with here and there some nobly eloquent passages ; contrasting very favourably, on the whole, with the strut and swagger of Dr French's laboured periods. There may be a few errors of detail, but the main arguments against the claim of the Heads are learned and accurate.

[1] This Statute directs that the election of *lectores*, *bedelli*, and other specified officers shall be conducted in the same manner as that of the Vice-Chancellor—i.e. by open poll after nomination by the Heads. It was Sedgwick's object to prove that these *lectores* were quite different from the *Professores*, whose offices had not been created when the Statute was framed.

The personal question is ably managed. No railing accusation is brought against Dr French, but the 'plain unvarnished tale' which Sedgwick tells leads irresistibly to a conclusion most unfavourable to his reputation as a man of honour. But, on the other hand, the whole pamphlet is far too long. This defect is partly due to the fact that it was written in the intervals of the author's lectures. As soon as the materials of one sheet were brought together, he tells us, they were sent to the printer ; and during their passage through the press, he was employed in preparing matter for the next sheet[1]. But this is a defect which detracted from all Sedgwick's writings, except his scientific papers. He never knew when to stop.

Dr French promptly published *Observations upon Professor Sedgwick's Reply* (21 January, 1824), but prudently refrained from comment on the personal question. His pamphlet is almost wholly devoted to the legal difficulty, which depended, in great measure, on the interpretation of the fortieth chapter of the Statutes of Elizabeth. Sedgwick replied (25 February, 1824), in another pamphlet of considerable length[2], confining himself, like his opponent, to law and precedent. Neither of these works need be examined in detail, and our account of the controversy shall be closed with a single quotation from Sedgwick's first pamphlet—partly as a specimen of his style, partly from the intrinsic value of what he says.

"Some one may, perhaps, contend, that the bustle of public elections but ill accords with the tranquil habits of this seat of science ; and that the question ought to be conceded to the Heads, out of regard to the peace of the University. Words of peace are always to be suspected when they are accompanied with acts of aggression. By conceding this question, we part with our own privileges without finding any remedy for the evil complained of. For it is notorious to the Members of the Senate, that no ordinary academical elections have been contested with more warmth, than those in which the Heads have nominated the two candidates.

" Had there existed any *flagrant abuse*—had there been a con-

---

[1] *A Reply to an Address to the Senate*, p. 79.
[2] It is called : *Remarks on the observations of Dr French: with an argument on the Law of Elections to offices created by the Senate.* 8vo. Camb. 1824.

Margin note: 1824. Æt. 39.

1824.
Æt. 39.

spiracy on the part of certain colleges, to exclude others from their fair share of academical distinctions, there might have been some plea for introducing new customs into the University. But in the present case, no abuse was even pretended; we were on the point of electing the very man, who was afterwards chosen by our opponents. And the lists of those who have filled our Professorships undeniably prove, that the Senate has, from time to time, selected out of its ranks the man who, by his zeal and his talents, was best qualified to promote the true interests of science, and to support the credit of our establishment.

"It was on this principle that Martyn, Watson, Milner, Wollaston, and Tennant, were elected; and on the same principles their successors have been, and will continue to be elected, as long as the privileges of the Senate are unextinguished.

"Had the Professorship of Mineralogy been the first office created by a Grace of the Senate, I should not have hesitated to pronounce an election by nomination, the worst form which was sanctioned by the usage of the University. It has all the evils of an open poll, with very little of the good. For it virtually gives the election to a few individuals, and what is worse, it gives it to them indirectly.

"Were these individuals led by their known habits of life, and their high official duties, to watch the progress and to examine the refinements of modern science; we might, perhaps, be content to surrender our privileges into their hands, and to repose with confidence on their wisdom. Collectively, they are entitled to all respect, as the Heads of our venerable establishments—as the guardians of our discipline—and as the directors of the studies of our younger members. Still more they are entitled to our veneration for their virtues, and for their talents, by which alone many of them have reached the greatest academical elevation. But this very elevation removes them from direct sympathy with the Senate, and imposes on them such high and important duties, that they have but little time for the elaborate investigations of Physiology, of Botany, of Chemistry, and of Mineralogy. Nay, some of them may even think, that these subjects are unfit for a course of public lectures—and that the Professors' chairs are nothing better than an academical incumbrance.

"Let the Senate look well to it, before, in any case, it surrenders the power of election into the hands of those who, to say the least of it, may be indifferent to the office, and therefore can have no deep interest in selecting an active candidate.

"I am not now warning the Senate against an ideal danger. My opponent has publicly told us, that *he thought the continuance of the Professorship of Mineralogy unnecessary.* I may tell him in reply, that the Senate thought differently—that the republic of science allows no such thing as official wisdom—and that his own opinion will be of little weight, unless it be founded on a deeper knowledge of the subject, than that which is possessed by his opponents. As for myself, I am well contented, on this question, to have acted with the majority.

"Individuals there are, at all times, who, not considering that   1824.
improvement is *innovation*, oppose themselves to every change, and   Æt. 39.
think every new appointment unnecessary. But the University of
Cambridge has not acted on such heartless suggestions during the
last century; and as long as her constitution remains unimpaired she
will never act upon them [1]."

One word more is necessary before we dismiss this tedious
affair. Three years after the publication of Sedgwick's last
pamphlet—in which a decision favourable to the views of
himself and his friends was confidently anticipated—the controversy was closed by an award of Sir John Richardson,
to whom the matter had been referred by the Senate. His
decision may be fairly described as a verdict for the defen-
dants—the Vice-Chancellor and the Heads—for he directed
that future elections to the Professorships of Anatomy,
Botany, and Mineralogy, should be conducted according to
the method prescribed in the 40th Chapter of the Statutes.

[1] *Reply*, etc., pp. 75—78.

CHAPTER VII.

(1822—1827.)

GEOLOGICAL EXPLORATION OF THE LAKE DISTRICT (1822—1824). CONTESTED ELECTION FOR UNIVERSITY (1822). DEATH OF HIS SISTER ISABELLA (1823). GEOLOGICAL PAPERS. WORK IN THE WOODWARDIAN MUSEUM (1823—1827). LECTURE TO LADIES. VISIT TO EDINBURGH WITH WHEWELL (1824). VISIT TO SUSSEX WITH DR FITTON (1825). CONTESTED ELECTION FOR UNIVERSITY. VISIT TO PARIS WITH WHEWELL (1826). ELECTED VICE PRESIDENT OF GEOLOGICAL SOCIETY. CONTESTED ELECTION FOR UNIVERSITY (1827). SOCIAL LIFE AT CAMBRIDGE. HYDE HALL. REVIEW OF SEDGWICK'S GEOLOGICAL WORK (1818—1827).

SEDGWICK'S first geological work in the north of England[1], briefly noticed in the last chapter, was succeeded by a thorough examination of the Lake District. "I spent the summers of 1822, 1823, and 1824," he says, "entirely among the Lake Mountains, and I made a detailed Geological Map of that rugged region—including a considerable portion of Westmoreland and Cumberland, and a small portion of Lancashire[2]." The scientific value of these explorations may be estimated from the papers read to the Geological Society between 1826 and 1828, and from the five letters addressed long afterwards to Wordsworth, of which the first three embody the results of the work done between 1822

---

[1] In that year, 1821, he began the researches into the relations of the Magnesian Limestone which were continued during 1822 and 1823. *Trans. Geol. Soc. Lond.* Ser. 2. iii. 37.

[2] To Archdeacon Musgrave, 5 October, 1856.

and 1824[1]. But of personal details the record is almost a blank. A brief but pleasant glimpse of Sedgwick at his work is afforded to us in one of Whewell's letters, written from Kendal in 1824: 'I got here on Thursday last, and next day saw Wordsworth at Rydal, and Southey at Keswick, by whom I was informed where to look for Sedgwick. I found him on Saturday at the base of Skiddaw, in company with Gwatkin[2], as I had expected[3],' but after this the writer passes on to other subjects. This dearth of information is the more provoking, as we know that many agreeable memories, both of adventures and of friends, clustered round these months in Lakeland.

1822 to 1824. Æt. 37— 39.

It was then that Sedgwick formed an intimate friendship with Wordsworth, at whose house he was always welcome, and who, to a certain extent, directed and assisted his explorations. Wordsworth has been credited with a cordial dislike for men of science, who looked upon Nature with other eyes than his; and the first of Sedgwick's letters opens with a sort of apology for writing on geology to one who had uttered "a poetic ban against my brethren of the hammer":

> He who with pocket-hammer smites the edge
> Of luckless rock or prominent stone, disguised
> In weather-stains or crusted o'er by Nature
> With her first growths, detaching by the stroke
> A chip or splinter, to resolve his doubts;
> And, with that ready answer satisfied,
> The substance classes by some barbarous name,
> And hurries on; or from the fragments picks
> His specimen, if but haply interveined
> With sparkling mineral, or should crystal cube
> Lurk in its cells—and thinks himself enriched,
> Wealthier, and doubtless wiser, than before![4]

---

[1] These three letters *On the Geology of the Lake District*, addressed by Sedgwick to Wordsworth in May, 1842, were published in *A complete Guide to the Lakes... with Mr Wordsworth's description of the scenery of the country*, etc., edited by the publisher, John Hudson of Kendal. A fourth letter was added in 1846, and a fifth in 1853.

[2] The Rev. Richard Gwatkin, Fellow of St John's College, B.A. 1814.

[3] Whewell's *Life*, p. 96.

[4] *The Excursion*, Book the Third.

This denunciation of a class did not prevent the poet from taking an interest in the pursuits of individual geologists; and the gratitude and admiration which Sedgwick felt for him can fortunately be recorded in his own words. In the third of the above letters he says: " Some of the happiest summers of my life were passed among the Cumbrian mountains, and some of the brightest days of those summers were spent in your society and guidance. Since then, alas, twenty years have rolled away; but I trust that many years of intellectual health may still be granted you; and that you may continue to throw your gleams of light through the mazes of human thought—to weave the brightest wreaths of poetic fancy—and to teach your fellow-men the pleasant ways of truth and goodness, of nature, and pure feeling;" and again, in the last of the series, written in 1853, when Wordsworth was no more, after some regretful musings on his own enfeebled powers, should he ever revisit Lakeland, he is led to speak of the friends of whom the district would remind him: " It was near the summit of Helvellyn that I first met Dalton[1]—a truth-loving man of rare simplicity of manners; who, with humble instruments and very humble means, ministered, without flinching, in the service of high philosophy, and by the strength of his own genius won for himself a name greatly honoured among all the civilized nations of the earth.

" It was, also, during my geological rambles in Cumberland that I first became acquainted with Southey, that I sometimes shared in the simple intellectual pleasures of his household, and profited by his boundless stores of knowledge. He was, to himself, a very hard task-master: but on rare occasions (as I learnt by happy experience) he could relax the labours of his study, and plan some joyful excursion among his neighbouring mountains.

"Most of all, during another visit to the Lakes, should I have to mourn the loss of Wordsworth; for he was so far a man of leisure as to make every natural object around him

[1] See above, p. 66.

subservient to the habitual workings of his own mind ; and he was ready for any good occasion that carried him among his well-loved mountains. Hence it was that he joined me in many a lusty excursion, and delighted me (amidst the dry and sometimes almost sterile details of my own study) with the outpourings of his manly sense, and with the beauteous and healthy images which were ever starting up within his mind during his communion with nature, and were embodied, at the moment, in his own majestic and glowing language."

Sedgwick frequently visited the Lakes again, sometimes for geological study, sometimes for the pleasure of looking at scenes in which he had taken so much delight, or of showing them to others. Many opportunities of recording his impressions of the district will therefore occur, and it might seem unnecessary to remove letters referring to it from their proper chronological position. On the whole, however, having regard to the dearth of contemporary information respecting the visits of 1822—1824, it seems best to print the two following letters in this place, as they give, incidentally, so many details respecting those years. Both were written for the instruction of geologists who were anxious to explore the Lake district for themselves.

To Rev. P. B. Brodie[1].

CAMBRIDGE, *September* 10, 1854.
My dear Brodie,

First of all, find out my old good friend Jonathan Otley, the author of the best guide to the Lakes that ever was written[2]. Tell him you are my friend, and that I wished you to call on him ; and you may read to him this letter. He will show you maps, &c. He knows the physical geology of Cumberland, and all the Lakeland, admirably well. He was the leader in all we know of the country. I wish, with all my heart, that my letters to Mr Wordsworth on the Geology of Lakeland had been printed in Otley's

---

[1] Rev. Peter Bellenger Brodie, Trinity College, B.A. 1838, M.A. 1842.

[2] *A descriptive guide to the English Lakes and adjacent Mountains:* with notices of the Botany, Mineralogy, and Geology of the district. By Jonathan Otley. Eighth edition. Keswick, 1849. In earlier editions (the second was published in 1825) it was called : *A concise description of the English Lakes, etc.*

Guide; but I promised Mr Wordsworth in 1822, before I knew Mr Jonathan Otley. Ask for a loan of my Letters to Mr Wordsworth; but they are printed in Hudson's Guide—see last edition, which contains a 5th Letter. *Secondly:* find out Charles Wright—a *guide* formerly; and *now*, I am told, a *guide director.* You must take what he says *cum grano salis,* for he is a bouncer. All Otley tells you, you may take for Gospel; for he only tells what he knows. He is a very clever truth-loving old man. Look at the mining operations at the back of Skiddaw. About Hesket Newmarket you have good Mountain Limestone, and a touch of the Old Red. N.B. Old Red Sandstone above Kirkby Lonsdale bridge, at bottom of Ulswater, near Shap Wells &c. &c. &c. If you visit it look for fish-scales. I had good eyes when I worked Lakeland; but at *that time* we knew not of the Old Red fishes; and I therefore never looked for them. No fossils in the Skiddaw slate, except a few graptolites and fucoids. Ruthven found them for me, and Otley will tell you the localities. If you could give me a list of the minerals turned out at the mines on both sides of Carrock Fell I should be obliged to you for it. It *might be of great use* to me. Also I should greatly thank you for a good account of the cleavage planes of the slates in Binsey, at the bottom of Bassenthwaite Lake. *Thirdly:* my old heart-of-oak friend John Ruthven lives at Kendal. See him by all means. He has all Westmoreland at his fingers' ends, and will tell you of all the fossil localities between the Coniston Limestone and the Old Red and Mountain Limestone of Kirkby Lonsdale. No fossils have, *as yet*, been seen in the slates &c. which alternate with the porphyries between the Skiddaw slate and the Coniston limestone; but if you cross them keep your eyes open; and possibly you may find some rare fossil. For when I crossed them *again and again* (30 years since) I was looking for sections rather than for fossils. And it is a good rule to keep a good look-out, and never to take for granted that no fossils are to be had. If Mr Gough[1] (the surgeon) be at Kendal, you ought to see him, but I think he is now away in bad health. You ought to see the Kendal Museum. I am President of the Society; and this letter will secure you an introduction and all needful attention.

There! I have done my best, in a rough way, to answer your questions, and I must now compltee my dress and prepare for morning Chapel.

Ever truly yours

A. SEDGWICK.

To Professor Harkness.

SCALBY near Scarborough, *August* 29, 1856.

My dear Sir,

Your letter has been long in reaching me, so I fear the information I can send you may come too late to be of any use.

---

[1] Thomas Gough, of Kendal, an intimate friend of Sedgwick's.

(1) I advise you to go to Kendal and to call on John Ruthven—the well-known collector of the northern palæozoic fossils. He knows the country well, and is the only person (so far as I know) who has found fossils in the Skiddaw slate. (2) You may procure *Hudson's Guide to the Lakes*; and in some letters published in an appendix to it you may see a general account of the several formations, tho' I am not sure that there is any notice of the Skiddaw slate fossils and their localities. (3) If old Jonathan Otley, author of an excellent little book, be still living (I saw him last year when he was turned ninety) he can give you good advice as to localities, and so can Charles Wright, one of the Keswick Guides, who went with me in some of my excursions in 1824. Since that year I have hardly looked at the Skiddaw slates. You should look at the new black-lead works somewhere behind Saddle Back, and see the manufactory at Keswick. I do not remember the name of the locality, though I saw it (in 1823) along with Mr Otley. These works are, I suspect, not in a vein, but in a variety of anthracitic slate. So they will give you the term of comparison you are looking for. I found black slates in the great Skiddaw Group, from which the dark carbonaceous colours were discharged by heat. Hence I concluded that such beds very probably would contain fossils; so I set Ruthven to work, and he found fossils—graptolites and fucoids—not far from the spots I pointed out to him. But he found no shells or crustaceans. Since then I have had some doubts about the age of the Skiddaw Group. It is of enormous thickness, and may well contain one or two groups of very distinct epochs both physically and palæontologically. (4) When you are seeking Skiddaw slate fossils I recommend you to take up your quarters at Scale Inn, at the foot of Crummock Lake. Hammer well the gritty rocks which appear in the several deep ravines which run up the mountains on the left side of the road from Scale Inn to Buttermere; they promise well for fossils. I never examined them for fossils in 1823 and 1824, because I foolishly thought that they were all below the region of animal life. At that time I had not quite learned to shake off the *Wernerian nonsense*[1] I had been taught. (5) Visit Black Coomb in the S. W. corner of Cumberland. It is of Skiddaw slate, brought up by enormous dislocations, and its ravines are of good promise. To the south it is overlaid by the green slate and porphyry zone—well marked, but of degenerate thickness; and over the green slate you have in the S. W. extremity of Cumberland the Coniston limestone, &c., and some appearances, in the cleavage planes, which I think defy the mere pressure theory. That there has been enormous compression, along with cleavage planes, no one can doubt, when the fossils are flattened and distorted. But they are not always distorted and flattened. You have to account for unflattened concretions, marking, though rarely, the average direction and dip of the

[1] In a letter to Lyell, written in 1845, Sedgwick speaks of himself as having been, in 1819, "eaten up with the Wernerian notions—ready to sacrifice my senses to that creed—a Wernerian slave".

1822 to
1824.
Æt. 37—
39.

cleavage planes. You have to account for the frequent change of cleavage dip when there is no change of conditions of pressure indicated in the sections; and you have to account for a second cleavage plane among beds that are by no means crystalline. (6) Visit Coniston, and look at the enormous dislocations &c. You have there (as also at Broughton in Furness, which you pass through on your way from Black Coomb to Coniston) the Coniston lime-stone, the Coniston flags, and the Coniston grits which form the boundary between a lower and an upper system—by whatever names you choose to call them. If these hints be of use I shall rejoice.

Yours very truly

A. SEDGWICK.

It will be readily conceived that a man so prominent as Sedgwick, and one endowed with so keen a sense of humour, became the subject of many jokes, both literary and artistic. His early visits to Lakeland recall one of the former, a humorous sketch, called *Joe and the Geologist.* The author has preserved a strict incognito, and mentions no names; but Sedgwick's numerous friends in the north recognised the accuracy of the portrait at once, and he himself laughed heartily over it, though he denied the accuracy of certain details, as, for instance, the white neckcloth and the "specks". "I never wore such things", he wrote, "while I was holding a hammer in Cumberland[1]." It should be mentioned that the tale, like other legends, is sometimes told with a different ending. This second, and probably later, version states that the geologist, before he had travelled many miles, discovered the fraud that had been perpetrated upon him, and travelled back, in furious anger, to catch Joe and make him tell what he had done with the contents of the leather bags. The boy took good care not to be found, but the stones he had thrown away were discovered in a heap by the wayside.

Ya het foorneun, when we war oa' gaily thrang at heàm, an oald gentleman mak' of a fellow com' in tul oor foald an' said, whyte

[1] To Rev. G. H. Ainger, 2 September, 1866. Mr Ainger, son of his old friend, had sent him a copy of the sketch. In writing to acknowledge it, Sedgwick says: "Thanks for your very amusing specimen of the Cummerland tongue, and the twit against the knights of the hammer."

natural, 'at he wantit somebody to gà wid him on't fells. We oa' stopt an' teuk a gud leuk at him afoor anybody spak; at last fadder said, middlin' sharp-like—(he ola's speaks that way when we're owte sae thrang, does fadder)—"We've summat else to deu here nor to gà rakin ower t'fells iv a fine day like this, wid nèabody kens whoa." T'gentleman was a queerish like oald chap, wid a sharp leuk oot, grey hair and a smo' feàce—drist i' black, wid a white neckcloth like a parson, an' a par of specks on t'top of a gay lang nwose 'at wasn't set varra fair atween his e'en, sooa 'at when he leuk't ebbem at yan through his specks he rayder turn't his feàce to t'ya side. He leuk't that way at fadder, gev a lal cheàrful bit of a laugh an' said, iv his oan mak' o' toke, 'at he dudn't want to hinder wark, but he wad give anybody 'at ken't t' fells weel, a matter o' five shillin' to gà wid him, an' carry two làl bags. "'Howay wid tha, Joe," sez fadder to me, "it's a croon mair nor iver thou was wūrth at heàm!" I meàd nèa words about it, but gat me-sel' a gud lūmp of a stick, an' away we set, t'oald lang nwos't man an' me, ebbem up t'deàl.

As we war' climmin' t'fell breist, he geh me two empty bags to carry, meàd o' ledder. Thinks I to my me-sel', "I'se gān to eddle me five shillin' middlin cannily." I niver thowte he wad finnd owte on t' fells to full his lal bags wid, but I was misteàn!

He tūrn't oot to be a far lisher oald chap nor a body wad ha' thowte, to leuk at his gray hair and his white hankecker an' his specks. He went lowpin' ower wet spots an' gūrt steàns, an' scrafflin across craggs an' screes, tul yan wad ha' sworn he was sūmmat a kin tul a Herdwick tip.

Efter a while he begon leukin' hard at oa't steàns an' craggs we com' at, an' than he teuk till breckan lūmps off them wid a queer lal hammer he hed wid him, an' stuffin t'bits intil t'bags 'at he geh me to carry. He fairly cap't me noo. I dudn't ken what to mak' o' sec a customer as t'is! At last I cudn't help axin him what meàd him cum sèa far up on t'fell to lait bits o' steàns when he may'd finnd sèa many doon i't deàls? He laugh't a gay bit, an' than went on knappin' away wid his lal hammer, an' said he was a jolly jist. Thinks I to me-sel' thou's a jolly jackass, but it maks nèa matter to me if thou no'but pays me t' five shillin' thou promish't ma.

Varra weel, he keep't on at this feckless wark tul gaily leàt at on i't efter-neun, an' be that time o' day he'd pang't beàth o't ledder pwokes as full as they wad hod wid bits o' steàn.

I've nit sèa offen hed a harder darrak efter t' sheep, owther at clippin time or soavin time, as I hed followin' that oald grey heidit chap an carryin' his ledder bags. But hooiver, we gat back tul oor house afoor neeght. Mūdder gev t' oald jolly jist, as he co't his-sel', some breed an' milk, an' efter he'd teàn that an toak't a lal bit wid fadder aboot sheep farming an' sec like, he pait me five shillin' like a man, an' than tel't ma he wad gie ma ūdder five shillin' if I wad bring his pwokes full o' steàns doon to Skeàl-hill be nine o'clock i't' mwornin'.

He set off to walk to Skeàl-hill just as it was growin dark; an'

neist mwornin', as seun as I'd gitten me poddish, I teuk t' seàm
rwoad wid his ledder bags ower me shoolder, thinkin' tul me-sel' 'at
yan may'd mak' a lal fortune oot o' thūr jolly jists if a lock mair on
them wad no'but come oor way.

It was anūdder het mwornin', an' I hedn't wok't far till I begon
to think that I was as gūrt a feul as t'oald jolly jist to carry brocken
steàns o't way to Skeàl-hill, when I may'd finnd plenty iv any rwoad
side, clwose to t' spot I was tackin' them tul. Sooa I shack't them
oot o' t' pwokes, an' then step't on a gay bit leeter widout them.

When I com nār to Skeàl-hill, I fūnd oald Aberram Atkisson
sittin on a steul breckan steàns to mend rwoads wid, an' I ax't him
if I med full my ledder pwokes frae his heap. Aberram was varra
kaim't an' tell't ma to tak them 'at wasn't brocken if I wantit steàns,
sooa I tell't him hoo it was an' oa' aboot it. T' oald maizlin was
like to toytle off his steul wid laughin', an' said me mūdder sud tak
gud care on ma, for I was ower sharp a chap to leeve varra lang i'
this warld; but I'd better full my pwokes as I liked, an' mak' on
wid them.

T' jolly jist hed just gitten his breakfast when I gat to Skeàl-hill,
an' they teuk ma intil t' parlour tul him. He gūrned oa't feàce
ower when I went in wid his bags, an' tel't me to set them doon in
a neuk, an' than ax't ma if I wad hev some breakfast. I said I'd
gittan me poddish, but I dudn't mind; sooa he tel't them to bring in
some mair coffee, an' eggs, an' ham, an' twoastit breed an' stuff, an'
I gat sec a breakfast as I niver seed i' my time, while t' oald
gentleman was gittan his-sel rūddy to gang off in a carriage 'at was
waitin at t' dooar for him.

When he com doon stairs he geh me tūdder five shillin' an'
paid for my breakfast, an' what he'd gittan his-sel. Than he tel't
ma to put t' ledder bags wid t' steàns in them on beside t' driver's
feet, an' in he gat, an' laugh't an' noddit, an away he went.

I niver owder seed nor heard mair of t' oald jolly jist, but I've
offen thowte ther mun be parlish few steàns i' his country, when he
was sooa pleas't at gittin two lāl ledder bags full for ten shillin', an'
sec a breakfast as that an'. It wad be a faymish job if fadder
could sell o' t' steàns iv oor fell at five shillin' a pwokeful—
wadn't it?

Sedgwick capped this imaginary narrative with an equally
amusing experience of his own:

"Two or three times I went with Mr Hunter (a statesman
at Mosedale) to break the syenites of Carrock Fell. On my
second visit I found his old fashioned chimney-piece decorated
with specimens of syenite. 'Do you think these curiosities',
I said. 'Not a bit', he replied, 'they are as common as
cow-muck. But I put 'em here, aboon the chimlay, to tell my

nebbers what mak o' things a Cambridge skoller will laëd 1822 to
his hors we.' But old Hunter played no tricks. He fed me
and my horse well; and he went with me and carried a great
sledge-hammer to break the hard syenites. The last time I
drove to Mosedale, he spied me before I reached his house,
and roared out: 'fain to see ye again; how do ye cum on wi
yer cobbles?'"

1822 to
1824.
Æt. 37—
39.

A suitable pendant to these anecdotes is a pen-and-ink
sketch of 't'oald jolly jist', just as Joe might have seen him
sitting in his carriage, with the bag of fossils at his feet. It is
believed to have been drawn by Mr J. E. Davis, and, if so,
belongs to a period long subsequent to that we are now con-

Sedgwick on a geological excursion, reduced from a
pen-and-ink sketch.

sidering. Fashions, however, did not alter rapidly in those
days, and it may well represent Sedgwick, hat (generally a
white one), coat, and all, as he appeared when exploring

Lakeland in 1822. It was on one of these expeditions that the following experience occurred, which shall be told, as Sedgwick used to tell it, in a dramatic form:

SCENE. *A room in a small wayside inn near Wastwater. Enter* Professor SEDGWICK, *dressed as in the above sketch, very hungry, calling for the landlady.*
S. What have you got to eat?
L. There's nothing in the house.
S. Nothing! What did you have today for dinner?
L. Potatoes and bacon.
S. Very well. You didn't eat it all, I suppose. Warm me up what's left.
Exit Landlady, returning presently with the remains of the potatoes and bacon, and a pot of ale. Sedgwick eats heartily.
S. (*having finished his dinner.*) What's to pay, missus?
L. Happen eight pence wouldn't hurt ye?
S. Nay, here's a shilling for ye.
Landlady takes the shilling, and produces four greasy pennies from her pocket, which she lays on the table.
S. (*pushing them back.*) Nay, nay, you may keep them.
L. (*after a long and earnest look at him.*) I'm thinking that you've seen better days.

On returning to Cambridge after the first of the above-mentioned tours Sedgwick was fully occupied for a time with lectures and geological work generally. But before the end of October a serious interruption occurred, in the shape of a contested election for the University[1]. It was not natural for him to keep long out of any political contest, and this particular occasion offered irresistible attractions. The burning question of the day was Catholic Emancipation, in favour of which, as related in the third chapter, he had already taken a prominent and decided line in the University. The excitement in the country was so great that a complete settlement of this important matter could not be much longer delayed; but the conservatives had no intention of yielding without an obstinate struggle, and a constituency such as that of Cambridge, composed in the main of clergymen, was easily roused to enthusiastic action by the cry that the Church and the Protest-

[1] John Henry Smyth, M.A. of Trinity College, who had been M.P. for the University since 1812, died 20 October, 1822.

ant ascendency were both in danger. Several candidates 1822.
presented themselves, but these were presently reduced to Æt. 37.
three : Mr Manners Sutton, Speaker of the House of Commons,
Lord Hervey, and Mr Robert Grant[1]. The success of the
Speaker, a conservative, was considered certain, when he felt
himself obliged to retire (2 November), in consequence of an
unexpected difficulty respecting his office. Two days afterwards Mr Scarlett[2] came forward. At this juncture it seemed
probable that the University would find itself in an anomalous
position. The three candidates now in the field were all in
favour of Catholic Emancipation, as was the sitting member,
Viscount Palmerston. Unless therefore a conservative could
be found, and returned, a body which annually petitioned the
House of Commons against the Catholic claims would
be represented by two members voting against its own
petition. Before long, however, Mr William John Bankes[3]
came forward, as determined an opponent of concession as
could be desired; and, before the day of election Mr Grant
retired. The three candidates left after these various changes
were all of Trinity College. Mr Scarlett, whom the whigs
seem to have specially adopted, and for whom Sedgwick and
his friends exerted themselves to the utmost, was already
a distinguished advocate, and had had three years experience
of Parliament as member for Peterborough. But he was not
popular, and besides, he had not come forward until most
votes were already pledged. Lord Hervey, who had proceeded to his degree as a nobleman only a few months before
the election, was called a whig, but could have had no recommendation whatever except his relationship to Lord Liverpool,
and this accident, it is whispered, caused several influential
whigs to support him 'for private and personal reasons.' As
one of the pasquinades of the day put it :

[1] Fellow of Magdalene College, third wrangler and second Chancellor's
Medallist in 1801. After a distinguished career at the bar and in parliament
he was knighted and made Governor of Bombay, where he died in 1838.

[2] James Scarlett, created Lord Abinger 1835, of Trinity College, B.A. 1790.

[3] B.A. 1808, M.A. 1811.

Hervey, pushed forth by Bury School
And backed by noble Liverpool,
First made his bow to Heads of Houses
And canvassed all their lovely spouses;
The Ladies smirked, the Doctors smiled :
" What? give a vote to a mere child ? "
" A child "—quoth Blomfield—" mark me, Sir,
He's nephew to the Minister[1]."

Mr Bankes, immortalised by Macaulay on a subsequent occasion as " our glorious, our Protestant, Bankes "—was a well-known, witty, popular, man of the world, and at that time specially interesting as a traveller in the little-known regions of the East. His personal canvass has been described as irresistible. " What could I do, Sir? He got me into the centre of the great pyramid, and then turned round and asked me for my vote," was an unwilling supporter's description of the way in which a promise had been extorted from him. These pleasantries might suit the study of a college dignitary; but for the main body of the electors he provided more substantial fare. He had no particular claims to represent the University, and therefore wisely presented himself as " an appendage to the anti-catholic idea[2]." His printed circular announced " the most steady and decided opposition to any measures tending to undermine or alter the established Church "; a well-selected phrase of no uncertain meaning, the value of which became evident at the close of the poll (27 November, 1822), when the numbers were : Bankes, 419; Hervey, 281 ; Scarlett, 219. Sedgwick's views on the contest and the result are summed up in the following passage from a letter written two months afterwards :

" You wanted to know something about our election. Bankes was principally brought in by the interest of the country clergymen, who came up from all parts of England to

[1] Lord Hervey was eldest son of the fifth Earl of Bristol, created first Marquess of Bristol 1826. His aunt married the second Earl of Liverpool. The family seat is at Ickworth, near Bury St Edmund's.

[2] This phrase occurs in a long and ably-written article on the election in *The Times,* 29 November, 1822.

vote for the anticatholic candidate. Undoubtedly all this was the operation of principle (though I think a mistaken one), because all the Government influence was exerted for Lord Hervey, the nephew of the Premier. The highest of our Cambridge high-church men (such as Rennell[1], Tatham[2], Calvert[3], Wood[4], etc. etc.) all went for Hervey, and thereby, in my humble opinion, did themselves no honor. If Lord Liverpool supported a relation, though favourable to Catholic concession, they ought not to have left their avowed principles to follow him. The whig candidate was not a popular one, and was not heartily supported by the staunch men of his own party. Our representative Bankes is certainly a very extraordinary man, and possesses a wonderful fund of entertaining anecdote. When an undergraduate he was half suspected of being a Papist: and he almost frightened Dr Ramsden[5] to death, by building in his rooms an altar at which he daily burned incense, and frequently had the singing-boys dressed in their surplices to chant services. For a long time, while in the East, he wore a long beard, and passed as a faithful follower of the law of Mahomet. I don't think we can depend on him as a man of business, though as a literary character, and a man of large fortune, he is a very proper person to represent us in parliament. For several years he had four artists in his pay in Asia Minor, and even now he has men employed in his service in Upper Egypt, excavating tombs and temples, etc[6]."

The year 1823 opened gloomily for Sedgwick. He was spending the Christmas vacation in Cambridge, arranging,

1822.
Æt. 37.

---

[1] Thomas Rennell, Fellow of King's College; B.A. 1810, Christian Advocate, 1816—21.

[2] Ralph Tatham, Fellow of St John's College; B.A. 1800, Public Orator 1809—1836, and Master 1839 to his death 19 January, 1857.

[3] Thomas Calvert, B.A. 1797, Fellow of St John's College, and Norrisian Professor of Divinity 1815—24.

[4] James Wood, B.A. 1782, Master of St John's College 1815—39.

[5] Richard Ramsden, one of the Senior Fellows of Trinity College; B.A. 1786, D.D. 1807.

[6] To Rev. W. Ainger, 1 February, 1823.

with Henslow's assistance, the collection which the latter had formed in Anglesea, when he was hastily summoned to Dent. His favourite sister Isabella had been for some time in a declining state of health, but no immediate danger was anticipated. Perhaps Sedgwick was not told the full truth. At last, towards the middle of January, he learnt that she was sinking fast.

"I left Cambridge without delay," he wrote to Ainger, "but in consequence of the great quantity of drifted snow, which detained us one day on the road, I did not reach home till Friday afternoon. Nor could I even then have completed my journey had I not left the coach behind, and pushed through the snow, for the last three stages, on post-horses. I did not reach home in time to see my poor sister, but I had the mournful satisfaction of accompanying her remains to the grave the day after my arrival. She was blessed with a quiet and affectionate temper which greatly endeared her to every one of us; and during her painful illness she exhibited a humble resignation to the will of God; bearing with patience her afflictions here, in the Christian hope of being received with favour by her Maker in a place where there is neither sorrow nor suffering. The shock produced by poor Bell's death had such an effect on our sister Jane that she was delivered of a daughter on the day following. She and the child, I am happy to say, are both doing well. The young one is to have the name of Margaret Isabella after my mother and sister, and I hope to take upon me the duties of sponsor before my return to Cambridge. My Father, who has now almost completed his 87th year, has borne his late affliction with that patience we all expected from him. His mind is better regulated than that of any man whom I have ever had the happiness of knowing; and, so far, he is enjoying, even in this world, the fruits of a well-spent life [1]".

[1] To Rev. W. Ainger, 1 February, 1823. The dates given in the letter shew that Sedgwick left Cambridge on Tuesday 21 January, and reached Dent on Friday 24 January.

Sedgwick says nothing about his own feelings in the above letter; but we know from other sources how bitterly he deplored the loss of a sister who had been the companion of his childhood, and for thirty years in after-life the object of the best affections of his heart[1]. To him—with his tender and affectionate nature—her almost sudden death was one of those calamities under which a strong man does not break down, but which he can never forget. Sedgwick's affection for his sister was transferred, so to speak, to the child-niece whose birth coincided with her death, and who became, as she grew up, his chosen friend and indispensable companion. After the death of her own father and mother, she resided with her uncle whenever it was possible to do so, and made his declining years happy by her tenderness and care. She might well have been, as he was fond of calling her, his own child.

After this long digression, which the sequence of events has rendered necessary, we must return to Sedgwick's geological work. In 1823 and 1824, as mentioned above, he continued his exploration of "the most intricate portions of Cumberland, Westmoreland, and Lancashire[2]," but he did not commit any of his conclusions to paper until 1831. In 1825 and 1826 he made no fresh geological explorations—unless we class under that head a very brief excursion in Sussex with Dr Fitton. During these four years, moreover, he worked out the information gathered in the period preceding his lengthened exploration of Lakeland, and was continually employed in writing papers[3], either for the Cambridge Philosophical Society, the Geological Society of London, or the *Annals of Philosophy.*

Nor, while engaged upon these works in his study, did he forget his Museum. His own reports, or those of the

---

[1] These words are used with reference to his own sister in a letter (dated 5 November, 1855) to Mr Lyell, whose sister had just died. See above, p. 53.

[2] Report to the Woodwardian Auditors, 1 May, 1823.

[3] A detailed list of Sedgwick's works, which we have tried to make complete, is given at the conclusion of this Biography.

1822-1825. inspectors, chronicle in each year some important work done,
Æt. 37-40. or some valuable specimens added. Among these additions
should be specially mentioned Mr Henslow's Anglesea
collection, arranged by himself, as before mentioned, during
the Christmas Vacation of 1822—23; a palæontological series
from the bone-caves of Yorkshire; and a cast of "one of the
finest fossils preserved in the Museum of the Jardin des
Plantes, Paris," presented by Mr Chantrey[1]. Unfortunately
the University was, for the time, but little the better for these
treasures, on account of want of space in the miserable room
in which the Woodwardian collections were then stowed—it
would be absurd to say displayed. This subject is dwelt
upon again and again in the reports of the Inspectors to
the Vice-Chancellor. We select, by way of illustration, a
single passage from their report for 1825:

"While we request your notice of the valuable additions which
continue to be made to this Collection by the indefatigable labours
of your learned Professor, and regret that the Museum should be
incapable of containing them, we cannot forbear expressing our
hopes that the result of the Syndicate appointed to treat for the
purchase of the buildings adjoining the Public Library will be
favourable to that enlargement of the Museum which has been so
long desired.
"Of the disadvantage arising from the present crowded state of
this place it would be needless for us to remind you; but we feel it
our duty to advert to an inconvenience which the Professor is
suffering from the necessity laid upon him of receiving in his private
rooms those specimens which have lately been received or collected
by himself."

Sedgwick's other occupations during these years are best
introduced by the following letters. Unfortunately none of
those written in 1824 have been preserved. His work in
Lakeland in the summer of 1823 prevented his presence at
the ceremony of laying the foundation-stone of The King's
Court of Trinity College—better known as The New Court—
which took place on Tuesday, 12 August, in that year. His

---

[1] Sedgwick's *Report to the Woodwardian Auditors*, 1 May, 1823. Unfortunately this is the only report by Sedgwick for those years that has been
preserved, but the series of those by the Inspectors is complete.

name, however, appears in the list of subscriptions as a donor of twenty-five guineas; and we learn, on the autho- rity of Professor Pryme, that he wished the name to be St Michael's Court[1], obviously in commemoration of Michael House, which had owned the ground on which the new buildings were to stand.

TRIN. COLL., *February* 19, 1825.

Dear Ainger,

I am really for once ashamed of myself, and acknowledge that I ought to have written to you some months since. One who has so often offended in the same way must needs be a merciful judge; I therefore venture to anticipate your forgiveness, and even to request that you will show your Christian temper by sending me an immediate answer. Pray tell me what you have all been about in the parsonage of St Bees. The young ones are now, I hope, all well. Give my very kindest remembrances to Mrs Ainger and every one of them.

Now for my own adventures. In about ten days after we parted I bent my way to Dent and spent a quiet week with my father. I then proceeded direct for Cambridge, and only reached my chambers about two days before I commenced my course of lectures. I had a very large class, and as usual was very busy during the term. Just when I thought my labours were happily terminated I found that the whole University was likely to be thrown into the greatest consternation by the sudden appearance of a kind of philosophical mania which broke out among the Cambridge Blues. Unfortunately for me their madness took a geological turn, so that I was obliged, out of pure compassion, to administer to them a sedative dose in the form of a three hours lecture. Peacock tells me that you were greatly scandalised at the news of this event; and that the electrical horror at this academical innovation caused your hair to stand erect, and your shovel to unfold itself.

[1] *Recollections*, p. 143.

A day or two after this act of homage to the Blues, Whewell and I started by the mail for Edinburgh. No words of mine can convey to you any notion of the pleasure which I experienced when I first saw this magnificent capital. The imposing flutter of the old town, which rises, in utter defiance of all regularity, along the sides of a steep declivity terminating in a perpendicular rock crowned by the battlements of the Castle; the beautiful symmetry and neatness of the new town; the happy grouping of great masses of building with natural features of gigantic magnitude; the beautiful glimpses of the Firth of Forth which from every elevated point is seen, like a great inland lake, winding between the shores of MidLothian and Fifeshire; these are the elements which go to the composition of a picture, at least in its kind, unrivalled in the whole world. I will say no more of dead things, but I will speak of the living. We had excellent introductions, and in two days after our arrival, were so completely in society that we had not a moment to call our own. We often went out to breakfast, and always found the tables covered with beefsteaks, ham and eggs, divers varieties of salt fish, marmalade, jellies &c. &c. In a corner of the table you might indeed see a tea-urn and coffee-pot; but these things are non-essentials in a Caledonian fast-breaking. Having out of such materials contrived to lay a good foundation, we sallied out, and spent the morning in running about the different lectures, examining the different institutions, making excursions &c. &c. At six o'clock we returned to some of our new friends, and had a second experience of Scotch hospitality, and a most sumptuous report I must give of it. Our labours did not always end here, as we not unfrequently went out to evening parties, where we met Belles, Beaux, Advocates, Savans, and Craniologists. In short we saw everybody and everything. Of the Savans, Leslie[1] and Brewster[2] are the

---

[1] Mr, afterwards Sir John, Leslie, then Professor of Natural Philosophy at Edinburgh. He died 1832.

[2] Mr, afterwards Sir David, Brewster: born 1781; died 1868.

most distinguished. The former is a short fat butcher-like figure with a red nose, and may be considered as a singular pachydermatous variety of the human species. He is, however, a man of very original powers, and possesses a great mass of curious information. Brewster is in many respects the converse of this. He is a thin gentlemanlike figure, and is so sensitive and thin-skinned that you cannot touch him without making him wince. The two philosophers hate each other most cordially. Jeffrey we met over and over again. He is on the whole a very agreeable man; but you may perceive, in most things he says, the tartness and causticity of the Edinburgh critic. Walter Scott was unfortunately away during the greater part of our visit: what we saw of him made us long for his better acquaintance. He talks exactly as he writes, and before you have been two minutes in his company he begins to tell good stories. Several of his portraits, and above all the bust by Chantrey, convey a most correct notion of his person. The advocates are a very agreeable set of men, not half so much the slaves of their profession, and on that account infinitely better informed on subjects of general interest, than our lawyers. But of all the people we met, the Craniologists afforded us the most amusement. They are perfectly sincere in their faith, tho' I confess I could only regard them as a set of crazy humourists. We met many of the Edinburgh Belles, Blue, Red, and White. The Blues, like the Blues of other countries, remind one of the Blue Boar. But among the Reds and Whites are many delightful persons, of whom I have no time to write.

On leaving Edinburgh we proceeded by the mail to Carlisle and Kendal. From Kendal I posted to Dent, and only remained a day or two, as I found by a letter that I was presented by the College to a small living near Cambridge[1], which I can hold with my Fellowship. A few hours after I reached Cambridge I went up to London to be instituted. Tomorrow

---

[1] Shudy Camps, a village in the S.E. corner of Cambridgeshire, 15 miles from Cambridge. The population, in 1831, was 418; the value of the vicarage, £146.

1825.
Æt. 40.

I read in. Such is the history of my life and adventures for the last five months. Now, my good Doctor, I have sent you a long letter which you must answer. Let me repeat my kindest remembrances to Mrs Ainger and your family.

Yours ever,

A. SEDGWICK.

P.S. I have accumulated so many materials that I must remain at home the greater part of this year to digest and write. I have no less than four memoirs on the stocks. One of these will run out almost into a volume. I fear I shall not have much time for sermons, but I have hired a curate.

TRIN. COLL., *August* 16, 1825.

Dear Ainger,

I reached College on Friday evening, and since that time have been employed in settling my last quarter's bills, reading Scott's last novels, and writing letters. It has long been a custom with me to answer all my friends' letters immediately on my return to College from my vagabondizing expeditions. The task must be performed some time, and in this way it fills up a day or two in which otherwise I might not be employed in anything, for after rambling about in the open air one always sets very reluctantly to work in a dull college room.

A day or two after we parted I proceeded towards the coast, and on my way passed thro' Canterbury, where we had the good luck to fall in with Metcalfe, who looks charmingly, and has a fine family about him. He conducted Dr Fitton and myself over Becket's tomb, and the other ecclesiastical buildings, which in general have more historical than architectural interest. On the whole I was rather disappointed with them. From Shakespeare's cliff we worked our way westward, examining the successive cliffs, hammer in hand, and making short excursions up the country wherever it seemed to promise anything good to our geological eyes.

The weather was beautiful, but so intolerably hot that Fitton
took fright, and ran home to take shelter under his wife's
petticoats. I had no wife to spread out her nether garments
over me, so I was compelled, like my old namesake, by the
sweat of my brow to go thro' my daily work. An account of
my labours, would, I know, be devoid of interest to the
uninitiated. Suffice it therefore to say that I sweated my
weary way as far as Bognor in Sussex, where the rocky cliffs
have entirely disappeared, and are succeeded by nothing but
sand and shingles, which offer but little scope for the exercise
of the hammer. This induced me to hire a boat, and make a
run direct for the Isle of Wight. The weather was delightful,
and the wind so favourable that we did not shift the sails
during the whole day. I shall never forget the glowing
beauty of the shores of the Isle of Wight as we swept up
the channel. The ships of war at Spithead were firing the
evening gun just as we reached the pier head at Ryde.

Next morning I pounded forward to Freshwater, and took
up my quarters with my brother, where I remained a month.
Between dinner-parties, water-parties with the ladies, and
geological expeditions to every corner of the Isle, I contrived
to pass the month most deliciously, and I left the place with
infinite regret this day week. As it blew a stiff breeze from
the right quarter I was induced to hire a boat, in which James
accompanied me to Portsmouth. At first we bounced over
the waves right merrily, but a heavy swell from the west
turned our mirth into sadness, and produced such internal
qualms that our stomachs almost came out thro' our teeth.
We therefore made for Cowes harbour, and "spliced our
main braces" with a glass or two of brandy, which acted like
oil on the troubled waters, and produced a dead calm in the
peristaltic regions. The rest of our voyage was performed
pleasantly enough.

The following evening we spent in the dock-yard with Dr
Inman[1], who kindly showed us everything in his power. I

---

[1] One of Mr Dawson's senior wranglers. See above, p. 65 *note*.

have seen this great naval arsenal two or three times before ; but I rejoice to say that I am as much alive to its interest as ever. My bedroom windows looked over the harbour, and the old Victory with an Admiral's flag at the main-top was at anchor within 200 yards of the house. As I rose very early to see my brother off, and the coach for London did not start till ten, I had an opportunity of hiring a boat, and rowing about the harbour for an hour or two, during which time I flew off into a fit of heroics which it would be impossible to describe in less than another sheet of paper. The state of the elements kept up this fit for the rest of the day; for we travelled to Town in the midst of claps of thunder and flashes of lightning. Near Petersfield a house which had been set on fire by lightning was blazing as we passed.

I only remained one day in Town, and here I am with plenty of employment for the next month. If I can finish a paper which I have on the stocks in a reasonable time I shall try to be at the York Meeting[1], and from thence I shall (D. V.) proceed by Leeds to Dent. My kindest remembrances to Mrs A. and my young friends.

Yours ever,
A. SEDGWICK.

LORD PALMERSTON'S COMMITTEE ROOM.
December 29, [1825].

Dear Ainger,

Strange things come to pass. I am now in the Committee room of a Johnian, a Tory, and a King's Minister; and I am going to give him a plumper. My motives are that he is our old Member, and a distinguished Member, and that I hate the other candidates—I mean with public and political hate, without private malice. Bankes is a fool, and was brought in last time by a set of old women, and whenever he rises makes the body he represents truly ridiculous. Copley is a clever fellow, but is not sincere, at least when

---

[1] The Musical Festival which took place at York, September 13—16.

I pass him I am sure I smell a rat. Goulburn is the idol of the Saints, a prime favourite of Simeon's, and a subscriber to missionary societies. Moreover he squints. Now, my good fellow, though I believe you have the liberality of a great Inquisitor, yet I think you will hardly vote against your own college, your own friends, and the cause of common sense.

1826.
Æt. 41.

Yours ever,

A. SEDGWICK.

P.S. If you don't give at least one vote to Lord Palmerston, I shall think you have rusted in the country, and lost your wits[1].

DENT, *February* 18, 1826.

Dear Ainger,

When we last parted I had no thought of finding my way so soon to Dent, but here I am at the corner of a breakfast-table in the old Parsonage, and the Pastor and his wife are making such a noise that my powers of attention must, I fear, be suspended, and my language incoherent. I will, however, do the best I can in making my way through three pages of this sheet. After I returned to College my whole time was taken up with a dull geological paper[2] which I was endeavouring to bring to a close, in order that it might appear in the *Annals of Philosophy* for next month. But my operations were interrupted by a letter from my sister, which informed me that my Father was much debilitated, and that he exhibited some symptoms of an incipient dropsy. I showed the letter to Haviland, and he advised me to come down, as a complaint of that kind would probably carry off a man of my Father's very advanced age in a few weeks. In consequence of this advice I met the Leeds coach at Alconbury Hill on Saturday last; spent the following day with my

---

[1] The poll-book shows that notwithstanding Sedgwick's efforts Dr Ainger voted for Sir J. S. Copley and Mr Bankes.

[2] *On the classification of the strata which appear on the Yorkshire coast.*

old pupil Charles Musgrave (who resides on his living near Leeds), and on Monday night reached my Father's house. He is, I am happy to say, very much better than I expected, and, on the whole, looks nearly as well as he did when I left him in the autumn. His legs are a good deal enlarged, but the disease makes very slow progress, and, thank God, he is quite free from pain. He is, however, languid and drowsy, and sometimes for a minute or two, even when awake, inattentive to what is about him. On the whole, however, there is no sign whatever of any sudden change, and if he should not get worse in course of next week I shall return to Cambridge and finish my lectures. Indeed I never expected that he would live over the year, and the only wish which his dearest friends have now any right to express is, that it may please God to preserve the faculties of his mind, and release him from a life, now only of labour and sorrow, without the additional burden of much bodily suffering.

After the illumination your mind received at Cambridge, not to mention a conversation one evening at your Father's house, you will, I am sure, rejoice to hear that Lord Palmerston's success at the next election is now quite certain. My best regards and love to all your family. My sister says she will cross the first page of my letter. She shall have her mind: but it is a dangerous thing to drive fresh sentences over my rugged text.

Yours ever,
A. SEDGWICK.

[TRIN. COLL., February, 1827.]

Dear Ainger,

Whewell and I left Cambridge the day after our commemoration (the 17th December), and went as far as Hyde Hall, where we spent the evening with Sir John Malcolm. Next morning we posted to town with the old General in time for breakfast; procured our passports, and went by the night coach to Dover. As soon as the tide served we embarked in a steamer, and in two hours and a half were at the pier head of Calais. In two more days the diligence conveyed us to the place of our destination. I intended to have written to you, and I thought about it every day I was at Paris, but I never had time. With the exception of two short letters to my Father which I regarded as a matter of positive duty, I did not write a line to any one.

My time was spent in the French capital delightfully, and I hope profitably. I attended public lectures, examined public institutions, and became in some measure acquainted with several men whom I before knew only by reputation. Many of the leading literary and scientific men give *soirées*, that is evening parties, once a week, at which any one may attend who has been introduced. Three of these I regularly attended: on Wednesday evenings at the old Marquis de Laplace's, on Thursday evenings at Professor Arago's at the Observatory, and on Saturday evenings at Baron Cuvier's. These parties were delightful. They assemble about nine, and break up about twelve. You meet there the first literary men of France, and you may talk or not as you like, for there is no restraint or ceremony whatsoever. Baron Humboldt is perhaps the most interesting character I have met, and I rejoice to think that I have in some measure formed his acquaintance. We have exchanged one or two letters, and I will endeavour to keep the ball up. He gave me up two mornings, which I considered a great compliment from one who is so much engaged. The day before I left Paris I called on old Laplace and had a long talk with him. He is thin and emaciated; but posessses great mental vigour for a man of 78 years. He asked a great many questions about Cambridge, and then began to talk of the *Catholic Question.* 'A Roman Catholic priest,' said he, 'cannot be a good man, for he is cut off from the rights of manhood, has no sympathy with other men, and only plots for the aggrandizement of his own order. He cannot be a good subject, for he acknowledges an authority which is external

1827.   and superior to the executive of his own country. You have
Æt. 42. these fellows down—keep them down—if you admit them
to power they will only endeavour to destroy those who lifted
them up'!! What do you think of this from a French Peer,
and a *nominal* Roman Catholic? I talked to him in French,
but I have translated what he said as literally as possible.
I found many more men in Paris in the same mind. I also
saw some good English society, and received very great
civilities from Bishop Luscombe[1]. Pray what is the exact
history of his consecration, and of his objects? I did not
exactly make them out. I have many other things to tell
you, if I had time and paper-room, but I must leave something till we meet. When must that be?

On my return from Paris to London I was three successive
nights on the road without being able to rest myself for a
single hour. The cold in France was horrible[2].

Give my affectionate remembrances to your family. Take
care of yourself[3]. I don't however think that writing a short
letter would do you much harm.

Yours most affectionately,

A. SEDGWICK.

It is provoking that this letter should be the only detailed
account of the six or seven weeks which Sedgwick spent in
Paris. He always spoke of the visit as having been not only
agreeable, but also extremely profitable. He learnt, at least
to some extent, what continental men of science were doing
and thinking—for at that time Paris was unquestionably the
scientific centre of Europe—and so paved the way for the

---

[1] An English clergyman consecrated (20 March, 1825) by the Primus of
Scotland, as 'a missionary Bishop for the superintendence of such of the English
clergy and congregations in France, Belgium, and Holland as were willing to
acknowledge his episcopate.' He also acted as chaplain to the British Embassy
at Paris.

[2] The Rev. Joseph Romilly, Fellow of Trinity College, whose diary will be
frequently quoted in subsequent years, notes under 5 February, 1827: 'Sedgwick
came alone in the malle-poste from Paris, with hay twisted round his legs.'

[3] Dr Ainger had just recovered from a severe illness.

researches into continental geology which he undertook soon afterwards. Moreover he added to the Woodwardian Museum "a considerable geological series from the neighbourhood of Paris, collected partly by his own hands, partly by the kindness of his friends, and partly by purchase.[1]"

The above letter may be supplemented by a few details derived from conversation :

" Laplace was a rather small man, with a white neck-tie, looking very like a parson, though he was reputed to be almost an atheist, as indeed was the case. He was then very old, and used an old man's privilege, retiring to bed at about nine o'clock. Arago was a fine-looking man, with a very fine wife, and a staunch republican. Laplace on the contrary was weak, and always shifting his politics according to the time. This led at last to such a quarrel between him and Arago that it was not usual for persons to attend the soirées of both. When Laplace was near his end Arago saw a man at his own soirée who usually went to his rival, and re-marked ' Ah ! he sees old Laplace is going, and so he has come to me.' It was usual for a visitor, when once introduced, to go regularly, and it was considered rude to cut many soirées consecutively. Laplace gave only tea and coffee, but Cuvier, after his soirée was over, would sit down with a few friends to tea and apple-pie.

" I believe that I was the last person (not of his own family) who ever saw Laplace[2]. I called on him before my departure, and sent in my card, as a formal leave-taking. To my surprise he received me, and I saw that he was in the humour for a good talk. He looked very ill, his voice was broken, sounding shrill, like a whistle, and his chin rested on his breast. He said he had desired above all things to visit Cambridge, the scene of Newton's discoveries, but that first want of means, and then the Revolution, had prevented him.

---

[1] *Report to the Woodwardian Auditors*, May, 1827.

[2] Pierre-Simon, Marquis de Laplace, was born 23 March 1749, and died 5 March, 1827, aged 78.

1827.
Æt. 42.

'Are you a clergyman?' he inquired. 'Yes.' 'A Protestant clergyman?' 'Yes.' 'Is there any objection to clergymen marrying?' 'On the contrary, a clergyman who does not do so is thought rather unwise.' Then, after some remarks about priests who could not marry, and therefore could not be good men in relation to humanity, and who obeyed a foreign despot and therefore could not be good citizens, he proceeded: 'I wish I could see your education at Cambridge. I am convinced that you are right in entrusting it to protestant clergymen. English clergymen can be good members of society and good citizens, and I have found out late in life that it is impossible to govern without the help of some religion.' He referred to Catholic Emancipation, then occupying the attention of Parliament, and said it would be an unwise measure: 'You have your foot on the Catholic priests, and you should keep it there!' This was called being 'liberal' in those days. After a long and earnest talk he said 'Good bye,' and that night or the next day I left Paris. On arriving in London I heard that he was dangerously ill. The news had probably travelled in the same coach as I had. He died soon afterwards[1]."

Sedgwick used to tell several other stories—about Cuvier and his daughter Clémentine—Humboldt—and other scientific men whose acquaintance he had the good fortune to make, and very amusing and characteristic they were; but unfortunately no one took the trouble to write them down.

TRIN. COLL., *March* 3, 1827.

Dear Ainger,

Your letter reached me this morning, and I do most heartily congratulate you on your new elevation in the Church[2]. When I mentioned the circumstance in Hall, every

---

[1] This account is derived from notes of a conversation held in Sedgwick's rooms on the evening of Sunday, 6 November, 1870, written down immediately afterwards by Dr Glaisher, Fellow of Trinity College, through whose kindness it is printed here.

[2] Dr Ainger had just been made Canon of Chester.

one who remembered you was quite delighted : and all agreed 1827.
that the manner of the appointment did the Bishop[1] the Æt. 42.
highest honour.   May you long live to enjoy this and still
higher dignities.   A thousand times greater than these things
is, however, the blessing of health, and I rejoice to hear that
now it is no longer withheld from you.   Give my affectionate
remembrances to your sister and all your children.   I hope
they do not forget me.   I must try to come down in course of
the year to rub up their memories.

On Saturday I finished my lectures and immediately came
to Sir John Malcolm's to spend a day or two.   I only returned
yesterday, and am now looking forward to a good spell of
hard work.   By the way, I was made Vice-President of the
London Geological Society at the last annual meeting[2].   But
this honour brings no grist.   There is no manger in my stall,
so that notwithstanding my V.P.G.S. at the tail of my
signature, I may die of hunger.   You plainly see I have
nothing to write about so, my dear fellow, let me once more
congratulate you and wish you a very good night.

Yours ever,

A. SEDGWICK.

The General Election of 1826, when Sedgwick was a
member of Lord Palmerston's committee, was fought on the
same general lines as the bye-election of 1825.   The conservative victory on that occasion had inspired hopes that the
second seat might be won for the anti-catholics, and every
nerve was strained to effect not only the return of Mr Bankes,
but the defeat of Lord Palmerston.   " This once liberal University," said a writer in *The Times*, " is seized at this moment
with such a violent horror of the Pope that in its panic
it forgets the services of an old and tried member, and would

---

[1] Charles James Blomfield, D.D., a former Fellow of Trinity College, was
Bishop of Chester 1824—1828.   He and Sedgwick were godfathers to Dr Ainger's
son.   [2] 7 February, 1827.

18—2

fling him away with as much unconcern as an old glove[1]." In addition to the sitting members, Mr Bankes and Lord Palmerston, two new candidates presented themselves: Sir John Singleton Copley, then Attorney-General, and Mr Goulburn. It soon became evident that Copley, from his brilliant legal reputation, added to an explicit declaration that he was, and always had been, "decidedly adverse" to the claims of the Catholics[2], was certain to be returned, and it became a question whether Bankes or Goulburn would do well to retire. As the former wrote: "by our both standing it is obvious to everybody that we are weakening the Anti-Catholic Interest, and frittering away the strength of our great cause[3]." Thereupon he authorised his Committee to institute "a fair comparison of strength, upon the understanding that whichever of the two should prove to be the weakest, should give way to the other," but the suggestion was declined, as Goulburn had pledged himself under any circumstances to go to the poll. This episode inspired a clever ballad[4], a few stanzas of which are still worth reading:

> Bankes is weak, and Goulburn too,
>   No one e'er the fact denied;
> Which is weakest of the two,
>   Cambridge can alone decide.
> Choose between them, Cambridge, pray,
> Which is weakest, Cambridge, say.
>
> Goulburn of the Pope afraid is,
>   Bankes as much afraid as he;
> Never yet did two old ladies
>   On this point so well agree.
> Choose between them, Cambridge, pray,
> Which is weakest, Cambridge, say.
>
> Each a different mode pursues,
>   Each the same conclusion reaches;
> Bankes is foolish in Reviews,
>   Goulburn foolish in his speeches.
> Choose between them, Cambridge, pray,
> Which is weakest, Cambridge, say.

[1] *The Times*, 13 June, 1826.   [2] Sir John Copley's address, 19 December, 1825.
[3] Mr Bankes to Mr Goulburn, 10 June, 1826.
[4] It is printed at length in *The Times*, 16 June, 1826.

Bankes, accustomed much to roam,
　Plays with truth a traveller's pranks;
Goulburn, though he stays at home,
　Travels thus as much as Bankes.
Choose between them, Cambridge, pray,
Which is weakest, Cambridge, say.

Sedgwick never did anything by halves, and the few letters from friends which he thought worth preserving show that he left them no peace till they came round to his views. The most liberal were evidently not a little scandalised—complained of being obliged to "pick the best out of a bad pack"—and the like; but in the end they voted as he had suggested. The result showed the unwisdom of divided counsels, for, though Copley headed the poll with 772 votes, Palmerston was second with 631, while Bankes and Goulburn had respectively 508 and 437[1]. The poll-book shows that Sedgwick ultimately voted for Copley as well as for Palmerston, as did a large proportion of the members of Trinity College.

Another matter connected with this election must be briefly noticed, as it made a great stir in the University, and Sedgwick's name appears in connection with it. For some years the expenses of non-resident electors had been defrayed by the candidate for whom they voted. It was notorious that the last election had cost a vast sum, and several stories were current of the way in which the liberality of the candidates had been abused. It soon became apparent that the practice would not be discontinued on the present occasion; indeed some leading members of one of the committees had been heard to say, with cynical frankness, that " whatever might have been the expenses of the last election, they were nothing compared with those which would probably be incurred at this." It was manifest that if such usage were not stopped, not only would grave scandals arise, but the choice of University representatives would be limited to men of large fortune. Under these circumstances an attempt was made to induce the four committees to agree in refusing to

---

[1] The poll was taken 13—16 June, 1826.

1826.   pay any expenses. The committees of Sir John Copley, Lord
Æt. 41. Palmerston, and Mr Goulburn agreed to do this, but that
of Mr Bankes declined even to discuss the subject. Some
further negotiations were entered into, but without effect, and
the subject would probably have dropped, had not Mr Lamb,
Master of Corpus Christi College, and Mr Henslow, drawn up
a " recommendation " to the following effect :

> "A very general expression of regret having manifested itself
> among the Members of the University at the practice of out-voters
> receiving their expenses from the respective Candidates for whom
> they voted; the undersigned resident Members of the Senate
> earnestly recommend that this practice should not be renewed at the
> ensuing election."

This document was signed by 102 members of the Senate,
among whom were seven Heads of Colleges, the Proctors, ten
Professors (including Sedgwick), several Tutors, and generally
most men of consideration in the University. It was therefore
easy to see on which side the opinion of residents, without
distinction of party, had been enlisted : but with non-residents
the case was far different, to judge by the literature which the
movement called forth. Nor was their wrath appeased by
the extraordinary conduct of two members of the Senate, who
on the day of election insisted that the oath against bribery
should be administered to each voter as he came to the Vice-
Chancellor's table[1].

It was not long before Sedgwick had another opportunity
of voting against Mr Bankes. The elevation of Sir J. S.
Copley to the peerage as Baron Lyndhurst, in April, 1827,
rendered one seat once more vacant. The candidates were
Mr Bankes, Mr Goulburn, and Sir N. C. Tindal, but Goulburn

---

[1] The history of this question will be found in : *Remarks upon the payment of
the expenses of out-voters at an University election: in a letter to the Vice-
Chancellor of Cambridge.* By the Rev. John Lamb, Master of Corpus Christi
College, and the Rev. J. S. Henslow, Professor of Botany. 8vo. Camb. 1826:
*Observations upon the payment of the expences of out-voters at an University
election, occasioned by remarks upon the same subject, by the Rev. the Master of
Corpus, and the Rev. Professor Henslow.* By a non-resident Master of Arts.
8vo. Camb. 1826; and several letters in *The Cambridge Chronicle,* 28 April—
16 June, in the same year.

retired before the day of election. Both the remaining candidates professed the same principles on the question in which
the electors were most interested, and therefore all they had
to do was to choose the best man. Tindal, a former Fellow of
Trinity College, was personally popular, and had a brilliant
forensic reputation; Bankes, now that the interest in his
travels had worn off, had lost the favour of residents, and in
his own college polled only 78 votes against 191 recorded for
his opponent. The country clergy, however, were still faithful
to him, and it was on this occasion that Macaulay wrote *The
Country Clergyman's Trip to Cambridge. An Election Ballad*[1].

> As I sate down to breakfast in state,
>   At my living of Tithing-cum-Boring,
> With Betty beside me to wait,
>   Came a rap that almost beat the door in.
> I laid down my basin of tea,
>   And Betty ceased spreading the toast,
> "As sure as a gun, Sir," said she,
>   "That must be the knock of the post."
>
> A letter—and free—bring it here—
>   I have no correspondent who franks.
> No! Yes! Can it be? Why, my dear,
>   'Tis our glorious, our Protestant, Bankes.
> "Dear Sir, as I know you desire
>   That the Church shall receive due protection
> I humbly presume to require
>   Your aid at the Cambridge election.
>
> It has lately been brought to my knowledge,
>   That the Ministers fully design
> To suppress each Cathedral and College
>   And eject every learned divine.
> To assist this detestable scheme
>   Three nuncios from Rome are come over,
> They left Calais on Monday by steam
>   And landed to dinner at Dover.

But, amusing as the whole poem is, we have no space for
further quotation. The result of the election was decisive:
Mr Bankes polled only 378 votes against 479 recorded for his
opponent, and, as *The Times* anticipated, he did not again
offer to represent the University.

[1] *Miscellaneous Writings of Lord Macaulay*, 8vo. Lond. 1860, ii. 413. It appeared
originally in *The Times*, 14 May, 1827. The Poll was taken 9—11 May, 1827.

Amidst the bustle and excitement of the election Sedgwick did not neglect his Museum. His next report (dated
1 May, 1827) records the acquisition of "a large collection
of very magnificent fossils, (chiefly the property of the late
Mr Parkinson, author of a work on the organic remains of a
former world[1]) which were purchased during last month at a
public auction;" and "a collection consisting of more than a
thousand specimens of fossil shells, collected in the Isle of
Wight by the Woodwardian Professor, and arranged by Mr
Sowerby of London."

It will be remembered that when Sedgwick first became a
Master of Arts he found the society of the Fellows somewhat
uncongenial. There is happily no evidence that such feelings
were of long duration, nor do his subsequent letters betray
any hint that he was otherwise than happy in his college
surroundings. Besides, in no place does society change so
rapidly as in a University; and in the years which intervened
between 1812 and 1827 the Fellowships at Trinity College
had been filled by a succession of very remarkable men. Most
of them, when their worth became known, attained to high
distinction in the University, in the Church, or in the world;
but, while they remained in Cambridge, they formed a society
whose social charm and intellectual brilliancy has never been
surpassed. They differed widely in tastes, in politics, and
in intellectual pursuits; but they were united by common
interests, by a common devotion to their College and their
University, and not only lived together harmoniously, but in
many instances formed intimate friendships. Some, like
Sheepshanks, Thirlwall, Macaulay, Airy, stayed for only a
short time; others, like Robert Wilson Evans, Peacock, Hare,
Thorp, gave many of their best years to College and University work; while Romilly and Whewell devoted their
whole lives to the same objects.

[1] *Organic Remains of a former World: an examination of the mineralized
remains of Vegetables and Animals of the antediluvian World, generally termed
extraneous Fossils.* By James Parkinson. 3 vols. 4to. Lond. 1804—11.

1827.
Æt. 42.

Among the Fellows here named Sedgwick's dearest friend
was undoubtedly Romilly, whose diary is nearly as full of
Sedgwick as it is of himself[1]. Next to him we would place
Whewell, whose insatiable love of knowledge, especially scientific knowledge, led him to add geology to the other departments of his omniscience. He therefore entered heartily
into Sedgwick's pursuits, and for many years they were
inseparable friends—not only seeing a great deal of each
other during term, but travelling together in vacation. Afterwards, when Whewell became Master, they drifted apart, and
their friendship was interrupted by more than one misunderstanding; but it is pleasant to be able to record that their
quarrels were not of long duration, and that in any serious
difficulty, or grave sorrow, Whewell always turned to the
ready sympathy of his old friend. But Sedgwick could be
approached from many other sides than geology; he was no
specialist, in the modern sense of the word, and those we
have mentioned had no difficulty in finding a large space of
common ground on which to build their friendship. He was
probably the most popular man in the college, and his rooms
the chief centre of attraction. Intimate friends were glad,
when their own work was over, to enjoy his original conversation, and not seldom his extravagant fun: while strangers
delighted to make the acquaintance of a learned Professor
who could talk on general subjects as well as they could
themselves, and who was always ready to lay aside his own
occupations for a while for the sake of their profit and
amusement. It is not too much to say that of the leading
men in Cambridge sixty years ago, no one made so lasting or
so favourable an impression on all who were brought into
contact with him as Sedgwick.

Genial as he was to all-comers, his special pleasure was to
entertain ladies and children, whom he amused in all manner
of quaint ways, and sent home with a store of memories that

---

[1] An account of the diary kept by the Rev. Joseph Romilly has been given in
the Preface.

never faded from their minds. One of the first of these
incidents which has been recorded befell in 1825. Mr
Leonard Horner, the geologist, had come with his family
to Cambridge, and the whole party, including the children,
breakfasted with Sedgwick. It was his first meeting with
Miss Frances Horner, afterwards Lady Bunbury, then a
pretty child of ten. When breakfast was over, he declared
that she should be made a Master of Arts. So she was
decorated with a cap and gown, which of course trailed
far behind her on the ground, and thus attired, ran across
the college grass-plots. Forty years afterwards Lady Bunbury reminded him of the incident, and quoted a passage
from her childish journal: " We got a most delicious breakfast—muffins, tarts, all sorts of nice things. Mr Sedgwick
was so kind as to give us some minerals." Another welcome, in which Sedgwick took part, equally warm-hearted,
though less boisterous, has been commemorated by Lord
Macaulay's sister, Lady Trevelyan. She went to Cambridge
in 1831 with her sister and brother: " On the evening that we
arrived we met at dinner Whewell, Sedgwick, Airy, and
Thirlwall : and how pleasant they were, and how much they
made of us happy girls, who were never tired of seeing, and
hearing, and admiring ! We breakfasted, lunched, and dined
with one or the other of the set during our stay, and walked
about the colleges all day with the whole train[1]."

What has been said of Sedgwick's college friends leads us
naturally to Hyde Hall, and its hospitable tenant, Sir John
Malcolm, who has been already alluded to in a letter. Sir
John, after a distinguished career as a soldier in India, and
a diplomatist in Persia, had returned to England in 1822,
and for a time fixed his abode at Hyde Hall, a large and
commodious mansion near Sawbridgeworth. During the five
years that he inhabited it it was the favourite resort of Hare,
Whewell, and Sedgwick, who have all, in different language,
sung the praises of the master and mistress, and of the society

[1] *Life of Lord Macaulay*, ed. 1881, p. 129.

they gathered round them. Hare, speaking of what conversation ought to be, described Hyde Hall as "a house in which I
hardly ever heard an evil word uttered against anyone. The
genial heart of cordial sympathy with which its illustrious
master sought out the good side in every person and thing,
seemed to communicate itself to all the members of his
family, and operated as a charm even upon his visitors[1];"
Whewell spoke of his acquaintance with the Malcolms as
one of the bright passages of his life[2]; and Sedgwick, as
might be expected, was even more enthusiastic and out-
spoken. After one of his early visits he wrote :

"Sir John has more of the elements of a great character
than any other man I have had the happiness of knowing.
As a mere author his rank is high, but with all this he is a
great oriental scholar, has been three times ambassador at the
Persian Court, has ruled empires with wisdom, and commanded victorious armies; and, what is of more consequence
in his own house, he is one of the most rationally convivial
men that ever sat at a table, or romped with a family of
smiling children[3]."

It is not improbable that in Sedgwick's eyes the smiling
children were not the least attraction that Hyde Hall had
to offer. He at once established an intimacy with all of
them; but his particular friend was the third daughter
Kate. He won her affection in the first instance by carrying
her on his back, and as she grew up he established a correspondence with her, which was carried on regularly until
his death. His letters were all carefully preserved, and it is
from this series that some of the gravest, as well as some
of the most amusing that he ever wrote will be selected.

In the next chapter we shall describe the geological work
which Sedgwick undertook in conjunction with Murchison,
first in Scotland, and then in Germany; but, before entering

[1] *Guesses at Truth*, ed. 1871, p. 528.
[2] *Life*, p. 239.
[3] To Rev. W. Ainger, 27 October, 1826.

upon this new field, it will be well to review what he had accomplished alone in the nine years which had elapsed since he was elected Woodwardian Professor.

When Sedgwick began to work, geology was still in its infancy. Until recently, theory, rather than induction based upon the observation of facts, had held undisputed sway; and, after the publication of such works as Woodward's *Theory of the Earth*, the rival opinions of the Wernerians and Huttonians had divided so-called geologists into opposing camps. While these profitless battles were proceeding, William Smith, whom Sedgwick rightly termed "the Father of English geology[1]," had shown that the proper sequence of the strata might be readily ascertained by observation of the fossils characteristic of each, and that by this means the composition of the crust of the earth might be arrived at—a pursuit likely to lead to more valuable results than theories of the forces by which that composition had been moulded. This discovery—the importance of which it is difficult to realise at the present day—worked a revolution. Theory was abandoned—the mineral composition of rocks, together with the whole science of mineralogy, ceased to be studied by geologists pure and simple—but, instead, a number of accurate and painstaking observers set to work in different parts of England to note the sequence of the strata, their relations to each other, and above all their characteristic fossils. Sedgwick became an ardent member of this band of explorers. His earlier papers, and some of his admissions in conversation or at lecture, show that, like many of his predecessors, he had once been a mineralogist, and a staunch Wernerian. In his first paper, for instance, several pages are devoted to an enumeration of the mineral constituents of the Cornish rocks; but this does not reappear in any subsequent treatise, and as for Werner, we find him dismissed with a jest: "For a long while I was troubled with water on the

---

[1] In his eloquent Address on handing him the first Wollaston Medal, 18 February, 1831.

brain, but light and heat have completely dissipated it;" and on another occasion he spoke of "the Wernerian nonsense I learnt in my youth." The first of these. utterances might be understood to imply an allegiance to the rival views of Hutton; but such was by no means the case. Throughout his geological life Sedgwick—thanks no doubt to his mathematical training—was no theorist. He held firmly to inductive observation, and, if he did advance a view, he took care to avoid the dangerous position of those who "view all things through the distorting medium of an hypothesis[1]."

It must not be thought, from what has been here advanced respecting the changes in geological methods, that geology had been made an easy study. The right road had been pointed out, but there remained many a tangled forest to pass through, many a steep mountain-side to scale. And, as we review Sedgwick's share in geological progress, it will be found that he always aimed at that which was most difficult, and that he did not often stop until he had reached a point beyond which, even at the present time, our data do not warrant us to advance.

Sedgwick's first paper, *On the Physical Structure of those Formations which are intimately associated with the Primitive Ridge of Devonshire and Cornwall*, recording the observations made during the summer of 1819, was read to the Cambridge Philosophical Society 20 March, 1820, and published before the end of that year.

He begins with a short notice of the New Red Sandstone, a deposit with which he was no doubt already familiar in the Eden Valley. He points out that these New Red conglomerates and sandstones are distinct from those which underlie the Mountain Limestone on the Banks of the Avon; but he does not deny that sandstones of this earlier age occur among the schistose rocks of Devonshire; thus, seventy years ago, suggesting, contrary to the opinion of all previous writers, the equivalence of some of the Devonian Rocks with the Upper

---

[1] Letter to Ed. *Annals of Philosophy*, 11 March, 1825. Vol. IX. p. 250.

Old Red. He collected fossils from the Devonian limestones, and from their character, and from a consideration of the stratigraphical position of the beds in which they occurred, he arrived at the conclusion that they might be referred to a formation distinct from the Mountain Limestone, and belonging to an earlier epoch.

It would be hard to find a better description of the physical geography and general structure of the part of Cornwall examined than that given by Sedgwick. He points out that the surface of the moors where the granite rocks predominate is covered " by granite boulders, the remains of larger masses of the same kind which have gradually disappeared, through the corrosive action of the elements," and explains the Loggan-stone as one of such spheroidally-weathered masses. Then, after describing the granite as a crystalline aggregate of quartz, felspar, and mica, he points out—with reference to some of the elvans which often resemble outwardly a fine sandstone—that "varieties, arising from the loss of one of these ingredients, or from the addition of some other mineral, are by no means uncommon;" notices and comments on the microliths in the felspar crystals, and the doubly-terminated crystals of quartz from other similar rocks; and, lastly, contests the opinion of De Luc that the great divisional planes of the granite represent lines of bedding. Having enumerated the different varieties of granite, he concludes in general that the granite of Cornwall "is a true granite, the oldest primitive rock of the Wernerian series"—an expression that does not convey any very clear idea to a modern geologist.

When he comes to the question of the relation of the granite to the overlying rocks, he is hampered by the then prevalent views of both Wernerians and Huttonians. The position Sedgwick takes up is this. The granite is the most ancient rock, because the killas, and other rocks, rest on it. The granite-mass and its veins cannot have been thrust into the killas because there is no displacement along the junction, such as would necessarily result from protrusion of the granite. The killas seems to have been deposited on the granite mass, and the dykes must be contemporaneous with the killas— that is, must represent some portion of it. If he had explained that the only way in which this was possible was by the dykes, or some of them, being portions of the killas assimilated to the granite by alteration along the more open divisional planes, he might have gained much support for his views at the present day. But he has left this not very clear, for it is impossible to imagine that he could have supposed that the killas was laid down among a number of preexisting protruding dykes.

His description of the mode of occurrence of the granite, the elvans, the killas, and all the associated phenomena, can hardly be improved as far as it goes.

The relations of the rocks known as the Old Red were not clearly understood seventy years ago—it can hardly be said that they are beyond controversy at the present time—and sufficient data had not yet been collected for establishing or refuting the various explanations suggested by Sedgwick. But, whatever may be thought of the results of the paper, everyone who reads it must be struck by the author's familiarity with the methods of field geology, with mineralogy, and with the general literature of the subject. It was written too with no external help, so far as we know, except what he might have obtained from Mr Gilby and his cousin, and before he had had the benefit of Mr Conybeare's experience. We are fully prepared to admit that, taken by itself, this paper throws considerable doubt on the truth of the story that Sedgwick knew no geology when he was appointed to the Woodwardian Chair in 1818; but, when we consider it by the light of the evidence which we have collected in favour of that view, it proves that by steady application a man of talent may be able to make observations of the first order in the field two years after commencing the study of the subject.

The memoranda which Sedgwick made on this expedition supplied him with materials for a second paper, *On the Physical Structure of the Lizard District*, read to the Cambridge Philosophical Society 2 April and 7 May, 1821. It discusses the relation of serpentine to the adjoining rocks, and, it should be noted, suggests that its origin may be metamorphic, but without expressing any definite opinion on the subject—a position he would have been justified in taking up even after much fuller investigation than he was able to bring to bear upon it—as may be inferred from the fact that the origin of serpentine is still considered a matter of such doubt and difficulty that at the International Geological Congress held at Bologna in 1881 it was the subject set apart for special consideration.

In 1821 Sedgwick published a *Syllabus of a Course of Lectures on Geology*, for the use of his pupils. The older palaeozoic rocks had not been yet worked out—he was himself the first to put them in order some ten years later— but he gives a classification of the sedimentary rocks which holds good in all essential points at the present day. It is very interesting to compare the syllabus of 1821 with the second edition of 1832.

Both commence with an introductory chapter, dealing with the history and progress of Geology—the distinction between Natural Philosophy and Natural History—ancient speculations on the Theory of the Earth—and the connection between Geology and other branches of Natural History. In the first syllabus it struck him as more important to emphasize the connection of Geology with Mineralogy, in the second their separation. In 1821 the views of Werner and Hutton were still subjects for difference of opinion, and consequently in the first syllabus prominence is given to the old cataclysmic theories, but in the second they are very briefly noticed under the heading "Ancient Theories," and the first chapter ends with "True mode of conducting geological speculations." This is succeeded by seven pages of notes which might form a rough index to Lyell's *Principles*. A section headed : " The great inequalities presented by the surface of the earth," is succeeded by another : " On the great agents by which the earth's surface is modified, and on the effects which have been produced by them during known periods." Among the examples of such agents it is interesting to find "Coral Reefs" enumerated, when it is remembered that in this very year Darwin began to work in South America, where, as he tells us[1], he first thought out his theory of their formation. The last division of this introductory portion is headed : " Ancient alluvion (" Diluvial detritus ")—including all superficial transported aqueous deposits which are un-connected with the present mechanical action of the waters." It is obvious that alluvium could not be considered without some notice of the organic remains contained in it; and, having regard to the very decided attitude taken up by Sedgwick afterwards in the discussions relating to the An-tiquity of Man, the following headings are worth transcription, as shewing that even at this period of his scientific life he had begun to pay attention to this question.

Organic remains. (1) In rolled masses derived from older formations. (2) Land and marine shells. (3) Bones of mammalia of extinct and living species, &c., &c. No human bones (?).
Description of some remarkable species.
Great local deposits of bones, formed before or during the period of the ancient detritus. Examples.
Ossiferous caverns, osseous breccias, etc.

Sedgwick always paid great attention to palæontology, and we find him recurring to the subject of organic remains in the same syllabus, Part II. (p. 13), concluding with " Importance of organic remains—in the identification of contemporaneous deposits—in determining the successive conditions of the earth." He fully recognised the value of palæontological evidence. As early as 1822 he insisted upon the importance of an intimate acquaintance with certain branches of natural history. " Without such knowledge it must be impossible to

[1] *Life of Charles Darwin*, i. 70. The theory was communicated to the Geological Society 31 May, 1837. *Proceedings*, ii. 552.

ascertain the physical circumstances under which our newer
strata have been deposited.   To complete the zoological
history of any one of these formations, many details are yet
wanting[1]."   He always carefully collected fossils, and referred
them to the best authorities he could find on each special group
for determination ; but, while he appealed to palæontological
evidence whenever he could, he always maintained that the
first thing was to get the rocks into the right order in the field.

Sedgwick's arrangement of stratified rocks in the two
syllabuses shews his gradual emancipation from the theories
of Werner.   In the first the term "transition rocks" stands as
the heading to a chapter; in the second he discusses, at the
end of a chapter: "Origin of the term *Transition*—with what
limitations it is applicable to the upper series—no clear line
of separation between the two."   Again: in the first he
worked up from the transition rocks through the stratified
series in ascending order.   It is true that while treating them
he pointed out the "great variety of fossil species", and taught
the "principles of classification of organic remains founded on
the classes of recent species", and dwelt upon the "connection
of fossil species with particular strata[2];" but it is a very
suggestive fact that in the second syllabus he reversed the
order, and worked down from the better known to the more
obscure, shewing that he now more fully realised the truth
that the history of the earth must be deduced from a study
of the operations of nature which we see going on around us
at the present time.   A sketch pasted into his own copy of
the second syllabus shews that among other illustrations he
cited in his lectures the floating island of Derwentwater, and
the sources of springs in the neighbourhood of Cambridge.
The execution is better than that of the sketches which
ornament some of his letters in later life, the rudeness of
which he used playfully to deplore.

[1] Letter *On the geology of the Isle of Wight*, 17 March, 1822.   *Annals of Philosophy*, N. S. iii. 329.
[2] *Syllabus*, ed. 1821, Chapter IV.   *Transition Rocks*, § 2.

It is pleasant to read a good practical paper founded on original observation in which the character of dykes is so well discussed as in the paper on the *Association of Trap Rocks with the Mountain Limestone Formation in High Teesdale* which Sedgwick read to the Cambridge Philosophical Society in 1823—24. He points out that dykes are of all ages, and refers them to an igneous origin in the following passage :

> "It is a matter of fact, which is independent of all theory, that an enormous mass of strata has been rent asunder; and it is probable that the rent has been prolonged to the extent of fifty or sixty miles.   If we exclude volcanic agency, what power in nature is there capable of producing such an effect?   By supposing such phenomena the effects of volcanic action, we bring into operation no causes but those which are known to exist, and are adequate to effects even more extensive than those which have been described."

In describing the columnar structure it did not escape his notice that the prisms were arranged at right angles to the cooling surfaces.   He mentions also the common mode of weathering into great balls by the exfoliation of successive layers from the joint faces.

Sedgwick contributed a paper to the *Annals of Philosophy* for April and July, 1825, on the *Origin of Alluvial and Diluvial Formations*, in which he distinguishes the older formations, which we should now call "drift", from the generally newer alluvial deposits.   He points out the anomalous position and irregular distribution of the boulders which occur in the drift; but that glacial conditions had once been prevalent in our island, and over extensive tracts throughout the northern part of our hemisphere, had not at that time been recognised.   There is still much difference of opinion as to whether the "drift" of certain districts is due to land-ice, or to icebergs; and, if we translate the views of those who hold the iceberg theory into the language of Sedgwick's time, we shall find that the observations which he records are not far wrong as to the direction of the currents which distributed the "drift". That he did not recognise the exact mode of transport is only to say that he had not that familiarity with glaciated districts which enabled Agassiz to suggest the agency of ice. He referred the accumulation of the diluvial detritus to the action of water; he concluded that the floods which produced it "swept over every part of England—that they were put in motion by no powers of nature with which we are acquainted —and that they took place during an epoch which was posterior to the deposition of all the regular strata of the earth."

As this question has acquired fresh importance from the recent invocation of a flood of waters as the sole agent in the deposition of the drift, the following passage, written it will be remembered sixty-three years ago, will be read with interest. Sedgwick maintained that while the evidence for the truths of revealed religion was of a totally different character from that by which physical laws are established, still the conclusions at which we arrive in the two cases should not be contradictory; and his statements on these questions are always fair and liberal, and far in advance of the age in which he lived.

"As we are unacquainted with the forces which put the diluvian waters in motion, we are also, with very limited exceptions, unable to determine the direction in which the currents have moved over the earth's surface. Many parts of the north of Europe seem to have been swept over by a great current which set in from the north. In some parts of Scotland there has been a great rush of water from the north-west. The details given above, show that the currents which have swept over different parts of England have not been confined to any given direction. It may, perhaps, be laid down as a general rule, that the diluvial gravel has been drifted down all the great inclined planes which the earth's surface presented to the retiring waters.

*   *   *   *   *   *   *

"The facts brought to light by the combined labours of the modern school of geologists, seem, as far as I comprehend them, completely to demonstrate the reality of a great diluvian catastrophe during a comparatively recent period in the natural history of the earth. In the preceding speculations I have carefully abstained from any allusion to the sacred records of the history of mankind; and I deny that Professor Buckland, or any other practical geologist of our time has *rashly attempted* to unite the speculations of his favourite science with the truths of revelation[1].

[1] Professor Buckland's *Reliquiae Diluvianae; or, Observations on the Organic Remains contained in Caves, Fissures, and Diluvial Gravel, and on other Geological Phenomena, attesting the Action of an Universal Deluge,* had been published in 1823.

"The authority of the sacred records has been established by a great mass of evidence at once conclusive and appropriate; but differing altogether in kind from the evidence of observation and experiment, by which alone physical truth can ever be established. It must, therefore, at once be rash and unphilosophical to look to the language of revelation for any direct proof of the truths of physical science. But truth must at all times be consistent with itself. The conclusions established on the authority of the sacred records may, therefore, consistently with the soundest philosophy, be compared with the conclusions established on the evidence of observation and experiment; and such conclusions, if fairly deduced, must necessarily be in accordance with each other. This principle has been acted on by Cuvier, and appears to be recognized in every part of the "*Reliquiæ Diluvianæ*". The application is obvious. The sacred records tell us—that a few thousand years ago "the fountains of the great deep were broken up"—and that the earth's surface was submerged by the waters of a general deluge; and the investigations of geology tend to prove that the accumulations of alluvial matter have not been going on many thousand years; and that they were preceded by a great catastrophe which has left traces of its operation in the *diluvial detritus* which is spread out over all the strata of the earth.

"Between these conclusions, derived from sources entirely independent of each other, there is, therefore, a general coincidence which it is impossible to overlook, and the importance of which it would be most unreasonable to deny. The coincidence has not been assumed hypothetically, but has been proved legitimately, by an immense number of direct observations conducted with indefatigable labour, and all tending to the establishment of the same general truth[1]."

At the end of the paper is an appendix, giving an account of some changes in the channels which drain the fen-land : an account full of interest to those familiar with the Humber and its tributaries, and all the phenomena of silting-up and warping. Sedgwick concludes as follows :

"If such extraordinary effects as those described in this note be produced by the accumulation of alluvial matter in course of a few hundred years, we may be well assured that the whole form of the neighbouring coast must have been greatly modified by the same causes acting without interruption, and without any modification from works of art, for 3000 or 4000 years."

The letter *On the Classification of the Strata which appear on the Yorkshire Coast*, which, like the last, was addressed to the editors of the *Annals of Philosophy*, 20 February, 1826,

[1] *Annals of Philosophy*, N. S. 1825, x. pp. 33—35.

must not be wholly passed over in silence. It is based on observations made in 1821, but which, after Sedgwick's fashion, were not worked out for five years. His object in writing it was to connect the phenomena observed in Yorkshire with those observed in other parts of England; and it contains, parenthetically, a good deal of information on the geology of Weymouth and its neighbourhood.

The value of the work done in Yorkshire appears again in the splendid monograph *On the Magnesian Limestone and lower Portions of the New Red Sandstone Series*, which was read to the Geological Society at various intervals between November, 1826, and March, 1828. Whatever turn geological research may take, this, with some other papers which followed it in quick succession, must always be referred to as standard works, which settled some of the disputed questions of English geology. It is at once broad and minute: broad in its generalisations—for it places in order a complex group of rocks which, until it was written, were in complete confusion; and minute in working out, through the whole of the district selected, from Nottingham to the southern extremity of Northumberland, the boundaries of the different formations, and their relations to each other. The labour which this research implies is almost incredible; the whole district appears to have been gone over, probably on foot, and compass in hand. Every quarry, cliff, and scar is made to bear its part in the general result. In carrying out these researches it must be remembered that hardly any help was at that time available. Mr Smith's geological map of Yorkshire had been published and is constantly referred to, but Sedgwick notes that "geological maps of the other counties through which the magnesian limestone passes were not published at the time the observations were made on which the greatest part of this paper is founded[1]." Moreover, the Ordnance Survey was not yet in existence. On this account it will be interesting, we think, to

---

[1] *On the Geological Relations and Internal Structure of the Magnesian Limestone, etc.*, p. 43, *note*.

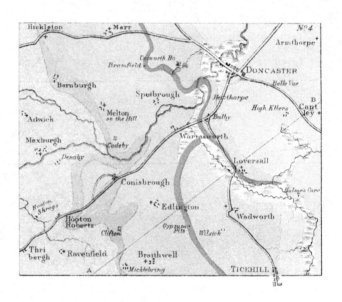

SECTION IN THE DIRECTION OF THE LINE, AB.(MAP Nº4)

MAP OF PART OF YORKSHIRE.

To illustrate Sedgwick's paper on the *Geological Relations and Structure
of the Magnesian Limestone.*
(*Trans. Geol. Soc. Lond.* Ser. II. Vol. iii. Plate 4.)

For a colour version of this map, see www.cambridge.org/9780521137706

reproduce one of the maps which Sedgwick drew and measured for himself. Difficult as his task must have been—for he could hardly have known much about practical surveying —it will be found that in accuracy of detail his unaided efforts have only been superseded by the elaborate work of the Geological Survey.

The paper opens with the following prefatory sentences:

"After the production of the rocks of the carboniferous order, the earth's surface appears to have been acted on by powerful disturbing forces, which, not only in the British Isles, but through the greater part of the European basin, produced a series of formations of very great extent and complexity of structure. These deposits, known in our own country by the name of new red sandstone and red marl, and, when considered on an extended scale, comprising all the formations between the coal-measures and the lias, notwithstanding their violent mechanical origin, have several characters in common, which enable us to connect them together, and, for general purposes of comparison, to regard them as one group. Great beds of conglomerate, coarse sand, and sandstone, frequently tinged with red oxyde of iron; and of red marl, associated with innumerable beds and masses of earthy salts, constitute, in many countries, the principal portion of the group we are considering. Many of these salts, though of almost constant occurrence among the rocks of this epoch, have been developed with so much irregularity, that the attempts to arrange them in distinct formations (when used for any purpose beyond local description) have sometimes, perhaps, served to retard rather than to advance our knowledge of the earth's history. The great calcareous beds which were produced during this period form, however, an exception to the last observation. They appear to have been chiefly developed in the upper and lower portions of the system we have been considering; and, though possessing some characters in common, are sufficiently distinguished by their position and their fossils to be separated into two distinct formations. The higher of these (the *muschel-kalk stein* of the continental geologists) has no representative in the series of rocks which have hitherto been observed in our island; the lower is represented by the great terrace of magnesian limestone which ranges from Nottingham to the mouth of the Tyne."

In the next place, after noticing the "general want of conformity to all the inferior formations" observable in the magnesian limestone series, he cautions his readers against pushing this kind of evidence too far:

"We have no right to assume, nor is there any reason to believe, that such disturbing forces acted either uniformly or simultaneously

throughout the world.   Formations which in one country are unconformable, may in another be parallel to each other, and so intimately
connected as to appear the production of one epoch."

Throughout the paper we find great care bestowed upon
the determination of the fossils found in the different deposits,
and a good many of the most characteristic are figured.
Sedgwick was of course very much excited by the discovery
of fossil fish in the marl-slate; and he hunted up the specimens and fragments of specimens which had found their way
into private collections all over the country.   When submitted to De Blainville it was decided that they were almost
identical with the celebrated fish from the copper-slate of
Germany.   At the end of the paper we find a short descriptive notice of three species, *Palæothrissum magnum*, *P. macrocephalum*, and *P. elegans*, illustrated by some excellent figures,
one of which is here reproduced, to shew the care with which
he elaborated his papers.   Sedgwick followed up the study of
fossil fish in subsequent researches, and succeeded in getting
Alexander Agassiz, the Cuvier of ichthyology, to come to
Cambridge and examine his collection.

The following classification of the rocks investigated, in
descending order, is given at the end of the paper:

1.   Upper red marl and gypsum.
2.   Upper red sandstone.
3.   Upper thin-bedded limestone.
4.   Lower red marl and gypsum.
5.   Yellow magnesian limestone.
6.   Marl and thin beds of magnesian limestone.
7.   Lower red sandstone.

Sedgwick's account of the lowest beds of the group, as
thus arranged, is not always clear, but it may be explained,
we think, in the following manner.

The Lower Magnesian Limestone passes down into red
or yellow sandy beds at Pontefract.   So, in the Eden Valley,
the *brockram* or basement conglomerate of the Poikilitic
series rests on red and yellow sandstones.   But these are of

PALÆOTHRISSUM MACROCEPHALUM.

(*Trans. Geol. Soc. Lond.* Ser. II. Vol. iii. Plate 9.)

totally different age. In the sandy beds of the roadside cliff near the great quarries just outside Pontefract *Schizodus obscurus* has been found. These beds undoubtedly belong to the Magnesian Limestone series. In the Eden Valley, on the other hand, it was found by boring near Appleby that the red colour did not extend farther into the rock than a little over one hundred feet. The colour was obviously a stain produced by infiltration from above. Again, there is, in the Woodwardian Museum, a collection of fossil plants from the north-west margin of the same area, obtained from red beds formerly referred to the Poikilitic series, but which contain nothing but carboniferous species. These facts obviously account for the difficulties which have arisen from some observers recording that the base of the Poikilitic series graduated into the carboniferous, while others saw a strong discordancy between the two. The confusion has been increased by an attempt to force the English classification into harmony with the as yet unestablished sequence of Germany, or the still less known deposits of Russia. On this point Sedgwick makes the following admirable remarks:

" Each country ought to be described without any accommodating hypothesis, according to the type after which it has been moulded. But, in comparing the unconnected deposits of remote countries, we must act on an opposite principle; learning to suppress all local phænomena, and to seize on those only which are coextensive with the objects we attempt to classify."

Whatever therefore may be convenient in respect of the Dyas of Germany, or the Permian of Russia, all attempts to bracket the Magnesian Limestone of England and its associated red marls with the carboniferous rocks, instead of making them the beginning of a new series, forming the base of the Secondary rocks, have been founded on stratigraphical mistakes, and tend to perpetuate an unnatural classification. Sedgwick's grouping of the New Red rocks of Britain, which brackets the whole Poikilitic series together, will undoubtedly stand the test of time.

CHAPTER VIII.

(1827—1831.)

The Geological Society. First acquaintance with Murchison. Tour with him in Scotland. Office of Senior Proctor (1827). Joint papers with Murchison. Summer in Cornwall. Dolcoath Mine. Visits Conybeare in South Wales (1828). Sedgwick President of Geological Society. Divinity Act. Mr Cavendish elected University representative. Summer in Germany and the Tyrol with Murchison. Joint paper on the eastern Alps (1829). Address to Geological Society. Summer in Northumberland. Contested election of President of Royal Society (1830). Addresses to Geological Society (1831).

Sedgwick had been elected a Fellow of the Geological Society of London in November, 1818. He could not have attended the meetings regularly—at any rate at first—but when he did go he thoroughly enjoyed them. In those early days of the Society's existence—it was but eleven years old in 1818—it was composed, as he has himself recorded, "of robust, joyous, and independent spirits, who toiled well in the field, and who did battle and cuffed opinions with much spirit and great good will. For they had one great object before them, the promotion of true knowledge; and not one of them was deeply committed to any system of opinions [1]." In such

---

[1] *A Synopsis of the Classification of the British Palæozoic Rocks*, p. xc.

an assemblage Sedgwick was sure to take a foremost place, and we are not surprised to learn that some of his "most honoured and cherished[1]" friendships originated in the Society. Before long the value of his cooperation, from a scientific point of view, came to be recognised; in 1824 he became a member of the Council; and in 1827 a Vice-President. His friend Dr Fitton, then President, in announcing his election to the latter office, assures him that he feels "no small gratification in the prospect" of having him as a colleague; and that he will be most happy "to receive any suggestions for the advancement of the subject or the welfare of the Society[2]." This language affords conclusive evidence of the favourable judgment which capable men of science had by this time passed on Sedgwick; before long the Society accepted him as a leader, and much of "the generous, unselfish, and truth-loving spirit that glowed throughout the whole body" was probably due to his influence.

Sedgwick soon became on excellent terms with the fathers of the Society; but it was with Roderick Impey Murchison, then one of the junior members, that he contracted the closest alliance. The origin of this alliance, like the origin of many great things, is shrouded in obscurity; but before we proceed much farther we shall find them taking long geological expeditions together, and collaborating in the *Transactions* of the Society. Murchison says that "from his buoyant and cheerful nature, as well as from his flow of soul and eloquence, Sedgwick at once won my heart, and a year only was destined to elapse before we became coadjutors in a survey of the Highlands, and afterwards of various parts of the Continent[3]." Mr Geikie, on the other hand, while fully admitting that his hero admired and respected Sedgwick, gives a more practical reason for the selection of him as a travelling-companion. Murchison was

---

[1] *Ibid.* p. xcii.
[2] From W. H. Fitton, M.D., 8 February, 1827.
[3] Geikie's *Life of Murchison*, i. 124.

then a beginner in geology, and in 1826 had found himself thoroughly puzzled by the problems presented to him in Arran and other parts of Scotland. He therefore "determined to return to the attack, bringing with him a geologist of ampler knowledge, and specially experienced in the complicated structure of the older rocks[1]." The geologist selected was Sedgwick, to whom he must have suggested a joint expedition to Scotland soon after his own return from it, for early in 1827 he writes: "Dinna forget the land of cakes, whiskey, and hospitality, where I hope to act as your aide-de-camp; which having been my old trade before I took to the road, enables me to say without presumption that I am not a bad caterer;" and Sedgwick replies (20 May):

"I am most anxious to accompany you to Scotland, and if possible will be ready by the time you mention. I think I told you that I am *out at elbows,* and that this sorry condition of my garments prevents me from going to Germany. You, it seems, have lately been *breech-less.* So on the whole we shall exhibit good specimens of denudation. I hope the country we pass through will be induced to sympathize with us."

Nothing happened to interfere with the proposed collaboration, and at the beginning of July, 1827, they started. The expedition was thoroughly successful. If Murchison gained an instructor, Sedgwick gained an agreeable companion, whom, before they got home, he called his intimate friend, though it may be doubted whether this friendship was based on a surer foundation than mutual convenience. They wrote affectionately to each other; Murchison signs himself "Rodericus," and Sedgwick replies with "Yours to the earth's centre," and so forth; but, with the exception of one remarkable letter in which Murchison confesses his religious belief, or rather disbelief, to "the man of my heart[2]," they wrote on geological subjects exclusively. Nor have we any evidence, either from personal recollection or from tradition, that Sedgwick ever spoke or wrote of Murchison in

[1] Geikie's *Life of Murchison,* i. 137.   [2] *Ibid.* i. 261.

the warm enthusiastic language he was in the habit of using 1827.
towards those whom he really loved. Murchison, on the other Æt. 42.
hand, sat for a time at Sedgwick's feet, and got twitted for his
hero-worship by those who adored less fervently. " Perhaps
you know," writes Lyell to Scrope, "that he idolises even
more than the Cantabs 'the first of men,' as Adam is usually
styled there[1]." It need not surprise us that such a friendship
came to an end as suddenly as it had begun; and that the
closing years of both were saddened by an estrangement
which each regarded as due to the vindictiveness of the other.

The special object of this first journey "was to ascertain,
if possible, the true relations of the red sandstones of Scotland[2]."
For this purpose it was agreed to commence with Arran, to
pass through Mull, Skye, Sutherlandshire, and Caithness, and
thence southwards by Elgin, Aberdeen, Forfarshire, Edinburgh, Dumfriesshire, Carlisle, and Newcastle, to York. The
progress of the travellers as far as Rosshire is agreeably
related in the two following letters to Whewell, who seems to
have had, at one time, some intention of joining them.

BRODICK, ARRAN, *July* 15, 1827.
Dear Whewell,
        Men who work fifteen hours a day cannot easily
find time to write long letters. I am stealing a minute at the
breakfast table in a pretty Highland village under Goatfell,
or Gengevane. We arrived at Edinburgh without any
accident; not even the adventure of a bout of sickness. I
slept like a top within three feet of the engine during the
three nights we were out, and during the day we were
geologising the coast from the deck of the steamer. I only
remained one day in Edinburgh, and heard many tender
inquiries after yourself.

We coached to Glasgow, where we remained all night,
and next morning we steamed to Lock Ranza on the
north point of Arran. Here we have been nine days, with un-

---

[1] *Life of Sir C. Lyell*, i. 310.   [2] Geikie's *Life of Murchison*, i. 139.

interrupted fine weather. We have fairly done the island, and have found a complete series of the Old Red, coal measures, and Young Red. The details have cost us some hard labour. I hoped to have run across to the coast of Ayr, but it was quite impossible. Tomorrow we are off for Bute—thence to Oban and Mull, where we shall remain about a week. Then we shall start for Portree in Skye. On this island we shall remain a week or ten days. If you have a notion of joining us, you may find your way by steamers from Glasgow to any of these places with as much ease as you travel from London to Liverpool; and indeed with much greater ease. All the lochs and arms of the sea are covered with steamers plying in all directions.

I am delighted with what I have seen of the Highlanders. They are good-humoured, high-minded, well-informed, racy, and dirty. The day before yesterday at Loch Ranza I asked a fine dark-eyed lass for a pair of slippers. She immediately pulled off her own shoes and offered them to me, saying: "I dinna want 'em. You may wear 'em yoursel while I clean your ain." On returning yesterday over the mountains we passed two fine lasses; one had a green veil, and the other a velvet reticule. Yet both were walking without shoes and stockings. Our guide is a fine old man whose mind is stored with traditions respecting all the families in the Highlands. He told Murchison many circumstances respecting his family history which quite astonished him. He (Mr M.) said that he did not believe any man in the world had known them except himself, and he had only learned them from an old manuscript of his grandfather's. My best regards to every body. Yours ever,

A. SEDGWICK.

ULLAPOOL, *August* 11, 1827.

Dear Whewell,

We are waiting at a small inn on the north-west coast of Ross, in hopes of better weather. Should it clear

up, in a quarter of an hour we are off. I have only time for
a few lines which may give you some notion of our intended
motions. Tomorrow and the next day we shall probably be
at Assynt ; Wednesday or Thursday we shall, if possible,
spend at Tongue (the seat of Lord Reay), from whose house
we mean to make an excursion to Cape Wrath. Afterwards
we turn to the north-east, and shall probably be at Thurso in
ten days from this time (on the 21st). From Thurso we
propose to coast by John o' Groat's, and to walk along the
shores of Caithness, Sutherland, and Rosshire, to Inverness.
This place we hope to reach in the second week of September.
If you have any disposition to join us we shall rejoice to see
you, and I can promise you a most hospitable reception from
the Highland Lairds. I will also promise you fair weather ;
for I am quite sure that all the rain must fall on the west
coast. We have so much here that the clouds can have none
to spare for the east coast.

I have neither time nor space for any description of what
we have been about ; but I will give you our track, and it may
serve as a text for some talk when we meet. We spent nine
active days at Arran, and from thence found our way through
the Kyles of Bute to Loch Fyne. We then crossed the Mull
of Cantyre, and coasted up to Oban and Ballachulish ; and
from the latter place we made an attack upon the primitive
chain, but were driven back by bad weather, and obliged to
take shelter in a steam-boat, which took us down Loch Linnhe
to Oban. From Oban we proceeded by the same conveyance
to Tobermory, on the north point of Mull. The weather then
cleared, and we had three or four glorious days at Staffa,
Iona, and the south coast of Mull. If ever you come among
the western Isles be sure to see the south coast of Mull.
The basaltic cliffs are most gorgeous. They are more than
equal to Staffa piled ten times upon itself. From Carsaig
Bay we crossed through the centre of Mull in a most dirty
condition ; the two shirts I had with me would have fetched
a good price from a tallow-chandler : but at Torliusk we

again embraced our portmanteaus, and found a stock of soap. After spending two days with Lord Compton we started by the Maid of Islay up the magnificent Sound of Skye, waiting for a few minutes by the way under the lofty Scuir of Eig. We landed at Portree, and proceeded forwards along the north coast of Skye to the utmost limit of the great basaltic chain. The character of the coast is this : a base of lias and oolite, sometimes forming a cliff five or six hundred feet high; over them great stacks of basalt rising several hundred feet above the horizontal beds. The tops of these form a kind of table-land, at the back of which is a great basaltic chain rising about a thousand feet above the plateau. From some points you take in at a single view from a boat, the stratified rocks, the superincumbent columns, and the bristling top of the chain, and the effect is glorious. This is a dreadful country for storms; you have such howling blasts along the coast that you might fancy that the gigantic columns of basalt were organ-pipes, and that the devil was blowing a tune through them. We examined the islands of Raasay, Scalpa, etc., and made some excursions to the south-west coast of Skye. Our attempt to cross the Cuchullins almost ended fatally. Lord Macdonald's forester was our guide; but in a dreadful storm of wind and rain, accompanied with the usual mists, he lost his way. After wandering many hours in a state of great misery we at length escaped from a labyrinth of precipices by the help of my needle, and found our way to a farm-house. Since leaving Skye we have been working our way, hammer in hand, up the west coast of Ross. The country is magnificent, and the weather, till this morning, has been highly favourable. Fish is so abundant that we have nothing to do but light a fire and put on a frying pan, and the salmon find their way into it without help. My paper is out. My best regards to all Trinitarians. Yours ever,

A. SEDGWICK.

P.S. Dear Hare, I will direct to you in case Whewell is

off. How goes on the Translation[1]? Murchison is walking
about with a cigar. He sends you his regards between the
puffs. Yours, A. S.

A third letter, written to Ainger three days later from
Assynt, where they were detained by bad weather, gives a few
additional particulars. Sedgwick was evidently thoroughly
happy. Everything was new, and strange, and delightful.
He enjoyed the geology, he enjoyed the scenery, we might
almost say that he enjoyed the bad weather and the rough
travelling. "Arran," he says, "is a geological epitome of
the whole world, and is, moreover, eminently picturesque. I
was greatly delighted with it as the first place in which I saw
the Highlanders in their native habitations. These indeed
are none of the best. Those of the lower orders have often
neither chimney nor window, and from the distance of two
hundred yards might be mistaken for peat-stacks." In Mull
they experienced some striking contrasts. One night they
"slept at a whiskey-shop, and breakfasted in the same room
with the pigs"; the next they dined with a laird, who gave
them venison and claret. Rosshire they found "very wild, no
roads, and what to us is of much greater moment, no bridges.
The mountain-streams have become torrents, and some of the
rivers impassable, so that we have been exposed to much
fatigue, delay, and vexation; and, what is worse than all, we
have in one or two instances, after walking twenty or thirty
miles, returned without effecting the purpose for which we
started. One day we crossed upwards of forty streams, some
of which took us up to the middle; and several times
yesterday our horses were nearly off their legs." Such were
some of the difficulties which a geologist had to encounter in
the Highlands sixty years since.

Mr Geikie, in his account of this journey, points out, as
indicative of Sedgwick's power of acute observation, that

---

[1] In 1828 Mr J. C. Hare and Mr Thirlwall published the first volume of their
translation of Niebuhr's *Geschichte Rom's*.

he had already recognized the peculiar structure of rocks
called "cleavage", as distinct from stratification. When the
travellers were examining the slate-quarries of Ballachulish
they fell in with two German geologists, K. von Oeynhausen,
and H. von Decken, whom Sedgwick tried in vain to convince
on this point. The argument was long, and the rain heavy,
but the Germans could not be made to see that there
was any difference between the two classes of phenomena.
This accidental meeting was the beginning of much agreeable intercourse, which was continued in Germany two
years later.

Having seen as much of the west and north-west of Scotland
as the weather would allow, the geologists explored the coast
of Caithness, and thence made their way down the east coast,
according to their original plan. On their way home Sedgwick
at least paid a visit to Lyell at Kinnordy[1], and then went for
three days to Dent, before returning to Cambridge for the
Michaelmas Term.

Trin. Coll. October 28, 1827.

Dear Murchison,

　　　I received your letter on Monday last, and was
greatly delighted with the account you have given of your
proceedings. In return I have no news to tell you. My
father I found in a state of extreme debility, and worn to a
perfect *anatomie vivante*, but still without pain, and I might
perhaps say in good health.

After remaining with him three days I posted up to Cambridge with all the expedition I could command, and only
arrived just in time to be enthralled in my new office. Behold
me now in a new character, strutting about and looking
dignified, with a cap, gown, cassock, and a huge pair of
bands; the terror of all academic evil-doers—in short a
finished moral scavenger. My time has been much taken
up with the petty details of my office, and in showing the

---

[1] *Life of Sir C. Lyell*, i. 199.

lions to divers Papas and Mammas who at this time of the year come up to the University with the rising hopes of their family. Dr Greville and Co.[1] spent two days in Cambridge on their way to the south coast, where the Doctor means to settle for the winter. Oeyenhausen and Decken have also been here for two days, and seemed much pleased with their visit. I wish I had been more perfectly disengaged for them; but Henslow helped me out. This week I have to make a Latin speech to the Senate, not one word of which is yet written; I mean to write a new syllabus of my lectures, which commence in about a week—in short my hands are as full as they well can be. I will, however, do the best I can for our joint-stock work, and indeed I should not be afraid of what is before me were it not for the weakness of my eyes. They have been much worse since we parted, and have almost entirely prevented me from doing anything by candle-light. For some days I have abstained from wine, and have dined in my own rooms in order to avoid the temptations of a College Hall. In consequence of strict regimen, and cooling medicines, I am now much better, but I am obliged to read and write as little as I can help. Fitton wishes me to write a kind of notice (I suppose in form of a letter) of what we have seen. What do you think of this? It might be put in, in our joint names, as the harbinger of our joint-stock papers. In the present state of my eyes, and with my 1001 engagements, I fear you will find me a bad helper. I can only promise to do my best, and this promise I make quite honestly.

I fear you will have a long bill against me, but I must cash up out of the profits of my new office. It was sheer poverty which drove me into harness. My eyes ache with what I have been writing; and from the way in which I have written I fear your eyes will be as bad as mine, before you make out my hieroglyphics. Give my kindest regards to Mrs

[1] Robert Kaye Greville, LL.D., of Edinburgh, author of *The Scottish Crypto-gamic Flora*, and other works.

20—2

1827.
Æt. 42.

1827.   Murchison, and my congratulations on the good work she
Æt. 42. has done during the summer[1].

Dear Murchison,

Yours ever,

A. SEDGWICK.

The laborious office of Senior Proctor, to which Sedgwick refers in the above letter, had been offered to him by
Trinity College; and moreover he had been specially solicited
by the Master, Dr Wordsworth, to accept it[2]. A certain
amount of unruliness, not to say dissipation, had made itself
apparent among the undergraduates, which could only be put
an end to by severe measures carried out with discretion and
strict impartiality. It was felt that Sedgwick, with his peculiar
geniality, and known sympathy with the younger members of
the University, had special qualifications for such a task.
Proctorial stories have a strong family resemblance, and therefore we need not repeat any of those which still survive
respecting his administration. We will only say that he
fully justified the hopes that had been formed of him. He
detected several evildoers, *flagrante delicto*, and had them removed from the University, without either losing his personal
popularity, or imperilling the dignity of his office. With
Sedgwick, however, even serious duties had their comic side,
and many were his jokes at his own expense. Let us quote,
as a specimen, the following description of his personal appearance, which he sent to Murchison at the end of a string of
reasons for not attending a particular meeting of the Geological Society: "You and Mrs Murchison would have laughed
had you seen me enter the Senate House this morning at
eight. I had a cap, bands, gown, and cassock—so far all was
regular—but under my silk petticoats appeared an enormous

---

[1] Mrs Murchison took a keen interest in her husband's pursuits, and frequently
gave him considerable help by collecting fossils, and making drawings.

[2] This statement is made on the authority of W. H. Thompson, D.D., late
Master of Trinity College.

pair of mud boots, and I had a great woollen ruff about my neck as big as the starched cravat of my Lord Bacon. At nine the Examiners adjourn to breakfast, and I provided a large pitcher of true *old man's milk*[1], which operated delightfully, and took the frost out of all our noses. We had kippered salmon, and many other good things, which will make my Proctorate quite celebrated; and the remembrance of it will, for ages unborn, continue to rise like a sweet odour in the nostrils of our *Alma Mater*. I am sitting at the top of the Senate House in an elevated arm chair, looking like an Inquisitor General, and scowling down on two hundred and fifty poor devils who are squeezing their brains to get out a few drops of mathematics[2]."

Would that one of the two hundred and fifty victims had occupied a few spare moments in executing a pen-and-ink sketch of the Grand Inquisitor, that posterity might have realised more vividly the truth of the above description! As it is, we must be content with a *silhouette* of Sedgwick taken during his year of office[3]. It gives a good general notion of his dress and figure, as he may have stood, watch in hand, to announce to those who were being examined, that the clock was about to strike, and that they must fold up their papers.

The joint labour of Sedgwick and Murchison in the field was to be succeeded by a joint labour at the desk,—the "joint-stock" work alluded to in the letter of October 28th—by which they were to put the results of their tour into shape for presentation to the Geological Society at as early a date as possible. It was decided to write two papers: the one on the structure of the Isle of Arran; the other on the Old Red Sandstone of the

---

[1] Sedgwick's name for the then popular combination of port wine, spice, etc., commonly known as "Bishop".

[2] To R. I. Murchison, 15 January, 1828. At that time the Senate House was not warmed.

[3] The *Cambridge Chronicle* for 29 February, 1828, records: "Monsieur Edouart, whose arrival we announced last week, has already met with a considerable patronage from the gentlemen of the University." The *silhouette* from which our copy was reduced was given to the Registry of the University by the Rev. J. Romilly.

1827.
Æt. 42.

1827.
Æt. 42.

north of Scotland. This proved a tedious and difficult matter. If the two writers could have retired together to some lonely spot, at a distance from all interruptions, their work might have proceeded smoothly and rapidly to its conclusion. This, however, they could not do, and therefore it had to be got

Sedgwick in 1828; reduced from a silhouette by Monsieur Auguste Edouart.

through by snatches, with the result that the summer of 1828 was far advanced before the second paper could be read to the Society. This long delay was quite unavoidable, partly from causes personal to the two writers, partly from the complexity of the subject with which they had to deal. Murchison, as will be seen presently, was not a little annoyed at it. He had no distractions to take him away from geology. He was ten years younger than Sedgwick; he was the fortunate possessor of a strong constitution; he was ambitious, and eager for

distinction in his new pursuit. Had he been left to himself
he would probably have written both papers in a very short
space of time. But it must be recollected that he had not
Sedgwick's experience; he had as yet published only one
paper—and that not a long one; and he could have had but
little idea of the labour of reducing extensive field-observa-
tions. It is perhaps fortunate for his reputation as a geologist
that he had to submit to an enforced delay in coming before
the public.

Sedgwick—as our readers know already—worked under
very different conditions; and it was only when he was away
from Cambridge that he could give undivided attention to
geology. In this year too he was more than usually occupied.
His duties as Proctor; his lectures; his museum; his parish;
his ailments; were all so many barriers to a speedy completion
of his share in the joint task. Moreover, his own paper on
the Magnesian Limestone, begun (as we have seen) in 1826,
had yet to be finished. This, however, he generously laid
aside, and proceeded as fast as he could with the paper on
Arran. The series of letters he wrote to Murchison while it
was proceeding is tolerably complete, and affords a more
graphic illustration than will again occur of the way in which
Sedgwick's work used to be hampered and impeded by incessant demands upon his time. The letters are valuable too
for another reason. They show how cordial the relations
between him and Murchison were at that period; and how
anxious he was to give full credit to his friend for his share in
the field-work. The first extract refers to what Murchison
calls "a tiff I had with our warm-hearted but hot-headed
President Fitton, who had suspected that I was not doing
justice to Sedgwick[1]."

TRIN. COLL. *November* 3, 1827.

"Your letter both vexed and surprised me...In one
respect I am almost certain that you are labouring under a

---

[1] Note by Murchison on the back of the letter.

false impression.   I have again looked at Dr Fitton's letter,
and I cannot persuade myself that anything unfair was intended towards yourself.   He merely wished to give me a
nudge on the elbow, knowing my habits of delay.   At all
events you may depend upon me, that nothing shall be done
by myself without consulting you.   Indeed, were I so disposed,
you have me on the hip; as all notes, sections, and indeed
everything from which our future papers must be compiled,
are in your possession.   You worked harder in many respects
than I did myself; and, till we reached the east coast, and
indeed there also, you were my geological guide.   I should
therefore be an ass indeed if I thought of anything beyond
what we meditated.   On the whole, perhaps it is better to put
in our papers one by one, and leave the results to the end,
which may be exhibited in the form of a *résumé generale.*   I
shall rejoice to see you here; perhaps I may be in town next
week to consult the cunning eye-man you mentioned.   My
eye is however much better.   I have been water-drinking,
and dephlogisticating, and certainly have reduced the inflammation.   My kindest regards to Mrs Murchison."

SENATE HOUSE, *Tuesday morning.*
[13 *November,* 1827].

"Last week my eyes were in such a state that it was
quite impossible for me to look over your penmanship[1].   Our
Professor of Physic ordered me to abstain from strong drink,
which I most religiously avoid, like a true Rechabite.   He
also recommended me to use animal food very sparingly ; and
not content with this he put twelve leeches on my left temple.
This treatment has produced a good effect, and if properly
followed up will, I hope, put out the fire of my eye.   I think I

---

[1] A sentence in the earlier part of the letter shows that Sedgwick here refers to
the MS of Murchison's *Supplementary remarks on the Strata of the Oolitic Series,
and the Rocks associated with them, in the Counties of Sutherland and Ross, and in
the Hebrides,* read to the Geological Society 16 November, 1827.   It is a continuation of the paper *On the Coal-field of Brora,* read 5 January and 2 February,
1827.

told you in my last that I don't look in a book by candle-   1828.
light, and as my mornings are sufficiently taken up with lec-   Æt. 43.
tures, and various other academic employments, I have truly
very little time for many things about which I wish to employ
myself. The whole of yesterday I was employed in attending
certain formal academic meetings. Today I had just commenced, under the genial influence of the tea-pot, to read over
your notes, when, to my great horror, I was summoned to the
cold marble pavement of our Senate House, to administer the
matriculation oaths to the freshmen of this year. The men
are brought up, about twenty at a time, and swear in volleys.
While a fresh set are signing our books, paying fees, and
priming for the next broadside, I dip my pen in the inkstand,
and try to write a word or two; in this way I have got so far
over the sheet at a rather hobbling pace; but still I make
way.

As soon as these swearing manoeuvres are over I will get
one of my friends to read over your paper to me, for I fear
my eye-sight will not bear the work, if I am obliged to put
it off till the evening. I will add tomorrow morning a few
notes if necessary, and forward it by "The Telegraph[1]" at
10 o'clock."

SENATE HOUSE, *Monday morning.*
[13 *January*, 1828.]

"It is not from want of inclination that I keep away from
town; but of all the days of the year 1828, Friday is the one
on which I shall be most completely nailed down by engagements. Our annual examination is going on, and does not
end till Friday night, when the Moderators, and other examiners, will meet at my rooms to give in the final result which I
shall have to publish in my capacity of Senior Proctor. On
Saturday the degrees are taken, and I shall again have to
officiate. I am truly sorry that I was not at the last meeting
from the account you give of it; besides, I was doing no good

---

[1] One of the coaches which then ran between Cambridge and London.

here. My cold is better, or rather I am now going on with a
second edition, which is not so voluminous as the first. My
eyes, I am sorry to say, are no better, and plague me sadly.
I set to work the evening of the day I last wrote to you and
finished the peroration of our *Arran* Paper. It is cram-full of
hypotheses, and truly may want defending; but you must
stand up for me. I really think we shall between us set the
coast-section at rest. I was, however, severely punished for
my exertions; for my eyes were so enraged at this treatment,
that they gave me no rest for nearly a week after. Since I
recovered I have written a few pages of the concluding part
of my Magnesian Limestone paper; and I can have something ready against next Friday fortnight, when I will, if
possible, attend *in propriâ personâ*. Can you contrive to find
a corner that evening for my gab? I truly hope Mr Pentland
will not have left London, as I want very much to talk with
him about the fish[1]. Don't prepare any map, as I have a
beautiful one by Decken, which he made at my rooms expressly for our *Arran* paper.

I wish I had been at your soirée to have had a fight with
Buckland; at the same time I can't help saying that the fight
against the footsteps is almost to destroy the evidence of our
senses; and this is going a long way. In plain truth I don't
in this case know any better argument than that clencher of
my uncle Toby, viz.—" By G— they are not footsteps[2]."

SENATE HOUSE, *Tuesday evening.*

[14 *January,* 1828].

Dear Murchison,

I am greatly obliged to you for your kind invitation,
and truly mortified that I have it not in my power to avail

---

[1] Mr J. B. Pentland was a travelled gentleman of scientific and antiquarian
tastes, who made himself useful to geologists by helping them to determine their
collections. He knew Cuvier, and took the fish in question to France for his
examination.

[2] Endorsed by Murchison: "Alludes to an experiment I made at a *soirée* with
live tortoises on paste."

myself of it. But my letter of yesterday will have informed
you how I am circumstanced. It is *literally impossible* for me
at this time to leave Cambridge. I fear you will think me a
sorry coadjutor; for all the work is left to yourself. This is
not as it ought to be; but I am at present almost a lame
soldier; at least, till my eyes are better, I shall only be fit for
invalid duty. You see I am trying to revenge myself for the
unlucky blow you gave me in Carsaig Bay[1]. It must be rather
a queer spot this weather. I wish I could transport myself
there for half an hour, and then come back again. The rust
of a spear is said to have healed the wounds inflicted by it.
Perhaps the sight of the Carsaig pitchstone would set my eyes
right. If you don't write sooner pray let me know how all
goes off on Friday night. I only ask this on the supposition
that you have five minutes to spare, and that you don't think
writing a bore.

The next thing for us to do is to give the structure of
Caithness, and the coast of the Murray Firth, and then to add
some general details on the conglomerates of the west coast.
The paper need not be very long; at the same time it cannot
be very short, for it will be a sort of *omnium gatherum*, in which
it will be necessary to speculate *de omnibus rebus et quibusdam
aliis.* And surely on such a text we may fairly be allowed to
preach our sand out. I started with a good pen, and through
the first page contrived to make a fair fist; but I am getting
worse and worse. My ink is thick, my brains are frozen, and
my time is up. So no more at present from yours till death

A. SEDGWICK.

The first part of the Arran paper was read 18 January,
1828, and concluded 1 February. Sedgwick then set to work
to finish his own paper on the Magnesian Limestone, but here
again ill-health stepped in, and March came before it could
be read, and even then it could hardly be called finished. But

---

[1] On the south coast of Mull.

1828.
Æt. 43.
for this, and his delay in beginning work on the second Scotch paper, he shall make his own excuses to Murchison.

TRIN. COLL. *Monday morning.*
[*February*, 1828.]

"Many thanks for your last parcel. I will use it as soon as I can, but I am out of sorts, and really *dare* not work at present. Since we parted I have had a short visit from an old and very unwelcome acquaintance, which has affected my arterial system, and produced a throbbing in my head which I must get rid of before I can fairly set to. Yesterday I did duty about sixteen miles from hence; and on my return I was frequently obliged to pull up, because my head would not bear the motion. The only radical cure for such feelings is exercise and abstinence, which I must practice forthwith. Your account of Mull delights me above measure. Mrs Murchison's picture is now framed, and looks magnificent. When will she do me the honour of coming to see it? My lectures begin on Wednesday next. Of course they will take up a considerable portion of my time."

March, 1828.

"I send you the end of the paper: and those parts which you may omit in reading I have marked at the side with a pencil line. You will probably on looking it over see other matter which may be left out. When you have finished, just say to the Society that the sections, and one or two small maps, are not finished; but that a description of them, together with a list of minerals and fossils in the formation, will be given in an appendix, not of course of a nature to be read. If I can get a drawing out of the hands of the man who was to do it for me I will send it up by the mail tonight directed to the Society's rooms, Bedford Street. If I send the abstract in a day or two I hope it will be in time[1]."

[1] The paper was read 7 March, 1828. The *Proceedings* of the Geological Society record under that date: "A sketch of the subjects contained in this paper was laid before the Society in 1826 (Nov. 17). They were resumed in a more systematic and detailed form during two meetings in 1827; and are now terminated by the observations read at the present meeting."

This work off Sedgwick's hands, his friend naturally
expected that he would find leisure to attack their second paper. But by this time he was immersed in the work of the Lent term, and moreover the President of the Geological Society had set apart the 16th May for the joint production— an unfortunate step when he had one so dilatory as Sedgwick to deal with, for it doubtless gave him a further excuse for delay.

[TRINITY COLLEGE, 12 *March*, 1828].

"I have received a letter from the President, who sends me the following programme of the order of papers to the end of the session. *March* 21. Dr Richardson. *April* 4. Blank. *May* 2. Buckland and Clift. *May* 16 and *June* 6. Murchison and Sedgwick. *June* 20. Tag, rag, and bobtail &c., &c. I am not sorry for this delay, except on your account. It will give us time enough, and I shall be much more disengaged. My Proctorial duties and lectures have pressed rather hard upon me, and left me for the last three weeks hardly a spare hour. This week I hoped to have made some progress with our joint papers, when to my dismay I found that I had, in right of my present dignity, to dance attendance upon my Lord Judge. I am off to church with him this morning, and then I go in the tail of his robe to hear him address the Grand Jury, &c. &c.; and as for my poor lectures they are for the time sent right about. But I hope to resume on Friday morning. By the way I gave our men a platoon fire about Scotland after one of the meetings of our Philosophical Society. I don't care one farthing how my paper went off when it was read, provided it read well when it is printed. It is necessarily dry, being so much in detail; but I *think* that the facts are important, and at least some of them are new. Pray did you get your friend Pentland to look at Mr Witham's big fish[1]? Does he consider it a *Palæothrissum?*

---

[1] A fossil fish from the Magnesian Limestone, sent by Henry Witham, Esq., of Edinburgh. He had already supplied Sedgwick with other specimens from this same locality. *Trans. Geol. Soc. Lond.* Ser. 2, iii. 116.

It seems to me to be a distinct species, but of the same genus with the *Palæothrissum magnum* of De Blainville.

My abstract is as short as possible, at least so I think. It will serve to convey a general notion of the whole paper, and that is what it ought to do. No abstracts were made of the parts read last year, because it was thought better that the whole of it should appear together. I have scribbled it in a hurry ; pray make any verbal correction you see fit.

Yours ever,

A. SEDGWICK.

TRIN. COLL. 15 *March*, 1828.

" I have not been lazy, but I have really had no time for our joint work. As soon as I get quit of the engagements of the term I can set to work in good earnest ; for my health, I am happy to say, is just now very good, and I think that my eyes are getting quite well. I will very carefully look over all the papers, and make a string of notes ; we can then divide our labours, and, though separate from each other, shall, I doubt not, get our separate columns into position, so that they may, at word of command, be made to deploy into line, and be ready for action. By the way, I am to blame to think of using military tropes to a soldier, as I shall thereby only show my own ignorance."

The term came to an end, but still Sedgwick was not ready. Murchison was then planning a journey to France —his first attempt at continental geology—and not unnaturally wished to see the paper completed before he set out. A stronger appeal than usual to Sedgwick elicited the following answer; a gentle rebuke which may be profitably laid to heart by anybody who advocates speed without making proper allowance for difficulties.

TRIN. COLL. *April* 7, 1828.

Dear Murchison,

You call upon me " for my own reputation, and your peace of mind, to make ready." I promise, if God spare my health, and preserve me of sane mind, to have all in good state before the reading; but to expect that our documents should exactly tally, so that we have only to stitch them together, is to expect impossibilities. One is making a key, and the other a lock, which never can fit till the wards are well rasped and filed. To rasp and file will be a part of my office, as well as to fit on a head and tail. All the specimens we mean to exhibit must be arranged before any good description can be given of the several sections. The general facts may be stated, but the skeletons must be clothed with flesh by the help of the specimens. I have tried my hand at the description of the Tarbet Ness[1] coast-section, but I cannot satisfy myself without the specimens—the subject seems to elude the grasp. I find the introduction, or *discours préliminaire*, excessively difficult—not from want of matter, but from having too much. If I could make up my mind what ought to be said I could take it at a canter.

My mind, ever since we parted, has been in a muddy state, for I have been living in a troubled atmosphere. A most painful case of ungentlemanlike profligacy has come under my official notice, and worried me almost out of my senses. For the soul of me I cannot take matters of this kind calmly. Till last Saturday night I had for a week hardly an hour of refreshing sleep. Three men have been expelled from Trin. Coll. Two or three of other colleges will be sent away from the University. These are the bitter fruits I have been gathering during the week. Thank God this harvest is over! and I hope we shall have no second crop of this kind during the season of my Proctorate. My head is, however, beginning to cool, and my sight to become more clear, so that I in some measure see my way through my work, and hope to lick it into form before we meet. Be therefore, my good fellow, in good cheer, and rejoice with me

---

[1] A promontory in Ross-shire, forming the north side of the Moray Firth. The section in question is described by Sedgwick and Murchison, *Trans. Geol. Soc.* Ser. 2, iii. 150. Plate 14, fig. 4.

that my wits have not been scared away for ever from their domicile.

During the few hours we spend together in Town we must devote one or two to the final arrangement of the specimens—both with a view to the Society and also to the systematic descriptions of the paper. As for the tail, it ought, like a spider's web, to be spun out of the body; it therefore can have no real existence before your work has assumed a substantial form. At present I can hardly form a guess about its length, curvature, or joints. When we have once determined what the head and body are to be there can then be no difficulty about it. This morning I have made a few notes upon the sandstones of the western coast; but I cannot make up my mind where to introduce them. At present I am disposed to throw them into the latter portion of the paper. I have no more time to tell you what I am doing, or what I am not doing, for in a minute or two the post closes. It is quite impossible for me to think of France for the present. My kindest regards to Mrs M.

Yours ever,
A. SEDGWICK.

Another matter, more agreeable than proctorial duties, had occupied Sedgwick during part of the Lent Term. The Lucasian Professorship of Mathematics had become vacant, and some of the friends of Mr Charles Babbage considered that he was the most proper person to fill it. The matter required very delicate handling; for Babbage was on the continent, and could not be communicated with. It was obviously impossible to announce him as a candidate unless it could be ascertained privately that his election would be certain. The electors were the Vice-Chancellor and Heads of Colleges—a body whose opinions it was not easy to discover. Babbage however was elected, and it appears that Sedgwick had had no small share in bringing about this result. Dr Fitton writes, 8 March, 1828:

I congratulate you very cordially on Babbage's election; which is not less creditable to the University than to him. And *you* certainly must have great satisfaction in feeling that your own efforts have so much contributed to *the spirit* that has produced this event.

The next letter describes the favourable reception of the long-expected paper, the first part of which Sedgwick read to the Geological Society on the 16th May.

CAMBRIDGE. 25 *June*, 1828.

My dear Murchison,

If you have thought me worth thinking about, I will venture to say that you will have accused me of breaking my promise. In this instance, however, I have not to plead guilty to any great offence; as I have been much harassed in mind, and somewhat also in body, by circumstances over which I have had no control. Our paper on the conglomerates increased to such a size that it was obviously too large to be taken in at one meeting. When all the details were left out, and almost every portion of the two coast-sections of Caithness, there was enough remaining to produce that peculiar oscillatory motion in Fitton's lower extremities which you have often marked on like occasions. All went off well, and ended with the dish of Caithness fish, which were beautifully cooked by Pentland, and much relished by the meeting. Greenough, Buckland, Conybeare, and all the first performers were upon the boards. The account of the conglomerates of the Murray Firth and the Old Red of the north-west coast, together with certain speculations and corollaries, were put off till the following meeting.

A most delightful party was next day organized at Greenough's. Pentland was about the middle of the week following to come from Oxford with Buckland, and Greenough at the same time was to start with Conybeare from London, and the party was then to bear down upon Cambridge, and spend three or four days with me. Our plans were, however, defeated by a melancholy event which we have long been

looking forward to. Before my return to College I received
the news of my Father's death, and in consequence hurried
down to the North with all the expedition I could command.
I felt a great pang at being separated from so old and dear a
friend; but the blow fell upon us as lightly as it could fall.
For he was in his 93rd year, and died without any pain
or illness whatsoever, of pure exhaustion; and retained his
intellects till within a few minutes of his dissolution. I never
knew a man of purer principles and warmer heart; and since
the time I was a boy I never have heard a word pass his lips
which implied a want of confidence, or was addressed to me
in anger. But enough of this—I have no right to obtrude my
own feelings on a subject like the one of which I have been
speaking.

I remained about a fortnight in the North, and returned
to London in time to attend the next meeting of June 6th.
Our paper was concluded, and Buckland had a short paper
on the fossils of the Isle of Portland. I had theory enough
for a long discussion, and fairly threw down the gauntlet to
old Mac[1]. No one, however, thought of taking it up for him.
In short, the meeting was thin, and the discussion meagre.
Greenough, however, spoke very handsomely of our labours....

I shall be extremely busy till the 4th of July, after which
I shall start for Cornwall, and join Whewell and Airy, who
are going to repeat their pendulum experiments. After they
are over we shall visit the granite veins, and make one or two
transverse sections. On my return I shall cross to South
Wales, and visit Conybeare, make arrangements for our joint
work (I think I told you we were going to scribble in company), and try to have a run through a part of North Wales.
I must be in Cambridge by the beginning of October to resign
the keys of my office, and I shall hail that day with rapture.
During the Commencement festivities the Duke of Gloucester

---

[1] John Macculloch, M.D. author of *A Description of the Western Islands of
Scotland, including the Isle of Man.* 8vo. Lond. 1819.

is coming down, so I fear I shall be half killed with hard-work and hard-eating. I wrote the other day for three bucks.

Believe me, Dear Murchison,

Yours to the earth's centre,

A. SEDGWICK.

In writing to Ainger, who knew well what the home-life at Dent had been, Sedgwick was naturally less reticent about his own feelings:

"In that humble but useful station in which God placed him, he has enjoyed an unusual share of health and happiness. If I could feel as I ought to do, I should rejoice and not mourn at this event, for surely no man could be better prepared for this great change. His mind was spared, and his kindly affections remained warm to the last; and by the operation of pure Christian principle he seems for years past to have triumphed over all the moral infirmities of his nature; so that he became an admirable example, and an endearing motive for virtuous life, to all those who were nearly connected with him. In this respect I feel as if I had sustained an irreparable loss. For years past I have never visited the old man without feeling better for it[1]."

Sedgwick has sketched, in a letter to one of his nieces, his father's personal appearance on that momentous evening when he himself came into the world: "He was then about fifty years of age, of robust frame, and of a rosy and cheerful countenance. He sat on the right side of the fire, wore his large well-powdered wig, his white cravat fixed behind with a large silver buckle, and he had a pair of large bright silver buckles to his shoes. The chair in which he sat was the very chair represented in the lithographic drawing taken about forty years after, which I dare say you have seen, so I need not describe it[2]." Mr Sedgwick had been blind for many years before his death, and his frame had shrunk a little as he

[1] To Rev. W. Ainger, 4 June, 1828.
[2] To Miss F. Hicks, 28 March, 1841.

1828.
Æt. 43.

grew older; but in other respects (save the wig), Mr Westall's drawing—a copy of which, slightly reduced, is here given—coincides exactly with the above description. In the year after his death a monument was erected to his memory on the south side of the chancel of Dent Church. The inscription, written by his son Adam, recounts the leading points of that character which has already been so graphically set before us by the same hand[1]. One sentence will fitly close this portion of our narrative: "He lived among his flock for fifty-four years, revered as their pastor and loved as their brother."

Sedgwick's allusion to a project for writing a geological work in conjunction with Mr W. D. Conybeare deserves more attention than such projects usually do; for though it was never really begun, yet for some years Sedgwick was always intending to begin it, and we believe that the needful preparation determined the direction of several of his geological tours. The scope of the proposed volume—a continuation of the *Outlines of the Geology of England and Wales*, published by Mr Conybeare in conjunction with Mr William Phillips in 1822—will be best explained by the following letter[2]:

BATH, *April* 24, 1828.
Dear Sedgwick,

Your letter gave me sincere pleasure. Nothing would be more agreeable to me than embarking in a joint voyage with you; and indeed nothing but some proposal of this kind would have held out to me the prospect of accomplishing a second volume.

The materials, as sketched out in my own mind, comprised these divisions. I. A description of the older rocks, throwing the transition and primitive classes together. The arrangement to be similar to that of the former books : first, a general account of the formations; secondly, the topographical detail of their distribution. All this part must principally devolve on you, and if more assistance could be had, more would be desirable—especially De la Beche, if he would undertake any portion of the unexamined districts, would be a very useful ally. We ought also to make a push to urge Aikin[3] to

---

[1] Chapter I. pp. 38—44.

[2] The letter has been slightly compressed in transcription.

[3] Mr Arthur Aikin, a distinguished mineralogist, and one of the founders of the Geological Society.

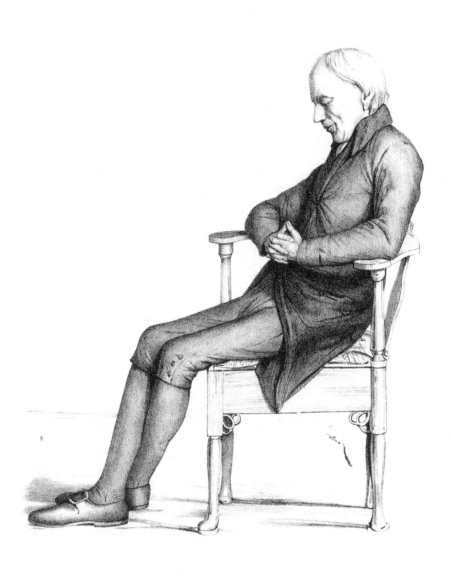

THE REV.ᴰ R. SEDGWICK. M.A.

Drawn by W. Westall. A.R.A.

To face page 324, *Vol. I.*

publish his Shropshire materials, which would be very important.
He means to do so in our *Transactions*, but wants stirring up.

I consider our materials at present as standing thus. The Cumbrian district you have done, the Cheviot you will do; of the insulated Midland districts enough has been said of Malvern; Charnwood requires doing, but might be accomplished in a fortnight; of Cornwall there exists a great quantity of scattered information, and your next visit may easily put it all together. Wales is the most unknown, and from all its local circumstances the most difficult. One ought, like Chalmers, to adopt the district plan. If Henslow would take the Caernarvonshire range, and indeed the whole ground north of Aikin's observations in Shropshire, you and I and De la Beche might easily accomplish the southern part; but I shall not be very efficient in the field, for I have not, from the demands of a large family, either time or funds for much touring.

II. The second division of my volume respects the collection of those phenomena which are perhaps more important as to the foundation of geological theory, comprising all the heads of my Introduction from § 6 to § 12. This would be more closet work than any other part of the subject, and I should sit down to it *con amore*, because I feel it easy to assemble such a mass of facts mutually illustrative of each other as I conceive must materially tend to establish on more positive bases the theory of our science.

III. The third division would be the corrections and additions to the former volume. Here I should principally depend on you.

You see therefore, from this general outline, that you would have a large half of the labour, and of course ought to have of the credit, such as it might be, which would result. But I fear that the work would hardly hold out a commensurate prospect of repayment in this way; for while you would deserve most, I should probably, from the earlier connection of my name with the work, get most. I do not know, however, that either of us can be more usefully employed for the advancement of our science, and I don't think we are either of us likely to quarrel for our slices of praise. Very sincerely yours,

W. D. CONYBEARE.

We have now reached a point at which it will be well to pause for a moment and examine the first joint work of Sedgwick and Murchison more closely.

The paper on Arran is a good example of the old stratigraphical methods—a well-kept diary of excursions made in a very interesting district; and, as an examination of the island from a new point of view by observers trained in other fields, it is a useful contribution to Scotch geology. The authors begin with a description of the sequence of the rocks observed; this is followed by a determination of the fossils by the best authorities on the subject; and from these data they attempt a correlation of the deposits with those of other areas already examined. It would be out of place here to go into details and criticise the succession inferred in each case, or to point out the corrections shown by later work to be necessary in the determination of their fossils. Such modifications are necessary from time time in the progress of all such descriptive work. This paper is a fine example of the way to set about the examination of a district, and is full of wise observations. The Islands of Scotland did not, however, offer new ground. That shrewd observer Macculloch[1] had been over it, and had clearly recorded the results of his work, though his credit was damaged by his too blind adherence to the tenets of the Wernerian school. He did not take in quite good part the work and criticism of Sedgwick and Murchison, for he was a man whose health and temperament made him impatient of contradiction, and inclined to resent as a personal injury any attempt to trespass upon ground which he had come to regard as his peculiar province. Sedgwick and Murchison recognised that in this outlying fragment of a continental area, which once extended further west, they might find a key to the phenomena observed on the mainland, and rightly thought that their work would "not only assist in completing the natural history of Arran," but would help "to fix the true epoch of all those interrupted fragments of secondary formations"[2] which are found along the West and North of Scotland.

The reader of these early papers must be cautioned that he will meet with some old-fashioned phrases, now changed, though not perhaps in all cases for the better. According to the old nomenclature the Primary Rocks included the Archaean, and, speaking generally, the great masses of crys-

[1] See above, p. 322, *note.*
[2] *Trans. Geol. Soc. Lond.* Ser. 2, iii. 22.

talline schist of unknown age[1]. Flanking these "Primitive ridges" were the rocks of intermediate character, in those days called "Transition Rocks," including the series which Sedgwick and Murchison afterwards made so well known under the names Cambrian and Silurian. Resting upon the upturned edges of these older rocks comes the Old Red Sandstone, which forms the base of what were then called the Secondary Rocks, and to this their attention was chiefly directed.

The local importance (for purposes of classification) of such an unconformity, as indicating lapse of time, is of course recognised, but the authors insist upon the fact that such phenomena are of limited geographical extent, and clearly state that they "do not think that a want of conformity is one of the elements which will much assist us in grouping together or in separating contemporaneous deposits in distant parts of the earth."[2]

The second paper is a continuation of the first, and, like it, refers chiefly to the Lower Secondary Rocks; that is, in the nomenclature of the day, to the Old Red Sandstone and overlying deposits.

They noticed among the older rocks the fan-shaped arrangement of which we have heard so much lately. They also distinguished the Old Red Sandstone from the Red Sandstone of Cambrian age, and drew attention to the fragments of older deposits which were found imbedded in the intrusive rocks. They considered that the Old Red beds had been accumulated between ancient ridges of crystalline schist, and pointed out that the basement beds were made up of fragments of the nearest Primary rocks. Above this lower conglomeratic stage they placed a middle flaggy stage, with fish remains. When wandering along the shore among these flags with their fish scales and bituminous patches, they

---

[1] It is worthy of note that the age and *genesis* of these rocks formed the principal subject of discussion at the International Geological Congress held this year in London. 　　　[2] *Ibid.* p. 33.

thought at first that some one had dropped tar here and there on the rocks. The occurrence of fish remains in these Caithness flags had been already recorded from one locality, but they found that they were far less uncommon than had been supposed, and traced the fish-bearing strata right across the country, and even into the Orkneys. Some of the specimens were referred to Cuvier, and his description is given *verbatim.* Some good figures, with a restoration by Cuvier, are published with the paper. Sedgwick's old work among the fish-bearing beds of the Magnesian Limestone must have made him take a special interest in this successful search for fossil fish of another and older type. In the middle flaggy stage they saw a connection with the Carboniferous System—a view not now accepted in the sense in which they understood it, namely that these rocks were the equivalents of rocks known as carboniferous further south. Their view has, however, an element of truth, in that the beds they were examining undoubtedly form a basement to the carboniferous rocks of Scotland, and exhibit the incoming of the characters by which they are distinguished. It was an interesting observation of theirs that the pholas-borings followed the calcareous bands everywhere along the shore.

They did not attempt to map the district in detail, but only proposed to indicate the distribution of the beds described. When we recollect this the map they give will compare not unfavourably with others published half a century later, after the country had been well worked out. They did not attempt to trace the "faults," but realised the probability of the occurrence of many lines of disturbance which had escaped detection in the rapid survey which they were making. They suspected their existence, among other reasons, because the thickness of the series would be so enormous were the beds in true geological sequence all along the sections examined.

Sedgwick's employments during the summer of 1828 are described in the following letter :

LONDON, *October* 8, 1828.

Dear Murchison,

Your letter of August 18th reached me in South Wales about ten days since. No one knew how to forward it from Cambridge before Whewell's return. He started it on a venture; and when I received it, it had been doubling in so many directions that it was blackened from one side to the other with addresses. When it did come it was most welcome, and made me almost envy you for the delightful work which you have gone through. Give my best regards and congratulations to Mrs M. and to Lyell, on the discoveries they have made, and on the dangers they have escaped. God preserve you all from the fury both of fire and water till you are by your own fire-sides in this murky capital; and I will contrive to join you as soon as I can find a moment's leisure, in order that I may have a *vivâ voce* narrative of the news you bring from the lower world. But what account have I to give of myself? Not I fear a very satisfactory one.

Immediately after the business of the Cambridge Commencement was over (during which festival I was figuring in processions, creating Doctors of Divinity, and going through many ancient monastic evolutions) I started pell-mell for Cornwall, and about the 8th of July contrived to join the pendulum party. I think you have heard of our expedition for the purpose of swinging pendulums at the bottom of the Dolcoath mine. Had I imagined what time the experiments would have taken I should certainly have kept far away from them ; we remained nearly two months at Camborne, during which time I indeed contrived to make a few interesting excursions, re-examined the principal junctions, and settled some of my notions : but all the work I did in the county might have been completed in ten days. After all we had a good deal of amusement out of the pendulums. Our ups and downs upon the ladders, which between the higher and the lower station amounted to more than fifty in number, and

extended to a length of nearly one-third of a mile; our young
attendant with a great belt stuffed with chronometers; the
dirt and the tallow; the uncertainty of the result; the
speculations of the mining-men and mining-women; these
were the materials out of which we extracted our share of
amusement. We had two beats, continued without interruption night and day for more than a week each—during
which two of Kater's pendulums were running against each
other for more than 600,000 beats. Whewell and Airy out of
these materials hoped to have reached a result against which
it would have been impossible to take any exception. I am,
however, sorry to say that in consequence of an unlooked for
fault in the instrument, which is called "*invariable*," the
successive results have been in some measure variable. We
have therefore after all only gained an approximation.

Whewell visited with me some of the finest junctions, and
has sketched some of the magnificent granite veins which are
found on the coast. I was very much surprised to observe,
what had before escaped me, that several of the metalliferous
deposits of Cornwall are true Stockworks. The great mass of
granite north of St Austell is traversed by innumerable *contemporaneous* veins, some of which bear oxide of tin, and
where they abound the metal is extracted, as far as I understand the case, exactly in the manner of the German Stockworks.

As soon as the pendulum-party broke up I turned my
face towards the east; just looked at the Exeter conglomerates;
then ran down to Ilfracombe, and crossed by a packet to
Swansea. From Swansea I ran down to Sully, the rectory
where our friend Conybeare has incarcerated himself. The
situation is, however, most delightful, and he has about him
the society of a charming family. I contrived to poke him
out of his den, and had a run of about three weeks with him
through a part of the South Wales coal-basin. It is a highly
interesting region, and exhibits the secondary rocks of the
older series in every variety of combination. After doubling

out of the coal-field I visited one or two friends with whom I have been eating and drinking to my heart's content, and I am now in admirable condition for the winter work at Cambridge. Yesterday I visited, for about an hour, the noble collection at the Bristol Institution, and last night I came by the mail to this place. I have all day been doing a great deal of little business, and among other persons contrived to see Greenough. He says that our Scotch paper wants rasping, and has reported to that effect; but he says that the authors must do it themselves. I have no doubt he is right, for it was a cobbled business. I will take it down and try my hand at docking and cropping. But really it must all be written over again, or we shall drive the printer's devils to despair. Tomorrow morning I return to my den in Trin. Coll. It is time for me to retire, and make up for my loss of sleep last night. As good a repose to you and Mrs Murchison as I am myself looking for.

Yours to the centre of the earth,

A. SEDGWICK.

1828.
Æt. 43.

The experiments at the copper-mine of Dolcoath, in which Sedgwick bore a somewhat reluctant part, had been commenced by Whewell and Airy in 1826. " The object was to determine the density of the earth, and the essential part of the process was to compare the time of vibration of a pendulum at the surface of the earth with the time of vibration of the same pendulum at a considerable depth below the surface. The experiment failed to lead to a satisfactory result[1]," because, as Sedgwick says, the pendulum could not be trusted; and also because, on each occasion, a serious accident occurred. In 1826 Whewell and Airy had conducted the experiment alone; but in 1828 it was thought

---

[1] *William Whewell*, by I. Todhunter, 8vo. Camb. 1876, i. 37. The only printed record of what Mr Todhunter calls "a very arduous experiment," is contained in an anonymous pamphlet, known to be by Mr Whewell: *Account of Experiments made at Dolcoath Mine, in Cornwall, in* 1826, *and* 1828, *for the purpose of determining the density of the Earth.* 8vo. Camb. 1828.

desirable to enlarge the party, so as to carry on the observations without intermission day and night.   By this means
some members of the party could take an occasional holiday,
and Sedgwick was probably absent when the accident of this
year—a subsidence of a portion of the mine—took place.
His account of the experiments, as given in the above letter,
may be supplemented by what he told Mr J. W. Salter:

"This mine had a great advantage for our purpose.
Besides being one of the deepest in Cornwall, it is overhung
by a steep hill 700 feet high, so that we got the means of
measurement to a greater extent than would have been
possible elsewhere.

"We went down in summer-time, and enjoyed ourselves
very much.   The weather was propitious; the company
excellent.   But the natives evidently thought us no better
than we should be, bringing, as we did, strange instruments,
and strange earnest faces to such a spot, and taking down
uncouth-looking packages and baskets to all the deepest and
most dangerous-looking places.   We often overheard their
remarks.   One morning I listened to two men who had
watched our descent the day before: 'I think they're no
good.   There must be something wicked about them—the
little one (that was Airy) especially.   I saw him stand with
his back to the Church, and make strange faces.'

"We gave them some cause for their suspicions.   Our
lamp-box, marked outside 'Deville, Strand,' stood well for
a formal address to his infernal majesty.   We were clambering down one day, when, to keep up the joke, I asked a
sturdy miner who was guiding us, 'How far is it to the
infernal regions?'   He was a match for me—for he replied—
'Let go the ladder, Sir, and you'll be there directly.'"

The party were most hospitably entertained by the neighbouring gentlemen—with some amusing results.   On one
occasion Sedgwick, Whewell, and Airy presented themselves
at the front door of a house, where they had been invited to
dine and sleep, in their working-dress.   The butler thought

that they were real miners, and had just exclaimed, somewhat
gruffly, "You go round to the back-door," when their host
came forward to greet them. At another house the host
himself is said to have mistaken the same party for agricul-
tural labourers in distress, and was just intimating to them, by
a shake of the head, and a wave of the hand, that it was no
use begging of *him*, when his friends revealed themselves by a
loud burst of laughter. We do not vouch for the absolute
accuracy of these stories. Sedgwick always found Cornwall
a land of humorous adventure, and other tales will have to
be related in connection with his subsequent visits to it.

In the autumn of this year, while Sedgwick was tranquilly
lecturing at Cambridge, Dr Fitton, whose term of office as President of the Geological Society was drawing to a close, came to
the conclusion that the Woodwardian Professor was the proper
person to succeed him. A certain amount of difficulty seems
to have been anticipated in persuading one so full of engagements elsewhere to accept an office which would entail regular
attendance in London at stated intervals; and the task of
sounding Sedgwick, and of obtaining, if possible, a favourable
reply was entrusted to Whewell. His letter is endorsed *To
be opened immediately*, an amusing indication of Sedgwick's
habitual carelessness with regard to his correspondence.

8 *October*, 1828.

Dear Sedgwick,

Fitton will come to you on Friday to try to persuade you
to be President. Pray do not refuse. Make it possible, somehow
or other, for the thing is every way in the highest degree desirable.
It is clear, from what he says, that he has spoken of it to so many
people in London, that it will be generally known that the offer has
been made you; and after Buckland had found it possible in his
case it will not be easy to make them comprehend that it is not
ungracious in you to reject the proposal. Fitton is very earnest on
the subject for the sake of the Society, and with great reason. He
says that having just received favors from government[1] it is very

---

[1] "The Society, at a special Meeting on the 18th of April, 1828, was informed
of the grant from the Lords Commissioners of His Majesty's Treasury, through the
mediation of the President and Council of the Royal Society, of apartments in
Somerset House." *Report of Council*, 20 February, 1828: *Proceedings*, i. 111.

desirable and important to have a person at your head who is sure to be independent and straight-forward. He adds too that a new era of the Society requires a leader who can fill his place with distinction. All this is very right—but *I* am very anxious that you should take the office for *our* sake and yours. It will undoubtedly give a degree of prominence and attraction to the science at Cambridge which you cannot give it in any other way, and will add weight and popularity to all your sayings and doings on the subject. Without this we are hardly on a level with the Oxford men, which we have a right to be, and which it is your business to make us.

You will tell me of your lectures, but I am persuaded they will gain more in effect, than they will lose by any curtailment or inconvenience. Then consider; this business will not interfere with the course of this term. You have often made your second term somewhat irregular; it *must* be possible by some contrivance of time or place to manage it again. Consider too that every such inconvenience is a practical argument for new rooms, and will I hope soon produce its impression.

I do not think the expense is a very formidable consideration. Fitton says he shall suggest to Gilbert[1] the advisableness of transferring his parties to Sunday night. This would make them a continuation of Fitton's, and might be very good.

Find the *will*, and make the *way*. I am sure you will not repent it, and it will be an excellent thing for all of us.

Ever yours,

W. WHEWELL.

These excellent arguments did not have the immediate effect intended by their writer; others had to try their hands at persuasion, before Sedgwick yielded. At last, 18 November, he wrote to Murchison: "My reluctance in accepting the office of President is by no means affected. I value the honour as I ought to do, and I should delight in it if I had all the accomplishments. But I am an absentee, and I am poor. These are sad drawbacks. My friends here, however, will not hear of a refusal. So, if you appoint me, I must promise to do my best. It will be a sad falling off after Fitton, who has done the thing magnificently." We do not know what verdict was passed by Sedgwick's contemporaries upon his performance of the duties of President; but it is clear, from various allusions in his correspondence, that he

[1] Davies Gilbert, Esq., then President of the Royal Society.

was not himself displeased with the work. He grumbled now and then, and vowed that he was "nearly ruined"; but when it was over he admitted that he had found the employment agreeable, and had liked the friends to whose society he had been introduced.

Christmas was spent, in company with Whewell, at Viscount Milton's house near Peterborough. At that time many of the Fellows of Trinity used to find their way to that hospitable mansion, and Whewell spent his Christmas there for many years in succession. On one of these occasions—possibly in 1828—he was asked if he would like to go out hunting. Of course he said "Yes". Mounted on a first-rate horse, well up to his weight, he inquired how he could see most of the run. "Keep close to Sebright (the huntsman)" was the reply. Whewell did as he was bid, and followed that splendid rider over everything. They had an unusually good run, over a difficult country, in the course of which Sebright took an especially stout and high fence. Looking round to see what had become of the stranger, he found him at his side, safe and sound. "That was a rasper, Sir", he exclaimed, in admiration at his pluck. "I did not observe that it was anything more than ordinary," answered Whewell. Sedgwick was either less inquisitive, or more prudent, and while the rest of the party were out hunting, rode quietly over to Whittlesea to have a chat with the Aingers. The Tory sympathies of his particular friend in that family were so strong that he could not be persuaded even to write to Sedgwick so long as he stayed with so pronounced a Whig as Viscount Milton[1]—an amusing illustration of the strength of political convictions, not to say prejudices, at that time.

The Lent Term of 1829 opened with an event which must have given Sedgwick unmitigated satisfaction. As soon as it became known that the Duke of Wellington and his colleagues intended to introduce a bill for the relief of Roman Catholics, the University decided to petition par-

---

[1] To Rev. W. Ainger, 20 May, 1829.

liament against it. In former years, as already related, the opposition which liberal members of the Senate offered to such petitions had been defeated; but on this occasion it achieved a signal success, and the Grace to affix the University seal to the petition was rejected by fifty-two votes to forty-three[1]. As the *Cambridge Chronicle* naively records: "the result appears to have been principally owing to the somewhat unexpected arrival of several members of the Inns of Court, who came down for the express purpose of voting upon the occasion; two Paddington coaches with full complements of inside and outside passengers arrived between one and two o'clock, and returned to London the same afternoon." The writer should have added that they dined in Trinity before they started. The fact was, that the Cambridge Liberals, chiefly Fellows of Trinity College, had written to their London friends; among whom Macaulay, then resident in the Temple, had energetically exerted himself in marshalling a number of barristers[2]. That Sedgwick would be among the promoters of these tactics might be guessed without evidence; but a contemporary ballad supplies distinct proof of his activity:

> Oh Sedgwick, Oh Peacock, Oh Whewell, Oh Romilly,
> I'll preach you a ballad, I'll sing you a homily;
> Come hear the prophetical words of a Daniel,
> They were uttered at Clare, they were heard at Emmanuel.

> When devils to Cambridge shall Paddington marry
> And St Pancras shall send an express to St Mary,
> When the Bank shall go down with four horses to meet her,
> Then down goes St Paul, and up goes St Peter.

> The cat's in the larder, the wolf's in the fold,
> The rat's in the garner, the thief's at the gold;
> Oh Journal, and Standard, and John Bull, and Age,
> The lawyers are come in the Paddington stage.

> Come down to the Senate, come up to the vote,
> From fen, and from dyke, and from ditch, and from moat;
> Come darker and blacker, and thicker and faster,
> Come web-footed parson, come well-landed master.

[1] Cooper's *Annals*, iv. 559. The Grace was offered to the Senate 11 February, 1829.

[2] Trevelyan's *Life of Lord Macaulay*, ed. 1881, p. 106.

Oh! were there no powers to check the Iscariots, 1829.
To hamstring their horses, to shatter their chariots? Æt. 44.
There sprung not a spring, and there split not a spoke,
Though the Journal protested the compact was broke.

All Cambridge crowds round them, both gentle and simple:
"Now are ye for Church, Sirs, or are ye from Temple?
What sort of beast are ye, or what kind of vermin?
Is it wig, is it mitre, is it lawn, is it ermine?"

"We come not for Church, and we come not for stall,
But we come for a dinner in Trinity hall;
We come not for King, if your commons you'll dish up,
We come not for Church, but we'll thank you for Bishop."

This victory in the restricted arena of the Senate House
proved only a foretaste of the pleasure which those who
sympathised with the Catholics had in store for them. In the
following month the House of Commons accepted Catholic
Emancipation by large majorities at each stage of the measure; nor did the Lords offer any serious opposition. Sedgwick was present when the Duke of Wellington introduced
the second reading in a speech which has become historical[1];
and listened with natural enthusiasm to the brilliant debate
that followed. "I have hardly yet come to my sober senses,"
he wrote on his return to Cambridge, "after the stimulus of
my last visit to London. Lord Grey's speech seems still to
be ringing in my ears[2]." Sedgwick's convictions, one would
have thought, hardly needed stimulating; but, possibly, the
general acceptance of principles which had hitherto been
held by a minority may have urged him to advocate with
even greater earnestness than heretofore the removal of
similar restrictions at Cambridge. At any rate we shall
find him, a few years later, taking a prominent part in the
great controversy respecting the admission of persons to
degrees without regard to their religious opinions.

On February 20, 1829, Sedgwick was formally installed
President of the Geological Society. At the anniversary

[1] To Mrs Norton, 5 September, 1863.
[2] To R. I. Murchison, April, 1829.

1829.
Æt. 44.
dinner, which then, as now, succeeded the general meeting, there was a full attendance of members, and the new President, according to Lyell, who was present, "quite astonished them. Among innumerable good hits, when proposing the toast of the Astronomical Society, and Herschel, their President, then about to be married, he said : 'May the house of Herschel be perpetuated, and like the Cassinis, be illustrious astronomers for three generations. May all the constellations wait upon him; may *Virgo* go before, and *Gemini* follow after[1].'"

How singularly pleasant the meetings of the Geological Society must have been when it was still a coterie of brilliant, enthusiastic men, who knew each other intimately ; and how mortifying it is that we should have to be content with far-off glimpses, and faint echoes of what they said and did! Would that we could recall, not merely Sedgwick's post-prandial fun, but his mode of delivering one of his scientific papers, or of handling the discussion which it was sure to elicit. Mr Geikie tells us that " by a few broad lines " he could "convey even to non-scientific hearers, a vivid notion of the geology of a wide region, or of a great geological formation. Embalmed in the Society's Transactions, the paper, as we read it now, bears about as much resemblance to what it must have been to those who heard it, as the dried leaves in a herbarium do to the plant which tossed its blossoms in the mountain-wind. Brimful of humour, and bristling with apposite anecdote, he could so place a dry scientific fact as to photograph it on the memory, while at the same time he linked it with something droll, or fanciful, or tender, so that it seemed, ever after, to wear a kind of human significance. No keener eye than his ever ranged over the rocks of England; and yet, while noting each feature of their structure or scenery, he delighted to carry

---

[1] *Life of Sir C. Lyell*, 8vo. Lond. 1881, i. 251. There were four Cassinis, not three, who were successively Astronomers Royal at Paris. The last, John Dominic Cassini, succeeded his father in 1784. Sedgwick had probably met him in Paris in 1826.

through his geological work an endless thread of fun and wit[1]."

While Sedgwick was President he did his best to attend the meetings of the Society with regularity, but, as he said when his two years of office were over, his had been "an interrupted service." The next letter enumerates the interruptions in a single month, April, 1829:

"My hands at present are sufficiently tied. I am in the first place reading the Fathers and School Divines by way of preparation for my Divinity Act, which I must keep on the 30th of this month. In the meantime I have a rascally examination to superintend which will nail me down for a whole week[2]. Lastly, we shall soon have a contested election, and they have already requested me to become chairman of the Committee which will sit at Cambridge. I shall not refuse if they come to the scratch, tho' it will be a tiresome business. It is a strange thing that good Christians can't keep out of troubled water[3]."

At that time Divinity Acts were held every fortnight during Term. Every Master of Arts of four years standing complete was obliged, under rather severe penalties, to be a Respondent, that is, to maintain a thesis against three Opponents. The proceedings were similar to those which preceded the Bachelor of Arts Degree, and therefore need no further explanation. If the regular days appointed for the keeping of Acts happened to be all engaged, a private Act was allowed, at which some Doctor in Divinity, other than the Regius Professor, might preside. A letter to Mill shows that Sedgwick had adopted the latter course.

TRIN. COLL. *April* 15, 1829.

"I expected from what you said, when we last met in the Athenaeum, that you would have been in Cambridge

1829.
Æt. 44.

---

[1] *Life of Murchison,* i. pp. 138, 195.

[2] The scholarship examination at Trinity College.

[3] To R. I. Murchison, without date, but endorsed by him "April, 1829."

22—2

before this.   My Act comes on on the 30th of this month.
If you cannot conveniently preside, Dr Lamb[1] has undertaken to perform the task for me.   Pray write to tell me
what you intend to do.   My questions are:

    1.   The Divinity of Christ.

    2.   A denial of the Millennium; perhaps in the words of
our expunged Article.

By *opposing* me, and pronouncing a *determination*, you
will get over two of your exercises, which will be some
advantage.   In regard to arguments, you may bring as many
or as few as you please.   In case of a private Act it is not
however customary to bring many.

Pray let me hear what is your final determination.   If you
can't come I must settle with Dr Lamb.   Though I have not
the honor of knowing her, I hope Mrs Mill will accept my
kindest wishes."

Murchison had found his foreign tour of the previous year
so instructive, that before he had been many weeks abroad
he had urged Sedgwick to come and do likewise.   Writing
from Nice, he describes what he had seen, and adds:

We left various things undone, consoling ourselves with the
parting reflection that such a case was to be worked out by Sedgwick
next year.   And here let me, by way of parenthesis, invoke the
philosophical spirit of inquiry which prevails at Cambridge, and urge
you, who are really almost our only *mathematical* champion, not to
let another year elapse without endeavouring to add to the stock of
your British Geology some of the continental materials.   Pray do it
before you marry and settle for life; pray even do it before you
bring forth that long-expected second volume on the Geology of
England and Wales[2].   Your comparisons will then have a strength
and freshness which will quite electrify us[3].

These arguments, enforced by conversation after his
return, had convinced Sedgwick, and he agreed to accompany Murchison on a second journey, so soon as he could

---

[1] John Lamb, D.D. Master of Corpus Christi College.

[2] The work which Sedgwick and Conybeare were supposed to be writing
together.   See above, p. 324.

[3] Murchison to Sedgwick, 18 August, 1828.

get away. They were to leave England towards the middle
or end of June, and explore the northern flanks of the Alps,
with the central parts of Germany, Bohemia, and Saxony—
"a glorious field for a knight of the hammer[1]." Meanwhile,
Sedgwick had plenty of work to do in presiding over the
Geological Society, and in putting the final touches to the
three papers read the previous year—"the rasping and
trimming of which, before they were finally delivered over
to the devils, was no small labour[2]." But other matters were
soon to interfere with his preparations for his journey. Before May was over Sir N. C. Tindal, one of the University
representatives in Parliament, was made Chief Justice of the
Court of Common Pleas, and his seat had of course to be
filled without delay. No burning question was agitating the
country, and it was hoped that some distinguished person
might be found, who would satisfy both parties, and a contest
be thus avoided. This hope proved delusive. Two candidates appeared in the field on the same day: Mr E. H.
Alderson, of Gonville and Caius College; and Mr George
Bankes, of Trinity Hall[3]. Alderson had been senior wrangler,
first Smith's prizeman, and senior Chancellor's medallist in
1809; Bankes had taken an ordinary law degree in 1825.
Alderson, moreover, besides his brilliant degree, could show
a distinguished career at the bar. Bankes had had some
parliamentary experience as member for the family borough
of Corfe Castle, and had filled a subordinate post in the
government of the Duke of Wellington, which he had resigned when Catholic Emancipation became a government
measure. After that measure had been passed, however, he
had resumed his place—a step which said but little for his
consistency. Meanwhile a feeling had gradually spread
through the University, that it would be well to elect Mr
Cavendish—now our honoured Chancellor—who in the

[1] To Rev. W. Ainger, 4 June, 1829.
[2] To the same, 20 May, 1829.
[3] Their circulars are dated 29 May, 1829.

previous January had been second wrangler and first Smith's prizeman; and besides, had won general admiration "by his superior talents, by his studious and reflective habits, and by the unimpeached regularity of his University life[1]." But an unexpected difficulty presented itself. Mr Cavendish, though himself willing to come forward, was for nearly a week prevented from entering the field by the head of his family, whose objections were only overruled by "a public address, signed by many distinguished members of the University." Mr Alderson then retired; and a canvass commenced for Mr Cavendish "unexampled for the energy and heartiness of those who were engaged in it[2]." That he was supported by the liberal party cannot be denied. The chairman of his Cambridge Committee was Dr Lamb, well-known for his liberal opinions, and in consequence one of the tory organs nicknamed him "Lamb's adopted"; but, on the other hand, many of the most decided tories voted for him. That Sedgwick was foremost in the fight will readily be believed. He "personally worked day and night so as almost to destroy his health[3]"; he marshalled the supporters who could be relied on; he stimulated the lukewarm; he exposed and controverted the tactics of the other side with a headstrong energy which in some cases lacked discretion. It was soon found that Mr Bankes had obtained numerous pledges of support before the resignation of Sir N. C. Tindal had been made public; and moreover the whole influence of the government was exerted on his behalf. "Not one member of the Senate," we are told, "who was placed directly, or indirectly, within a minister's influence, escaped an official canvass[4]." Not only did his brother, once member for the University, write to his former supporters; but Mr Goulburn, now Chan-

---

[1] From *A letter to a Member of the Senate of the University of Cambridge*, by Robert Grant, M.A., M.P., Fellow of Magdalene College.

[2] These quotations are from an article in *The Times*, 19 June, 1829, which is known to have been written by Sedgwick.

[3] To Dean Monk, 1 November, 1829.

[4] *The Times, ut supra.*

cellor of the Exchequer, wrote to many resident members of
the Senate on his behalf.   These tactics were controverted by
Sedgwick in a letter to Goulburn which was printed and
circulated in the University.   It is signed *A resident member
of the Senate*, but there could never have been any doubt
about the authorship.   It was evidently written under the
influence of strong excitement, and had it been merely an
ephemeral composition dashed off to serve the purpose of
the hour, the obvious course would have been to leave it in
the oblivion into which it has long since fallen ; but it is so
vigorously written, and throws so much light on the University politics of that day, that it has been decided, after
much hesitation, to reproduce it.

To the Right Honourable Henry Goulburn.

Sir,

I expected before this time to have seen you at Cambridge ;
and when I at length found that a reluctance to meet your old
partisans, or some other motive well understood by yourself,
prevented you from again being a candidate for the vacant seat
in the University, I took for granted that you would at least know
what was due to your former supporters, and that you would preserve
a dignified neutrality.   In this expectation I have been disappointed.
You have condescended to become the bustling advocate of Mr G.
Bankes ; and in a position so unnatural, the result has been what
you ought to have anticipated.   Your letters from Downing-street to
certain resident members of the Senate, have done no good to the
cause of which you have so unexpectedly become the advocate, and
have been received only with expressions of contempt and resent-
ment.   Your personal elevation prevents you from hearing at all
times the language of truth ; but it is well that it should sometimes
be spoken, and so loudly too, that even those who sit in high places
should not find themselves exalted above its influence.   I beg leave,
Sir, to remind you that you have twice been a candidate for the
representation of our University,—that you stood forward as the
champion of the Protestant ascendancy,—and that you received on
that ground the support of many high-minded and honourable men.
There is a vulgar proverb of very obvious application, which may,
perhaps, explain the rancorous bitterness which existed between
some of your friends and the party which supported Mr W. Bankes.
Your person and your conduct were assailed by that party with
long-sustained invectives, conveyed in language such as English
gentlemen are not often in the habit of giving or of receiving.   You
received these assaults with exemplary calmness, and endured them

with a patience which was the admiration of your friends, and was
worthy of the high religious ground which you had then taken.   I
may, however, remind you, that although our religion commands us
to forgive our enemies, it never enjoins us to be the patrons of those
by whom we have been vilified, or the champions of those whose
principles are in open hostility with our own.   Mr W. Bankes is not,
indeed, in the field ; but his brother professes (how consistently is
not now the question) to be the representative of the same party,
and of the same opinions, and on this ground alone comes forward
to ask for our support.

During the last session of Parliament your opinions underwent
one of those sudden revolutions which, whether they happen in the
physical or the moral world, astonish and confound us.   I was not
among the number of those who made an unfavourable analysis of
your motives.   I believe you, and I still wish to believe you, sincere.
If you were not sincere in the line of conduct which you have
recently adopted during the agitation of one of the gravest questions
which ever came before Parliament, you must be content to find
your name written in the list of those men who barter themselves
and their faculties for office and emolument, and who hold to no
principle with a grasp which does not relax at the approach of a
vulgar temptation.   Of this baseness I dare not and I cannot
accuse you ; but if you escape from this imputation, I would tell
you, in the language of our schools, that you are still on the horns
of a dilemma ; and I would ask you, in the name of the Senate, by
what new metamorphosis you are become the champion of Mr G.
Bankes, whose only pretension,—I repeat it, whose only pretension
is, that he opposed you and your colleagues on that great question
to which I have alluded, and who, had he succeeded, would have
contributed to thrust you out from that office which you now fill
through the kindness of your Sovereign.   If you were in the right,
Mr Bankes was in the wrong ; and for this wrong he is to have your
support, and appear among us backed and recommended by your
autographs from Downing-street.   But Mr G. Bankes has returned
to the party which he once vilified and opposed, and he must be
treated with the affection of a brother, because he also now wears
the semblance of an apostate.   Whether I am right in this conjecture
I know not ; but I do know that the members of the Senate are
justly indignant at any direct interference with the freedom of their
elections ; that they believe themselves to be the proper judges of
who is the best person to represent them, and that they are not yet
reduced so low as to supplicate at a Government office for the
nomination of a candidate.

In what you have done, you have not appreciated our character
or our sentiments.   The resident members of the Senate, by their
votes on a late occasion, did good service to the Government; and
by way of return for this, you now endeavour to force upon us a
representative who does not himself stand upon the high ground of
political consistency,—who is almost unknown to us,—who is not a

member of our Senate,—who is decorated with no academic honours,
—whose name is associated with no pleasant recollections,—and
who, by his only public acts connected with our body, encouraged
and vindicated a combination of the Undergraduates, avowedly
made in a violation of all discipline, and in contradiction to our
existing authorities[1]. I do not wish to speak harshly of Mr G.
Bankes, because I think that he does not deserve it : but I am bold
to say that he comes forward with no high pretensions, and that he
has done nothing to entitle him to the honour of being thrust upon
us by all the forcing power of Government influence. If he consults
his own honour and the dignity of the University, he will immediately withdraw : and you, Sir, if you have any regard for your own
consistency, and the good opinion of those distinguished and honourable men by whom you were once supported, ought to be among the
very first to recommend this measure to him.

Mr Cavendish is this moment arrived amongst us. He is urged
forward by no party and no faction. He was put in nomination
(without his own knowledge or participation, and against the wishes
of the highest members of his family) by many distinguished resident
members of the Senate, who, however they may differ on other
questions, think it for the honour of our establishment that on this
they should be united. They come forward to support Mr Cavendish
because he is a young man of modest and amiable temper, and of
unsullied life,—because, during the years he lived among us, he
conformed himself in the purest and highest sense to the true spirit
of our institutions,—because he has proved, by his academic distinctions in literature and in science, that he possesses talents of no
ordinary kind, and habits of application which even in early life
have resisted extraordinary temptations. I have now lived more
than twenty years in the University. I can assert without the risk
of contradiction, that during this long lapse of time, no young
nobleman has appeared amongst us who could have been brought
before the Senate with such high and unsullied pretensions. If,
from his youth, he has been hitherto prevented from exhibiting his
powers as a senator, at least he has been saved from error during
times of no ordinary difficulty, and comes before us without any
tarnish of inconsistency. He has reaped his first laurels amongst
us ; they sit fresh upon him, and they will wear well, and they will
for ever be associated with the ardent recollections of early life.
Under these circumstances, we may safely count upon his lasting
attachment to us and to our venerable institutions, and upon that
consistent and dignified exercise of his great talents which will be for
his honour and for our own.

With such qualities, I cannot for an instant doubt the success of
Mr Cavendish. All the high aristocracy, belong to what party they

---

[1] Mr George Bankes presented to the House of Commons, 23 March, 1829, a
petition signed by about 600 Bachelors and Undergraduates, against any further
concessions to the Roman Catholics. Cooper's *Annals*, iv. 560.

may, are interested in his success: for he stands forward as the representative and the ornament of their order, and has assisted to keep alive in a great public body that constitutional respect for dignity and for rank, in the absence of which the highest privileges would lose all their grace and much of their importance.

I cannot believe that the illustrious individual who is now at the head of administration can have given his sanction to a canvass from the Treasury, which is so plainly against the best interests of his own order. Be this as it may, the University of Cambridge can and will judge for themselves; and are, notwithstanding your humble opinion of them, placed far above the reach of improper influence, however high the quarter from which it may descend.

I have now performed the task I have undertaken. I could have wished to have had more time for its performance, but I hope I have made myself understood. I believe I have fairly represented the motives of a great body of the Senate, and the feelings which your unexpected canvass has excited. I therefore leave this homely expression of truth to its proper influence; and, notwithstanding the strange revolutions I have witnessed in the conduct of others, I venture still to subscribe myself, with great respect, Sir,

Your most obedient and humble servant,
A RESIDENT MEMBER OF THE SENATE.
CAMBRIDGE, *June* 3, 1829.

A few extracts from letters to private friends show better than any description Sedgwick's feverish condition during the first fortnight of June—divided as he was between the election, preparations for his journey, and his duty to the Geological Society.

To R. I. Murchison, Esq.

TRIN. COLL. *Tuesday Morning,*
2 *June,* 1829.

"We are up to the ears in politics. Bankes has started for the University, and we have pitted Cavendish against him. I hope to God we shall succeed. I shall take care to be up in time for the Council on Friday; but in our present disturbed state I don't know that I can be with you sooner. I am sorry for it; as I should have rejoiced to meet your party of Wednesday. This hurly burly at this time is unfortunate, but I can't help it. Pray can you do us any service? If you know any voter, or any one who can influence a vote, at him by all means. Our cause is good in both ways. Cavendish is a man who would do us great honor — Bankes is nobody, and wishes to ride upon the shoulders of the *ultras*. *No Popery* was a grand stalking-horse; but I hope it has now broken its knees, and will not carry weight."

During this visit to London Sedgwick tried to secure Mill, who had returned from India for a short holiday, and was staying at an hotel. Mill was out when he called, but he introduced himself to Mrs Mill, and left with her the following note, endorsed, "From Adam Sedgwick, Professor, canvassing for Cavendish. Written with his heart's blood." The pollbook shows that the appeal was successful.

GEOLOGICAL SOCIETY.

Dear Mill,

By your love of virtue—of Trinity College—and of literature, and of science, come and vote for Cavendish. He has committed no political sin. If this will not do—by your friendship for myself and for the other residents who were the academic companions of your early life, do not vote against him. I wish I had time to see you, but I am engaged here from three till half-past eleven.

Yours ever,

A. SEDGWICK.

To Rev. William Ainger.

TRIN. COLL., *June* 4, 1829.

"I dare not canvass you for Cavendish, because I know that you see things with eyes so different from mine that we hardly on some matters can find a starting-point from which we may begin an argument. He is one of the most amiable and accomplished young noblemen whom we have had among us in this last century, and has been started by men of all parties, purely on his personal merits. I dare say you don't like his name, and think it sounds Whiggish. Bankes has no

1829.
Æt. 44.

merit that I know of except that he *pretended* to go out on the Catholic question. It was all mockery, he never *was out* or he would not *now be in.* The cast-off rags of *No Popery* won't cover his nakedness. But enough of this. If Cavendish come in the University will have an honourable rest. All I ask is that you do not come against us."

To R. I. Murchison, Esq.

COMMITTEE ROOM, *June* 11.

"I have not one thought, word, or deed, except for Cavendish. Pray do what you can in the way of preparation for our tour. The election ends on Thursday week. On Friday morning following I shall come up to Town to attend the meeting of the Geological Society. But I fear it will be impossible for me to be ready by the Wednesday following. By Saturday I should be able to start. Coddington[1] means to accompany us up the Rhine. His German will be of great use to us. I don't think he will go very far. Whewell will if possible join us in the Thüringerwald. Excuse this hurry."

To the same.

CAMBRIDGE, 15 *June*, 1829.

"Good God! you will have to do everything for me. To start at six o'clock in the morning of Friday, immediately after the election, and to be off on Wednesday! But I will do my best. Pray look out some papers for me to read, e.g. Boué's &c.; enquire about maps, and other geological necessaries. If these be in readiness, I hope I may be ready myself by Wednesday. How abominably unprepared I shall be. My mind will be like white paper—ready for any impressions. Be it so. The election is horribly inconvenient. I am sitting on the edge of a razor. At 8 o'clock tomorrow we start. Both sides are confident."

The Poll closed on Thursday, 18 June, when Mr Cavendish was elected by 609 votes to 462—a majority of 137. When

---

[1] Rev. Henry Coddington, M.A., Fellow of Trinity College.

the result was declared an undergraduate in the gallery shouted: "Farewell, a long farewell to all—the Bankeses!" The day ended, according to Mr Romilly, with "a huge dinner" in the Hall of Trinity College, at which Sedgwick "spoke finely."

The next morning Sedgwick was off to London, hurried through his preparations, and started with Murchison for the continent on the appointed day. The route followed up to the middle of September is described in the next letter[1].

GMUNDEN NEAR SALZBURG, *September* 14, 1829.

My dear Ainger,

I have for some hours been twirling my thumbs and watching the weather; but there is no longer a gleam of hope. The spirits, under such circumstances, undergo a kind of recoil. When a man cannot move his body forward, he casts his thoughts backward, and thinks of those who are behind him. If this letter deserves thanks, you must thank the weather and not me. In some respects I am still to be envied. While I wield a pen in my right hand, I hold a German pipe in my left, and the images of past scenes are floating before my mind's eye among the fumes of Hungarian tobacco. If the elements were less turbid, I should have before me one of the most lovely lakes in the world, backed by peaks of the Saltzburg Alps. I am too much a man of business to write much; but I always intended to send you one sheet full of such matter as I could scrape together. So here I take up my parable.

I left England on June 24th, steamed to Rotterdam, and after a delay of a few hours not ill employed in that truly Dutch city, continued my journey up the Rhine, by the same conveyance, to Bonn. There we halted, and took in a quantity of geological ballast from sundry German Professors. We again embarked, and landed at Andernach; and made an excursion on foot up the country to visit some very interesting

---

[1] Our account of this tour should be compared with that in Geikie's *Life of Murchison*, i. 157—162.

1829.   extinct volcanoes. We traced lava-currents to their craters,
Æt. 44. and travelled for miles upon pumice, scoria, and ashes. I was
bewildered and confounded at the sight, for these fires have
never smoked within the records of mankind. We then
travelled along the lovely banks of the Rhine to Maintz, and
crossed to Frankfort, where Murchison and myself purchased
a carriage; and since that time we have been travelling by
post, always excepting excursions over hill and dale, above
ground and under, in places where horses have never trodden.
Our first excursion was to Cassel, a beautiful capital; from
thence we walked over some of the Hessian mountains, and
met our carriage on the south edge of the kingdom of
Hanover. We halted one day at Göttingen, and were above
all measure delighted with old Professor Blumenbach. Thence
we posted to the Hartz mountains. They detained us some
days; but I will not torment you with geology. We then
posted to Eisleben, famous for fossil fish and copper, and
after angling, or more properly haggling, for these fish, we
went to Halle, and again rested one day, and smoked with
German Professors. From Halle we posted across the sandy
plains of Prussia to Berlin. It is a fine modern capital, but
is devoid of any venerable monument of former times; and,
after the first flash which astonishes you, ceases to give any
pleasure. We found some very well-informed persons there,
who gave us the kind of information we wanted, and after a
halt of four days we started for Saxony. I do wish that I could
take you by the skirts of your coat and place you upon the
Bastei, a perpendicular rock on the banks of the Elbe, and I
would show you one of the finest views in the world. From
thence we would walk over the field of battle, pause at the
spot where Moreau fell (marked now by a small granite
pillar)[1]; thence we would track our way through the defiles
of the Bohemian mountains; sleep at a small inn close by

---

[1] Moreau, the celebrated French republican general who joined the Allies after
Napoleon's defeat in Russia, was fatally wounded at the battle of Dresden,
27 August, 1813.

Culm, where Vandamme was defeated[1]; and next day we would visit Töplitz, a broiling hot city where the streets during the season are filled with German Barons and Counts, Bohemian and Polish princes, and where you might walk in the public rooms cheek by jowl with the King of Prussia. All this I cannot do ; I must therefore be content to tell you that this was my track : that from Töplitz I went to Prague, one of the most magnificent cities I ever beheld ; and from thence posted through southern Bohemia and Moravia, a dull and dismal long journey, to Vienna. We were rather unfortunate, as the most eminent men of science, at least in our way, were gone out of the city. A few Professors were, however, left ; and our ambassador very politely invited us to his country-house, and we spent a delightful day with him. His house stands upon the site of the old Turkish camp occupied during the siege of Vienna. It commands a view of the Danube, the fatal plains of Aspern and Wagram, the city, the Hungarian mountains, and the Styrian precipices which form the eastern termination of the Alpine chain.

On leaving Vienna we took the road towards Trieste, and, after crossing a corner of the Alps, descended by the banks of the Mur, and spent about ten days in lower Styria. It is full of interest, moral and physical; a most lovely country peopled by a most beautiful race, who are simple and kind-hearted beyond anything I have ever seen. We had some excellent introductions, and saw everything we wanted, with one exception. We had letters to the Emperor's brother, the Archduke John, the Governor of Styria, and he was unfortunately absent. He is one of the most extraordinary men in Europe; accomplished as a man of science, kind-hearted, liberal, and of extraordinary simplicity of manners. He was unfortunate in the wars against Napoleon, and, perhaps in some disgust with a court life, retired to his government; adopted a simple style of living; visited every corner of his extensive province, and almost every family,

1829.
Æt. 44.

---

[1] At the second battle near Culm, 30 August, 1813.

often travelling on foot without a single servant. He has established museums and scientific institutions, encouraged everything good and liberal, and has gained such influence that in case of need he could raise up the whole population by a motion of his finger.

We crossed the desolate mountains of Carinthia, and at length reached their southern limit and found ourselves in one moment looking over the blue waters of the Adriatic. I dare not attempt to describe my sensations when the rocky shores of Idria, the plains of Italy, and the great wall of the Alps burst upon the view; I should fly into heroics which would be out of keeping for a geologist who travels with stones in his pockets, and is therefore kept from soaring. We spent a day at Trieste, among surly English captains, sleepy Dutchmen, and Levant merchants in oriental dresses. We crossed the sultry plains of Italy among olive-groves and vineyards, and then plunged into the defiles of the Taglia-mento, and, after wandering several days among the southern flanks of the Alps, crossed the axis of the chain at the Tauern, a pass which is at the elevation of about 6500 feet above the level of the sea. We then descended by some gorgeous defiles to Salzburg, and here I am, as I before said. My pipe is out and my eyes are nearly out....

I am now going to zigzag along the north flank of the Alps, and sometime in October hope to hammer my way to Paris. In the meantime I wish you a very good night. My kind regards to your sister, and my love to your children.

Yours most affectionately,

A. SEDGWICK.

On entering the Tyrol from Italy they fortunately fell in with the Archduke. Already, though they had not seen him, they had profited by his presence in the country; for, writes Sedgwick, "wherever we went in the valleys of Styria with our hammers, we were set down at once as odd fellows who were friends of the Archduke John[1]."

"We first saw him at a little village of the higher Alps called Bad Gastein; and in five minutes found ourselves as much at home with him as if we had known him twenty years. He had received the letters we intended to have presented to him, and therefore knew our objects of search and who we were. He proposed an excursion to the glacier and waterfalls [at Nassfeld], to which we of course joyfully assented; and added most courteously that it would give us an occasion of talking of many things by the way, and this would be the only opportunity, as the day following he was going to cross the great chain on foot to visit a friend in the southern Carinthian Alps. We started in a machine with two seats, but in every other respects like a Dent's shandery-dan; and, after going as far as this machine would go, we scaled the precipices on foot, and traversed one of those magnificent amphitheatres of ice and snow of which no written language can convey any adequate description. We descended in the evening to a little tidy alehouse [at the village of Böckstein] where we supped upon trout and bottled ale, and we finally tracked our way by the light of a lantern to the village from which we started.

"Everything we had heard of this excellent man was more than realised. He is sensible, liberal to a degree which offends the despotic Emperor, accomplished as a man of science, of most amiable temper, and wonderful simplicity of manners. The moment the girls of the little alehouse knew who he was, I thought they would have gone into fits through joy. They seized his hands, kissed them a hundred times, and, if he would have allowed them, would have gone down on their knees before him. He talked with great freedom; spoke of the partition of Poland as an iniquity which he feared would some day bring down a great national punishment, and frankly pointed out many existing evils in

1829.
Æt. 44.

---

[1] To Rev. John Sedgwick, 10 August, 1829.

the system of government. I before told you of the way in which he passes his time among the people whom he governs. Everybody seems happy under him, and all institutions flourish. There was one subject on which I dared not speak to him, and that was religion. He is a catholic, and assuredly is a liberal one. Where the people wanted ministers and chapels he has built them, in several parts of Styria. All this is right, for the country is entirely catholic; but about matters of faith he did not seem inclined to speak.

"The morning after our excursion I called on him just as he was about to start. He was dressed in worsted stockings, hob-nailed shoes, and jacket, with a little green hat and feather, the costume of Styria. Three men in a kind of uniform, with rifles, followed him, for the purpose of shooting chamois, wild-deer, or other animals they might meet with. His parting was like the rest of his manner; simple, kind-hearted, and unceremonious. We saw him start, and ascend towards the higher Alps on foot[1]."

The travellers next made a rapid exploration of the Salz-kammergut, with which Sedgwick was delighted. "The whole region," he wrote, "is of exquisite beauty, and the inhabitants have the same honest, simple, kind-hearted character which I praised and admired so much among the Styrians[2]." Letters from the Archduke gave them ready access to the salt-mines; and at Berchtesgaden they came in for the close of a grand hunting-party given by the King of Bavaria to some foreign princes.

---

[1] To Rev. John Sedgwick, 31 August, 1829. The Archduke John was the sixth son (born 20 January, 1782) of the Emperor Leopold II. He commanded the Austrians at Hohenlinden, where he was defeated by Moreau (3 December, 1800), and his subsequent military career was equally unfortunate. Throughout his whole life he took great interest in the Tyrolese and Styrians. The former were incited by him to the unsuccessful revolt under Hofer (1809). In after-life he resided in Styria, chiefly at Gratz, where he married the daughter of a postmaster. He had no official post in Styria, but employed himself, as Sedgwick says, in the improvement of the people. In 1848 he became vicar-general of the Empire, an office which he held for only a few months. He died in 1859.

[2] To Rev. John Sedgwick, 26 September, 1829.

"The sport was nearly over before we arrived, but we saw the company, and the manner of the chase, which was all we wanted. The scene altogether reminded me of some of Sir Walter Scott's finest descriptions, but was upon a scale more grand than Scotland could ever boast of. For some weeks before the visit of the royal party many hundred persons are employed in driving the deer, chamois, and other wild animals into a particular part of the Bavarian forests just under the snowy Alps. They form two great lines on the opposite extremes of the great forests, at the distance of twenty or thirty miles from each other, and by means of dogs, horns, etc. drive the affrighted beasts towards the central region. It is so contrived that these central forests are under a long succession of precipices, through which there is no escape to the Alpine summits, except by a few ravines and narrow gorges. On an appointed day the king and his attendants place themselves in these ravines accompanied by soldiers armed with rifles; and on an appointed signal thousands of persons, some employed by government, others led by curiosity and love of the sport, rush into the forests, and drive out the wild inhabitants from their hiding-places. The poor animals, thus beset on all sides, become frantic, and rush out of the forest, sometimes singly, sometimes in large herds, and, having no means of escape except through the narrow defiles I have mentioned, scores of them are brought to the ground before they can get out of the reach of the riflemen. The different ranks and costumes of the assembled multitude, the shouts of the hunters, the echoes of the guns among the great precipices of the Alps—produce a combination of circumstances well fitted to excite the imagination[1]."

From Salzburg they went to Munich, and thence by Ulm, Stuttgart, and the Lake of Constance to Strasburg. A letter from the latter place to Whewell, who had been disporting himself in Switzerland with Mr Coddington, enters so

[1] To Rev. John Sedgwick, from Stein on the Lake of Constance, 26 September, 1829.

much more into geological detail than any of the other letters of this year, that, at the risk of some repetition, we will give a few extracts from it. After describing their visit to the Hartz, Eisleben, Halle, etc. Sedgwick proceeds:

"This is the focus of Wernerian geology, and to my infinite surprise it is the most decidedly volcanic secondary country I ever saw. The granite bursts through on one side, sends out veins, and along the whole eastern flank the secondaries are highly inclined and often absolutely vertical. Near Goslar they are absolutely heels over head....

" In Styria we found a great deal of good tertiary geology. Our Styrian tertiaries led us down into the edge of Hungary, from which we doubled to the great road, and beat our way down to Trieste. Dull geology, but the finest caverns in the world.... From Trieste we crossed the plains of Italy to the Tagliamento, by which we entered a great gorge in the Julian Alps. We emerged from these gorges at Bleiberg, and began to ascend the primary axis. To our great surprise found the oldest rocks of the calcareous zone full of gryphites, and not older than our lias, though crystalline as white as sugar! We crossed the axis at the top of the great Tauern Alp amidst mica-schists and crystalline marbles, serpentines, etc., etc., and, what do you think? in this series we found beds top-full of encrinites. I could hardly believe my eyes. Thence down the high road to Werfen, from which place we again doubled, and ascended to the primary axis by a parallel valley.... On our return to Werfen we set off to Salzburg, and afterwards threaded our way among the links of the great southern calcareous zone. And how shall I describe the wonders we here saw? The tertiary deposits resting on the outskirts of this calcareous zone are thicker than all our secondary formations put together. For scores of miles they are in a vertical position. In many places the Alps, in rising through them, have lifted great rags of them into the regions of snow. Some of these rags are 3000 or 4000 feet thick, and stuck on like great poultices on the bruised pates of the older

rocks. From Salzburg to Innspruck. Thence once more
over the calcareous chain—top-full of fish, and stinking of
fish-oil, which in many places trickles out like tar. From the
fish-beds to a bed at Munich. Pictures and antiques one
day—off to the great tertiaries on the Bavarian flanks—so
to the Lake of Constance. Two noble sections linking in
our work with the tertiaries of Switzerland. Thence to
Oeningen, Murchison's fox-cover[1]. Thence to the Danube—
Ulm. N.B. Freshwater hills all around the city. From Ulm
we visited the famous field of Blenheim on our way to
Solenhofen ; a wonderful place for lithographic stone and
fossil fish. From this dépôt we crossed the Jura limestone,
through some beautiful freshwater basins, to Stuttgart, and so
down the Neckar to Heidelberg. This outline will give you
some notion of what we have been about. I think we have
done some good work. I am anxious to be home again, but
we must go by Paris. We have some work by the way, and
may not be there before the 17th or 18th. My kindest
regards to all who regard me[2]."

Lyell tells us that Sedgwick returned "full of magnificent
views ; throws overboard all the diluvian hypothesis; is vexed
he ever lost time about such a complete humbug ; says he
lost two years by having also started as a Wernerian, etc.[3]"
He did not himself admit that his conversion was so complete
as this report of his conversation would imply ; but no doubt
his views had been greatly modified and extended by what
he had seen on the continent, and by his intercourse with
foreign geologists.

On this occasion Murchison had no cause of complaint
against Sedgwick on the ground of delay in getting their
joint work ready for publication. In about a fortnight after

[1] In the previous year Murchison had obtained from this celebrated quarry a
unique skeleton of a fossil fox (*Galecynus oeningensis*) now in the British Museum.
He described it in the *Transactions* of the Geol. Soc. iii. 277. Compare also his
Life, i. 154.

[2] To Rev. W. Whewell, 10 October, 1829.

[3] *Life of Sir C. Lyell*, i. 256.

1829.
Æt. 44.

their return (6 November, 1829) their first paper *On the Tertiary Deposits of the Vale of Gosau in the Salzburg Alps*, was read to the Society; and, wonderful to relate, in spite of Sedgwick's occupations and ailments, which appear to have been unusually severe, it was succeeded, at the two following meetings (20 November and 4 December) by a second, *On the Tertiary Formations which range along the Flanks of the Salzburg and Bavarian Alps.* The method of setting about the work, and the value of the results, do not call for much comment. It was not a new and unexplored district, such as Sedgwick loved, and yet it was an area where great problems were suggested, and it formed a fine field for a holiday tour. Murchison, as was his very useful custom, "got the subject up" before starting; he read what had been written on the district, corresponded with the authors and authorities upon it; and, thus furnished, the colleagues started to examine for themselves, and to criticise the interpretation of the geological structure of the country given by Boué and others.

In the summary given by Sedgwick in his presidential address to the Geological Society, we have as clear an account as we can desire of what was proposed and what was done. It was a question of identification and correlation, and Sedgwick and Murchison were among the great host of explorers and authors who have treated of the bands of calcareous and arenaceous rocks, with nummulites in the newer beds, and hippurites in the older, which flank the Alpine ridges from the Rhone to the Danube. They did not collect materials for a minute classification—indeed it would have been impossible for them to do so—but they gave a good account of the district, with much new work; and they brought the whole subject before English geologists for the first time.

The principal points established are thus stated by Sedgwick:

" We have shewn that several transverse sections from the central axis of the Alps to the basin of the Upper Danube

would present a succession of phenomena in very near accordance with those of other transverse sections from the same
axis to the tertiary formations at the other base of the chain
in the North of Italy. On both sides of this chain, after
passing over the great secondary calcareous zones, we meet
with the lower tertiary strata—always highly inclined, sometimes vertical, and occasionally conformable to the beds
of the older system. We contend that this remarkable
symmetry confirms the hypothesis of a recent elevation
of the Eastern Alps; and makes it probable, independently
of arguments derived from organic remains, that the tertiary
deposits of the Sub-Apennine regions and of the basin of the
Upper Danube belong to one period of formation.

"Thick masses of strata full of organic remains, and often
occurring at low levels near the northern foot of the chain,
are sometimes also found (e.g. in the valley of Gosau) in
unconformable positions, caught up among the serrated peaks
of the Alps, four or five thousand feet above the level of the
sea. Such a disjunction of corresponding strata is inexplicable on any hypothesis which rejects the theory of elevation.
We have concluded, chiefly on zoological evidence, that the
unconformable beds of Gosau are more recent than the chalk.
We believe that they contain neither ammonites nor belemnites, nor any other known species of secondary fossils; and
on the whole we regard them as a term of that unknown
series of formations which may hereafter close up the chasm
between the lowest beds of the Paris basin and the chalk.

"We have pointed out the limits of the old chain of the
Salzburg and Bavarian Alps, and traced the direction of its
valleys anterior to the tertiary epoch: and we have described
a great deposit of lignite far up the valley of the Inn, containing fresh water and marine shells, which seem to connect it
with the period of the London clay. We have further shewn
that there are within the basin of the Upper Danube two or
three higher zones of lignite separated from each other by
sedimentary deposits of enormous thickness.

1829.   " The tertiary system of Bavaria is shewn to pass into,
Æt. 44. and to be identical with, the *molasse* and *nagelflue* of Switzerland.   The higher part of this series must therefore be of the
same age with some of the formations of the Sub-Appennines.
We have proved that enormous masses of sandstone and
conglomerate many thousand feet in thickness, stretching
from the base of the Alps to the plains of the Danube, are
chiefly derived from the degradation of the neighbouring
chain—that many of these masses cannot be distinguished
from the newest detritus which lies scattered on the surface
of the earth—that in their prolongation into Switzerland
they sometimes contain bones of mammalia—that they are
regularly stratified, and alternate with beds containing marine
shells—and that they cannot have been caused by any
transient inundation.

" Finally, we point out the probable effect of *debâcles*
which took place when the basin was deserted by the sea.
We shew that the excavations produced by the retiring
waters have been augmented by the bursting of successive
lakes, of which we found traces in all the upland valleys
of Bavaria; and that these excavations have been since
carried on by the erosive power of the streams which roll
down from the sides of the Alps to the plains of the Danube[1]."

To read an elaborate paper is one thing; to make it fit
for publication is another; and it was soon found that some
very hard work had yet to be done.   When the abstracts
appeared in the *Proceedings* of the Society, the views therein
advanced—especially those relating to the Valley of Gosau—
were combated both in England and abroad, notably by
Dr Ami Boué.   It became therefore necessary to test conclusions by a second visit to the ground.   Sedgwick, apparently, had no wish to leave home again so soon; and the
task therefore devolved on Murchison, who devoted the
summer of 1830 to it, with complete success.   On his return

[1] *Address to the Geological Society,* 19 February, 1830, p. 9. *Proceedings,*
i. 193.

he read a separate memoir to the Geological Society, in 1830.
which the old conclusions were fortified with fresh facts. Æt. 45.
After this the Council of the Society decided that it would
be more instructive, and save repetition and correction, if the
whole subject were treated in a single memoir. This was no
doubt a wise decision, but it entailed the re-writing of both
memoirs, so as to weld them properly together. A good deal
of this labour fell to Sedgwick's share, and occupied him,
conjointly with other work, for several months in 1831. The
volume in which the paper appears in its final form was not
published until 1835.

At the beginning of 1830 it became Sedgwick's duty to
deliver, as President of the Geological Society, the customary
address at the Anniversary Meeting. Perfunctory work of
this sort rarely repays careful analysis. Questions which the
author weighs in the balance of a good-natured criticism
have long since been settled, or forgotten; and praise or
blame when delivered from the Chair is apt to lose in
sincerity as much as it gains in authority. Sedgwick—
always honest and straight-forward—avoids these defects as
far as the nature of the case permits. He passes in review
what had been accomplished during the previous year; the
papers by Lyell and Murchison; by Murchison and himself;
by Dr Fitton, and others; and these works lead him to
discuss the action of river-currents, with his own views there-
on; the true sub-divisions of the tertiary strata; the importance of the study of organic remains, etc.; so that the speech
is lifted above the accidents of the moment, and possesses
permanent value. These diverse subjects are not only clearly
and eloquently treated, but what in other hands would have
been a dry discussion is enlivened with graphic similes—
humorous touches—inspiriting appeals to unwearied labour,
especially in the field of English Geology—and hints at the
true method of correlating facts, and establishing a correct
induction from them—in a manner well worthy of the
"mathematical geologist." The concluding paragraphs are

1830.
Æt. 45.
devoted to a piece of criticism as severe as anything that Sedgwick ever penned. A member of the Society, Andrew Ure, M.D. had lately published *A New System of Geology*, "in which the great revolutions of the earth and of animated nature are reconciled at once to modern science and to sacred history." This "monument of folly," as Sedgwick calls it, is pulled to pieces without mercy, and some of its worst blunders exposed. Into these we need not enter—indeed the subject would not have been alluded to at all had it not given occasion to a passage so noble, and of such general application, that we cannot resist the pleasure of quoting it.

Laws for the government of intellectual beings, and laws by which material things are held together, have not one common element to connect them. And to seek for the exposition of the phenomena of the natural world among the records of the moral destinies of mankind, would be as unwise, as to look for rules of moral government among the laws of chemical combination. From the unnatural union of things so utterly incongruous, there has from time to time sprung up in this country a deformed progeny of heretical and fantastical conclusions, by which sober philosophy has been put to open shame, and sometimes even the charities of life have been exposed to violation.

No opinion can be heretical but that which is not true. Conflicting falsehoods we can comprehend, but truths can never war against each other. I affirm, therefore, that we have nothing to fear from the results of our inquiries, provided they be followed in the laborious but secure road of honest induction. In this way we may rest assured that we shall never arrive at conclusions opposed to any truth, either physical or moral, from whatsoever source that truth may be derived: nay rather, (as in all truth there is a common essence) that new discoveries will ever lend support and illustration to things which are already known, by giving us a larger insight into the universal harmonies of nature[1].

The first half of 1830 was fully occupied with the preparation of the above address (delivered 19 February)—with journeys to London on the business of the Society, and other matters which his recognised position as one of the first of English geologists imposed upon him—as, for instance, examination before a Committee of the House of Commons on the Coal Measures, and attendance at "a Committee appointed

[1] *Address*, p. 23.

to direct a Survey of the Thames[1]." In the midst of all this he found time to congratulate his friend Dean Monk on being promoted from the Deanery of Peterborough to the Bishopric of Gloucester. "We all rejoice at this event," he wrote, "from feelings of personal regard, founded in a long experience both of your unwearied kindness and of your great services rendered to ourselves and our Society while you were one of our resident members: we all rejoice on public grounds, for we see in yourself an instance of honourable distinction, founded, as it ought to be, not on party interest, but on high literary claims[2]."

1830.
Æt. 45.

May brought the usual preparations for the Woodwardian audit. This year he had plenty of additions to record: "specimens from the extinct volcanoes near Bonn;" from nearly every district visited in 1829, including "a very fine series of organic remains from Solenhofen," and several geological maps purchased at Berlin. As usual he laments "that want of room prevents him from having the pleasure of exhibiting many of these additions," but "hopes that before long he shall have an opportunity of unpacking them, and of arranging them in a Museum of the University[3]"—a sanguine aspiration which was not realised for more than eleven years after this report was written.

Murchison had started early in the summer to "riddle the Alps in all directions," as he said; and he kept Sedgwick constantly informed of all that he was doing—of the help he obtained from continental geologists, and of the new facts he was ascertaining in support of their common position. His letters, interesting and valuable as they are, are too technical for a biography; and indeed, would be hardly intelligible without the maps and sections to which they refer. Of Sedgwick's letters to him one only has been preserved— perhaps only one was written. It is occupied chiefly with a

[1] To R. I. Murchison, May, 1830.
[2] To Dean Monk, 14 February, 1830.
[3] *Report to the Woodwardian Auditors*, 1 May, 1830.

1830.
Æt. 45.

very severe criticism on the abstract of their joint papers which Murchison had sent to the *Annals of Philosophy*, and which Sedgwick had fortunately intercepted before it was printed off. Murchison calls the letter "very cross," and declares that Sedgwick must have been "in a mathematically exact humour" when he wrote it; but he concedes all that was wanted when he speaks of the paper in question as "my very careless abstract[1]." This matter despatched, Sedgwick proceeds to tell him that he had stayed in London for some time, "being in daily expectation of the commencement of a canvass for a new contest in the University;" that "fortunately the storm had blown over," and that he was then "in all the press and confusion of a man who is going to desert his quarters. Tomorrow morning I am off on my way to Northumberland. I shall halt at Newcastle, and try to lay in a stock of information; and then, as soon as possible, bear down upon the Cheviots. If I knew how to hit you I would send a fly-leaf after you when I had properly pounded the porphyries. I am delighted with your account of the Low Countries; you have done some excellent work there....Lyell has been off about three weeks, and has not been since heard of. His book[2] has a hard delivery; it is not yet out. This is very vexatious, as I wanted to take it with me. The King's death; speeches in the House of Commons; the fear of a contested election; these have been the chief topics.... Give my best regards to Mrs Murchison. How does she bear the fatigues of your campaign[3]?"

A letter to Mrs Murchison (21 October) records a few particulars of the summer's work, which had proved somewhat disappointing: "Among the Cheviot hills I worked hard for about ten days; and I did some good work in a small way on the Scotch borders. Soon after I was attacked with indisposition; and in a great measure driven off the

[1] From R. I. Murchison, Ischl, 15 August, 1830.
[2] The first volume of his *Principles of Geology*.
[3] To R. I. Murchison, Cambridge, 15 July, 1830.

field. After making one or two vain efforts among the
Cumberland mountains I finally took shelter in my native
valley." There he was detained by a long spell of bad
weather: "all the powers of the air," he says, "were in league
against me." He was therefore obliged to give up a projected
excursion to North Wales and Ireland, and content himself
with Ingleborough. In that district he "ransacked the hills
from the ridge of Stainmoor to the heart of Craven." By the
end of October, as usual, he was back at Cambridge.

Towards the close of this year certain Fellows of the Royal
Society let it be known that, in their opinion, His Royal
Highness the Duke of Sussex was a fit and proper person to
be President; that he was willing, not to say anxious, to accept
the office; and that he did so with His Majesty's approval.
Thereupon the actual President, Mr Davies Gilbert, retired.
The scientific section of the Fellows, indignant at what they
regarded as an interference with their independence on the part
of the Court, persuaded Mr Herschel to allow himself to be
nominated in opposition to the Duke. Murchison took an
active part in getting up the requisition to Herschel, which was
signed by Sedgwick and most of his Cambridge friends. He
had been for some years anxious to see Herschel President[1],
but, being out of London, did not work himself up to the
white heat of anger that seems to have been there prevalent.
At the same time he did what he could to stop the Duke's
pretensions in his usual straightforward fashion. He wrote
to Murchison (21 November): "I *did sign* a paper requesting
Herschel to come forward. What this *new paper* is I don't
exactly know, and I don't intend to take any more public

[1] In 1827, when he had heard of a suggestion to elect Sir R. Peel, he wrote to
Murchison (25 November): "The republic of science will indeed be degraded if
the Council of the Royal Society is to become a mere political junta, and we are
to sit under a man who *condescends* to be our patron. The Institute of France
have not yet learned to degrade themselves by placing in the chair a man who has
no other recommendation than that of having been a King's Minister. Why don't
some of you propose Herschel? He is by far the first man of science in London,
and would do the work admirably." Sedgwick became F.R.S. in 1820.

steps. But I intend to take a private step, and a very strong one. By this post I shall write to the Duke of Sussex, and explain my views to him very plainly. I shall then have liberated my conscience. In case of a contest I cannot come up. Whewell and some others will come up—Coddington, Willis, &c. Whewell wishes me to impress upon you the very great importance of inducing Herschel to come forward as a candidate. Many men will *tail off* if they have an excuse, and Herschel's unwillingness is a good apology for weak minds. Indeed it is no bad apology for any one."

At that time the Duke was a frequent visitor to Cambridge, where he greatly enjoyed the hospitalities profusely laid at his feet by the Fellows of Trinity College, two of whom, Sedgwick and G. A. Browne, had been appointed his chaplains[1]. On these occasions he would breakfast with one, dine with another, sup with a third, and in general behave as if he were one of themselves. He was evidently not displeased with Sedgwick's boldness, and wrote in reply: " I thank you for your candour, and whether our opinions may differ on this or any other subject, I know how to respect the talents as well as the motives of any individual"—but he persisted in his candidature, and was elected by 119 votes to 111.

At the beginning of January, 1831, Sedgwick read to the Geological Society his first paper *On the General Structure of the Lake Mountains of the North of England*, the materials for which had been collected several years before. For the present, we merely note the fact, reserving for a subsequent chapter an account of his views on the geology of Lakeland.

In the following month he concluded his two years of office as President of the Geological Society; but, before retiring from the chair (18 February, 1831), he had a singularly agreeable duty to perform. Three years before, Thomas Hyde Wollaston, M.D. had transferred one thousand pounds to the Society, the dividends on which were to be applied after his death, " in promoting researches concerning the

[1] Sedgwick's patent is dated 10 May, 1819.

mineral structure of the earth, or in rewarding those by whom
such researches may hereafter be made." Dr Wollaston died
22 December, 1828, just a fortnight after executing the above
transfer. The first year's income was devoted to the purchase
of a die, designed by Chantrey, "bearing the impress of the
head of Dr Wollaston"; in order that a commemorative gold
medal, value ten guineas, might form part of the annual
donation; and it was not until early in 1831 that the Council
met to decide upon the first award. They resolved unanimously that the medal should "be given to Mr William Smith,
in consideration of his being a great original discoverer in
English Geology; and especially for his having been the first,
in this country, to discover and to teach the identification
of strata, and to determine their succession, by means of
their imbedded fossils."

The conscience of any other President than Sedgwick
would probably have been satisfied by declaring this award
in a few well-selected phrases at the begining or the end of
his own address; for, he might well have argued, the merits
of "Strata Smith" are by this time fully recognised by
geologists. But Sedgwick was too generous, too warm-
hearted, to adopt so selfish a course. "I for one," he said
"can speak with gratitude of the practical lessons I have
received from Mr Smith: it was by tracking his footsteps,
with his maps in my hand, through Wiltshire and the
neighbouring counties, where he had trodden nearly thirty
years before, that I first learnt the subdivisions of our oolitic
series, and apprehended the meaning of those arbitrary and
somewhat uncouth terms, which we derive from him as our
master, and which have long become engrafted into the
conventional language of English Geologists." He determined
to publish to the world, with all the authority of the position
he then held, the wonderful story of that humble land-
surveyor who, in the course of his professional work, had
discovered the key, if we may so speak, to the geological
cipher. The result is a rapid, but singularly clear and

1831.
Æt. 46.

appreciative sketch of a remarkable life, interspersed with some nobly eloquent passages, the effect of which, as originally delivered, with all the force of Sedgwick's energy and enthusiasm, may be measured by that which they still produce upon a reader. Towards the close of his speech he demanded from his hearers their approbation of the Council's award : " I would appeal " he said " to those intelligent men who form the strength and ornament of this Society, whether there was any place for doubt or hesitation ? whether we were not compelled, by every motive which the judgment can approve, and the heart can sanction, to perform this act of filial duty, before we thought of the claims of any other man, and to place our first honours on the brow of the Father of English Geology.

" If in the pride of our present strength, we were disposed to forget our origin, our very speech would bewray us ; for we use the language which he taught in the infancy of our science. If we, by our united efforts, are chiseling the ornaments, and slowly raising up the pinnacles of one of the temples of Nature, it was he who gave the plan, and laid the foundations, and erected a portion of the solid walls, by the unassisted labour of his hands[1]."

In the evening of the same day Sedgwick delivered his own address as President. As in that of the previous year, he reviews the progress of stratigraphical geology, with even more than his former felicity of treatment, and clearness of exposition ; while the publication of Herschel's paper *On the Astronomical Causes which may influence Geological Phænomena*, of the first volume of Lyell's *Principles of Geology*, and of the papers contributed by M. Elie de Beaumont to the *Annales des Sciences Naturelles*, gave him an opportunity of dealing with the fundamental theories of the science.

Of Lyell's work he spoke with genuine admiration. " Were I to tell him," he said, " of the instruction I received

[1] *Proceedings of the Geological Society*, i. 110, 270—279. *Memoirs of William Smith, LL.D.*, by John Phillips, 8vo. Lond. 1844.

from every chapter of his work, and of the delight with which
I rose from the perusal of the whole, I might seem to flatter
rather than to speak the language of sober criticism;" but,
when the criticism came, it struck at the very foundation of
the author's theory. He "could not but regret" that Lyell
seemed to stand forward as "the champion of a great leading
doctrine of the Huttonian hypothesis", and that "in the
language of an advocate he sometimes forgets the character
of an historian." Sedgwick had not time to deal with the
whole of even the single volume then published; but
addressed himself, in the main, to the theoretical portion.
The following paragraphs are, we think, of especial value,
as the truth of the doctrine of Uniformity has again been
called in question.

According to the principles of Mr Lyell, the physical operations
now going on, are not only the type, but the measure of intensity, of
the physical powers acting on the earth at all anterior periods : and
all we now see around us is only the last link in the great chain of
phænomena, arising out of a uniform causation, of which we can
trace no beginning, and of which we can see no prospect of the end.
And in all this, there is much that is beautiful and true. For we all
allow, that the primary laws of nature are immutable—that all we
now see is subordinate to those immutable laws—and that we can
only judge of effects which are past, by the effects we behold in
progress....But to assume that the secondary combinations arising
out of the primary laws of matter, have been the same in all periods
of the earth, is an unwarrantable hypothesis with no *a priori*
probability, and only to be maintained by an appeal to geological
phænomena.

If the principles I am combating be true, the earth's surface
ought to present an indefinite succession of similar phænomena.
But as far as I have consulted the book of nature, I would invert
the negative in this proposition, and affirm, that the earth's surface
presents a definite succession of dissimilar phænomena. If this be
true, and we are all agreed that it is; and if it be also true, that
we know nothing of second causes, but by the effects they have
produced; then "the undeviating uniformity of secondary causes",
the "uniform order of physical events", "the invariable constancy
in the order of nature", and other phrases of like kind, are to
me, as far as regards the phænomena of geology, words almost
without meaning. They may serve to enunciate the propositions of an hypothesis; but they do not describe the true order
of nature.

1831.   We are not surprised that Lyell should have written:
Æt. 46. "Sedgwick's attack is the severest, and I shall put forth my
strength against him in [1]" the second volume.

But, cautious as Sedgwick was in expressing agreement
with Lyell's theory, he accepted, almost without hesitation,
the startling views of M. Elie de Beaumont on the elevation
of mountain-chains, and exhausted the vocabulary of praise
on his "noble generalisations," "admirable researches", and
so forth—because "his conclusions are not based upon any
a priori reasoning, but on the evidence of facts; and also,
because, in part, they are in accordance with my own
observations." It is only just, however, to mention, that
Sedgwick warned his hearers that even these generalisations
had been "already pushed too far."

This careful study of M. de Beaumont's work led Sedgwick
to one important conclusion, which has by no means lost its
interest at the present time, namely: "that the vast masses
of diluvial gravel, scattered almost over the surface of the
earth, do not belong to one violent and transitory period."
And then he had the courage to proceed as follows:

It was indeed a most unwarranted conclusion, when we assumed
the contemporaneity of all the superficial gravel on the earth. We
saw the clearest traces of diluvial action, and we had, in our sacred
histories, the record of a general deluge. On this double testimony
it was, that we gave a unity to a vast succession of phænomena, not
one of which we perfectly comprehended, and under the name
diluvium, classed them all together.
To seek the light of physical truth by reasoning of this kind, is,
in the language of Bacon, to seek the living among the dead, and
will ever end in erroneous induction. Our errors were, however,
natural, and after the same kind which led many excellent observers
of a former century to refer all the secondary formations of geology
to the Noachian deluge. Having been myself a believer, and, to
the best of my power, a propagator of what I now regard as a
philosophic heresy, and having more than once been quoted for
opinions I do not now maintain, I think it right, as one of my last
acts before I quit this Chair, thus publicly to read my recantation.
We ought, indeed, to have paused before we first adopted the
diluvian theory, and referred all our old superficial gravel to the

---

[1] *Life of Sir Charles Lyell*, i. 318.

action of the Mosaic flood. For of man, and the works of his 1831. hands, we have not yet found a single trace among the remnants of Æt. 46. a former world entombed in these ancient deposits. In classing together distant unknown formations under one name; in giving them a simultaneous origin, and in determining their date, not by the organic remains we had discovered, but by those we expected hypothetically hereafter to discover, in them; we have given one more example of the passion with which the mind fastens upon general conclusions, and of the readiness with which it leaves the consideration of unconnected truths.[1]

It is strange that so cautious a writer as Sedgwick should have written the last paragraph, even in 1831. But he lived long enough to make a second recantation, and to admit that Man had appeared upon the earth at a period long anterior to that for which he had previously contended.

---

[1] An interesting account of the way in which Sedgwick was led to change his views on diluvium occurs in a letter to Murchison, dated 17 November, 1831. "If I have been converted in part from the diluvian theory (which by the way I never held to the same extent with Buckland, as you may see if you read the last page of the only paper I ever wrote on the subject) it was...by my own gradual improved experience, and by communicating with those about me. Perhaps I may date my change of mind (at least in part) from our journey in the Highlands, where there are so many indications of *local diluvial* operations....Humboldt ridiculed [the doctrine] beyond measure when I met him in Paris. Prévost lectured against it."

CHAPTER IX.

(1831—1834).

THE REFORM BILL. CONTESTED ELECTION FOR UNIVERSITY.
GEOLOGICAL PAPERS. TOUR IN WALES WITH CHARLES DARWIN
(1831). DECLINES THE LIVING OF EAST FARLEIGH. MRS
SOMERVILLE'S VISIT TO CAMBRIDGE. BRITISH ASSOCIATION
AT OXFORD. SUMMER AND AUTUMN IN WALES. PRESIDENT
OF CAMBRIDGE PHILOSOPHICAL SOCIETY. DISCOURSE ON THE
STUDIES OF THE UNIVERSITY (1832). BRITISH ASSOCIATION
AT CAMBRIDGE. THE BEVERLEY CONTROVERSY (1833). DIS-
LOCATES RIGHT WRIST. PETITION AGAINST TESTS. BRITISH
ASSOCIATION AT EDINBURGH. MADE PREBENDARY OF NORWICH
(1834).

WHEN Sedgwick ceased to be President of the Geological
Society, and, as he put it, "exchanged dignity for liberty,"
he probably looked forward to a spell of leisure, during which
he might work at geology without interruption. Such hopes,
however, if he ever entertained them seriously, were doomed to
disappointment; and we shall find that during 1831 and the
three succeeding years he was more than ever absorbed in
University and College occupations.

In less than a fortnight after the annual general meeting of
the Geological Society at which Sedgwick delivered his
farewell address, Lord John Russell introduced the Reform
Bill into the House of Commons. A man of letters or of
science, whichever side he took in the controversy which

thenceforth divided England, might well have exclaimed,
' O now for ever farewell the tranquil mind, farewell content';
and therefore it need not surprise us that Sedgwick should
complain shortly afterwards: "I am sadly out of sorts, and
involved in politics, which are dividing old friends, and
playing the devil amongst us[1]." More than half a century has
elapsed since these words were written, and few are left in
Cambridge who can remember the state of feeling which
justified them. But there are still some who can recall
the after-effects, and can truthfully describe the Reform Bill
as the nightmare of their childhood. It appeared to them,
from the way in which they heard their elders speak of it, to
have been a maleficent influence—an embodiment of the
spirit of evil—which had brought discord into a peaceful
society, and had left to the next generation an inheritance
of sundered friendships and bitter feuds. When a child
inquired, "Who is that?" he was not unfrequently answered,
"That is Mr So and So; we used to be very intimate before
the Reform Bill, but we never speak now." But nothing of
this sort could be said about Sedgwick. He was an ardent
reformer; but his high personal character, his great popularity,
and his uniform kindliness and good humour towards those
who differed from him, enabled him to pass through the ordeal
unscathed. It has been often said that he and Mr Pryme
were the only liberals in the University who took a prominent
part in favour of reform, and yet neither made an enemy nor
lost a friend.

Soon after the introduction of the Reform Bill, the
University sent a petition to the House of Commons against
it. The promoters were evidently by no means sure of their
ground, for a contemporary fly-sheet records that the petition
"was carried through with a haste extremely indecorous and
reprehensible; the notice to the Senate having been barely of
the legal extent, and given on a day on which the post did not

---

[1] To R. I. Murchison, 24 March, 1831.

leave Cambridge; the previous meetings of the Heads having
been held with the shortest possible notice; some of their
deliberations having taken place on a Sunday; and the petition
having been agreed to by them in the two hours preceding
the congregation at which it was offered at the Senate[1]."
Under these circumstances it was of course impossible to give
notice to non-residents; and the grace to affix the University
Seal to the document, though opposed, was carried in each
of the houses into which the Senate was then divided.

The petition having been carried, Sedgwick tells us that
he and his friends are setting to work "to remedy the evil
of it as well as we can, by getting a *Declaration* up in favour
of the sitting candidates[2]." This *Declaration* was supported
by some of "the original promoters of the recent petition
to the House of Commons against certain provisions in
the Reform Bill;" and was probably suggested by the
knowledge that the success of the petition would be followed
by an attempt to unseat Lord Palmerston and Mr Cavendish,
in the event of a dissolution of Parliament[3]. The *Declaration*
is dated 23 March, and is signed by thirty-six members
of the Senate, among whom, in addition to Sedgwick,
are Smyth, Cumming, Whewell, Henslow, Airy, Worsley,
J. C. Hare, and Thirlwall. It was promptly succeeded (28
March) by a counter-declaration, the signatories to which,
forty-two in number, bound themselves "to promote the
return of two representatives entertaining more moderate

---

[1] *Reasons for regretting the University Petition of March* 21. This fly-sheet
is dated "Trin. Coll. March 23, 1831," and is believed to have been written
by the Rev. W. Whewell. Notice was given on Saturday, 19 March, for a Congregation to be held on Monday, 21 March, at 11 A.M. A letter from Cambridge
printed in *The Times* of 23 March says: "If two days' notice had been given, we
should have had twenty or thirty Masters of Arts from London, who would have
thrown the petition out, as we did in the case of the Catholic Petition [in 1829]."
See above, p. 336.

[2] To R. I. Murchison, 24 March, 1831.

[3] Sedgwick says, in the last-quoted letter: "The petition will, I fear, create a
contest at Cambridge, in case of a dissolution, and such an event would be almost
the death of me."

views than those of the present representatives of the Uni- 1831.
versity upon the vital question of Parliamentary Reform." Æt. 46.
It became evident, therefore, that, whenever a dissolution
should occur, there would be a contest, and probably a
severe one, for the honour of representing the University
in Parliament.

When the dissolution took place (22 April) the conservatives brought forward Mr Goulburn and Mr William
Yates Peel, who declared themselves, in the most unqualified
language, opposed to the Bill, though not averse to the
consideration of reform in general. Neither said a word on
any subject except reform—an omission of which Sedgwick
was not slow to take advantage. He printed a short address
(2 May), signed, like his letter to Mr Goulburn, *A Resident
Member of the Senate.* After pointing out that the two
candidates had rested their claims to election solely on
their opposition to the Reform Bill, he proceeds:

And why, may I ask, have they reduced the question within
these narrow limits? because that they are well aware they have
nothing else to produce as a recommendation to any part of the
University, except their pledges in opposition to the Bill brought
forward by Ministers.

How far these pledges are likely to be fulfilled, I will not now
stop to inquire; but I wish to put this question to every Member of
the Senate, "whether it is consistent with the dignity of the University to select two Members to represent them in Parliament,
on account of the Vote which they may give upon a particular
question, without considering their general qualifications for defining
the interests, and maintaining the character, of this Learned Body?"

With regard to the measure of Reform, the result of the
Elections which have already taken place, must convince every
impartial mind that its success is no longer doubtful. The ground,
therefore, upon which the new Candidates lay claim to your favour,
is fast crumbling beneath their feet, and when the Reform Bill has
passed, I defy the most zealous Anti-Reformer to point out a single
advantage which the University can derive from this change in its
Representatives.

This appeal to what might happen in the future met, we
may well suppose, with but little consideration in those stormy
times. A large majority of the electors were satisfied with

the knowledge that Lord Palmerston and Mr Cavendish were both reformers. The former, as Secretary of State for Foreign Affairs, was a member of the Government, the latter had voted for the Bill. The non-residents flocked to the Poll, and at its close Mr Goulburn had polled 805 votes, Mr Peel 804, Mr Cavendish 630, and Lord Palmerston 610[1]. When the election was over, Sedgwick wrote : "I was extremely fatigued with last week's work, and mortified at the result more than I can find words to express; and it certainly does not take away from the painful feelings when I reflect that the defeat was courted by the vacillation of our own party[2]."

Amidst this excitement Sedgwick found time to write a second paper in continuation of his work on the Cumbrian Mountains—a *Description of a Series of Longitudinal and Transverse Sections through a Portion of the Carboniferous Chain beween Penigent and Kirkby Stephen*, read to the Geological Society in March, 1831. With this—and his usual share in the Scholarship Examination at Trinity College, he was fully occupied till the middle of April, when he set to work on the revision of the paper on the Eastern Alps, which, as previously explained, had now to be recast, and almost re-written. He had hoped to get this work rapidly off his hands, and to be ready to start for Wales at the beginning of the Long Vacation. But, when June came, he was out of sorts, his time had been cut up by a succession of visitors, and, as he went on with his work, he found it increase, rather than diminish.

Early in June Murchison started to commence those investigations into the then little known Transition Rocks, as they were called, which ended in the publication of *The Silurian System*. Sedgwick had been invited to accompany

---

[1] The poll was taken 3—6 May, 1831. An analysis at the end of the Poll-book shews that the number of voters was 1450—a strong proof of the excitement of the hour—for it exceeded by 157 the largest number polled on any previous occasion.

[2] To Mrs Murchison, 11 May, 1831.

him, but his plans were made, and he declined to alter them. As events turned out it was an unfortunate decision. Had he said "Yes" instead of "No," how different the future of the two men might have been! They would have commenced their exploration of Wales from the same point, instead of from opposite sides of the principality; they would have worked out the proper sequence of the rocks together instead of separately; Siluria might never have had an existence independent of Cambria; and no misunderstanding need ever have arisen between the two explorers.

The next letter describes Sedgwick's difficulties with the paper, and at the commencement of his survey of North Wales.

LLANLLYFNI, near CAERNARVON,
September 13*th*, 1831.

Dear Murchison,

Had the elements been more favourable you might have waited long for an answer. But I was driven by stress of weather to Caernarvon on Saturday, and on Sunday morning found a packet of letters waiting for me, and yours among the rest. Yesterday was fine and I did some work here; today everything is wrapped in mist, and the rain is falling in buckets. I did not get from Cambridge before the 1st of August. The Alpine paper was infinitely more troublesome to reduce than I expected. The fossils, I took for granted, would fall into their right places, and, as their determination was not a part of my labour, I hoped simply to have the trouble of writing out the lists. You may therefore judge of my vexation when I tell you that I was stopped by the fossils over and over again; that I had two journeys up to London, and that Lonsdale had one down to Cambridge arising out of them. This was not as it ought to have been. The lists ought to have been settled for better for worse sooner. And, after all, the result is far from satisfactory. I last year actually bullied Boué about the lowest strata in Styria, assuming that the fossils

were those of the London clay. Now it turns out, after the final
revision, that in the lowest Styrian clays there is not a single
London Clay fossil ascertained. There is one with a query,
and that is all : in fact, I do not believe now that the London
Clay is found in Styria. But enough of this.

I spent one day at Dudley and two days at Shrewsbury,
and finally entered North Wales on the 5th August. As the
Prince of the Air would have it, I was almost drowned in a
thunderstorm the very morning I commenced my labours.
As the greywacké hills continued in cloud I crossed to the
vale of the Clwyd, hoping at least to do some work among
the secondaries[1]. It would have delighted me to have
attacked the Mold district, but I knew that I had no time,
so I confined myself to the vale. The day following was
beautiful, and I worked my way down to Denbigh. Next
day I made a traverse, and descended to St Asaph, thence
in my gig to Conway. The Old Red all round by Orm
Head &c. &c. is a pure fiction. At least I can't see a trace
of it. There is not a particle of it between Denbigh and
the Isle of Anglesea. There are, however, some red beds
(which may pass for Old Red for want of better) in a ravine
west of Ruthin, and in one or two places near Llangollen
under the Mountain Limestone escarpment. The band of
limestone on the east side of the vale of Clwyd is not, I
believe, continuous—there are, if I mistake not, several
interruptions in it. I spent some days in the Isle of Anglesea
in the hopes of learning my lesson for Snowdonia. Henslow's
paper[2] is excellent, but the lesson is worth next to nothing;
for Anglesea is almost as distinct in structure from Snowdonia,
as if they had been separated by the Atlantic sea rather than
the straits of Menai. I have now been at real hard work,
cracking the rocks of Caernarvonshire for rather more than
three weeks, and can report progress. My health is much
better, but I am liable to rheumatic attacks at night, after

---

[1] This term includes the carboniferous rocks, as was usual at that time.
[2] See above, p. 234.

the fatigues of the day, and truly fatiguing work it is to climb these mountains ; but nothing can be done without. Already I have been upon all the most elevated summits in this county. This is pretty well considering the many interruptions from mist and rain. If my health continue, and my limbs are not jostled out of their sockets, I shall remain here as long as the weather will let me, or at least till I have finished this county. I can then quit the country with a good conscience, and if I live till next year, can come back to the Principality in good hope of finishing my work in another summer.

Under these circumstances York is quite out of the question. I should be a traitor to quit my post, now that I am keeping watch among the mountains. It would be very delightful to meet the philosophers, and commence deipnosophist, but it would be very bad philosophy in the long run. You may tell Mr Vernon[1] that keeping away is a great act of self-denial on my part, and that I am in fact doing their work by staying away. I shall rejoice to meet you and Mrs M. at Cambridge on your return ; you may then tell me all about it. If you write, address me still at Caernarvon. I consider it my head quarters, though I may not be there again for a fortnight or three weeks....I have no room for a Snowdonian transverse section. The structure is on the whole regular, and the *strike* longitudinal ; I have nearly completed one *base line* to work upon; the rest must be done by traverses. My best regards to Mrs M.

Yours ever, A. SEDGWICK.

For two or three weeks, at the commencement of this tour, Sedgwick was accompanied by Charles Darwin, then a young man of twenty-two. It is provoking that neither should have written down his impressions of the other at the time ; for

----

[1] The Rev. William Vernon (afterwards Vernon Harcourt) third son of the Archbishop of York, zealously promoted the first meeting of the Association. Geikie's *Life of Murchison*, i. 185.

it is evident that from this time forward Sedgwick took a keen interest in him. In 1835, while Darwin was absent on board *The Beagle*, Sedgwick wrote to Dr Butler of Shrewsbury: "His [Dr Darwin's] son is doing admirable work in South America, and has already sent home a collection above all price. It was the best thing in the world for him that he went out on the voyage of discovery. There was some risk of his turning out an idle man, but his character will be now fixed, and if God spares his life he will have a great name among the Naturalists of Europe[1]." In after life, though they differed widely, Sedgwick always spoke of his geological pupil, as he may be termed, with cordiality and kindness; and Darwin, replying to a note received from Sedgwick not very long before his death, could write: "I am pleased that you remember my attending you in your excursions in 1831. To me, it was a memorable event in my life: I felt it a great honour, and it stimulated me to work, and made me appreciate the noble science of geology." In 1875, in answer to an inquiry from Professor Hughes, Darwin wrote down all he could remember about the tour of 1831.

DOWN, BECKENHAM, KENT,
May 24, 1875.

My dear Sir,

　　　I understand from my son that you wish to hear about my short geological tour with Professor Sedgwick in North Wales during the summer of 1831; but it is so long ago that I can tell you very little.

　　As I desired to learn something about Geology, Professor Henslow asked Sedgwick to allow me to accompany him on his tour, and he assented to this in the readiest and kindest manner. He came to my father's house at Shrewsbury, and I remember how spirited and amusing his conversation was during the whole evening; but he talked so much about his health and uncomfortable feelings that my father, who was a doctor, thought that he was a confirmed hypochondriac.

　　We started next morning, and after a day or two he sent me across the country in a line parallel to his course, telling me to collect specimens of the rocks, and to note the stratification. In

[1] To Dr S. Butler, 7 November, 1835.

the evening he discussed what I had seen; and this of course now suspect that it was done merely for the sake of teaching me, and not for anything of value which I could have told him. I remember one little incident. We left Conway early in the morning, and for the first two or three miles of our walk he was gloomy, and hardly spoke a word. He then suddenly burst forth: "I know that the d—d fellow never gave her the sixpence. I'll go back at once;" and turned round to return to Conway. I was amazed, for I never heard before, or since, anything like an oath from him. On inquiry I found that he was convinced that the waiter had not given to the chambermaid the sixpence which he had left for her. He had no reason whatever, excepting that he thought the waiter 'an ill-looking fellow.' On my hinting that he could hardly accuse a man of theft on such grounds, he consented to proceed, but for some time he grumbled and growled. At last his brow cleared, and we had a delightful day, and he was as energetic as on all former occasions in climbing the mountains. We spent nearly a whole day in Cwm Idwal examining the rocks carefully, as he was very desirous to find fossils.

I have often thought of this day as a good instance of how easy it is for any one to overlook new phenomena, however conspicuous they may be. The valley is glaciated in the plainest manner, the rocks being mammillated, deeply scored, with many perched boulders, and well-defined moraines; yet none of these phenomena were observed by Professor Sedgwick, nor of course by me. Nevertheless they are so plain, that, as I saw in 1842, the presence of a glacier filling the valley would have rendered the evidence less distinct[1].

Shortly afterwards I left Professor Sedgwick, and struck across the country in another direction, and reported by letter what I saw. In his answer he discussed my ignorant remarks in his usual generous and frank manner. I am sorry to say that I can tell you nothing more about our little tour.

I find that I have kept only one letter from Professor Sedgwick, which he wrote after receiving a copy of my *Origin of Species*[2]. His judgement naturally does not seem to me quite a fair one, but I think that the letter is characteristic of the man, and you are at liberty to publish it if you should so desire.

Believe me, my dear Sir,

Yours sincerely,

CHARLES DARWIN.

[1] These phenomena are described in a paper by Darwin in the *Philosophical Magazine* for 1842, xxi. 180. See *Life and Letters of Charles Darwin*, i. 57.

[2] This letter, written in 1859, will appear when we come to speak of Sedgwick's attitude towards Darwin's great work. It has already been printed in Darwin's *Life*, ii. 247.

Sedgwick got home sooner than he had intended, driven out of Wales by bad weather. His letter to Murchison, announcing his return to Cambridge, indicates a certain amount of disappointment.

TRIN. COLL. *October* 20, 1831.

" I came here the night before last. The weather became so bad that I was driven out of Caernarvonshire before I had quite finished my work; but, God willing, I hope to be in North Wales next year before the expiration of the first week in May, and with five months before me I shall perhaps be able to see my way through the greater part of the Principality. If I live to finish the survey I shall then have terminated my seventh or eighth summer devoted exclusively to the details of the old crusty rocks of the primary system[1]. What a horrible fraction of a geological life sacrificed to the most toilsome and irksome investigations belonging to our science! When I finished Cumberland I hoped some one else would have done North Wales—but I have been disappointed. *N'importe.* I am now in for it, and must go on!

Many thanks for your account of the York meeting. I suppose I must enrol myself one of your body corporate, though I shall certainly not be able to attend the next meeting at Oxford. But we will talk of it when we meet."

Before the Michaelmas Term was over it fell to Sedgwick's lot to urge the claims of his friend W. H. Mill to be chosen the first Boden Professor of Sanscrit at Oxford. Mill had written to him from Calcutta expressing his anxiety to obtain an office which, while it took him away from a noxious climate, would enable him to "prosecute the Indian studies I most like in an academic retirement which, though not the abode of my mother, is that of her venerable sister"; and in a style not unlike Sedgwick's own, proceeded to claim his good offices:

---

[1] He told Lyell that their investigation was like "rubbing yourself against a grindstone." *Life of Sir C. Lyell,* i. 367.

1832.
Æt. 47.

"Now I do not suppose, my dear Professor, that your *ad eundem*[1]
degree gives you a vote in Convocation; nevertheless I do canvass
you, and implore you, if I have appeared to you one on whose
behalf such things may be done or attempted without unjust
partiality, and acceptation of persons such as Scripture and sound
reason do condemn—by the memory of old times, and our several
meetings and crossings in France, Switzerland, and Alsace—by the
canvassing scenes of 1818 and 1829, diverse though they be in many
material respects from this—that you will wisely bethink yourself
how you may befriend me in this affair. Perhaps your friends
Professor Buckland (to whom you introduced me at Somerset House),
and Mr Lyell, and other scientific men of the other University might
be induced to lend me their powerful aid by your mentioning my
name to them[2]".

Sedgwick did as Mill suggested, and at once wrote to
Professor Buckland, who entered warmly into his views. His
letter "transcribed in a fair hand"—a preliminary step which
was doubtless necessary if it was to have any effect upon the
University—was placed in the hands of the Vice-Chancellor,
together with the testimonials of the other candidates; and at
first Buckland seemed sanguine of success; but, when the day
of election came, an Oxonian was preferred to a member of
another University, and the Professorship was conferred upon
Horace Hayman Wilson.

In March 1832, Lord Chancellor Brougham invited Sedgwick to accept the living of East Farleigh in Kent. The
offer was probably prompted by personal feeling as much as
by the wish to reward a political supporter, for Brougham
had sought Sedgwick's cooperation in his schemes for the
diffusion of knowledge, and we believe that the correspondence had led to more than one interview. Sedgwick declined,
apparently without hesitation; a refusal which was much
deplored by several of his friends, and especially by Lyell,
who records in his diary (16 March) his own conviction that,
were Sedgwick to leave Cambridge and marry, "he would be
much happier, and would eventually do much more for

---

[1] Sedgwick, Peacock, Whewell, Airy, Henslow were all admitted *ad eundem*
gradum at Oxford, 17 June, 1830.
[2] From Dr Mill, 6 June, 1831.

geology[1]." Lyell states that it was Murchison who advised Sedgwick to decline; but, had this been the case, Sedgwick could hardly have written the following letter:

To R. I. Murchison, Esq.

"I returned this morning from the little living of Shudy Camps, from which I used to derive £40 or £50 *per annum*, but which now is worse than nothing. Under such circumstances you will think it strange that I have been mad enough to refuse the living about which you wrote, but so it is. I cannot accept it without my Professorship being vacant, and breaking off my work in the middle. I do not think this would be to my honour, or that it would add to my happiness. Many thanks for your kind note. I have written to Le Marchant[2], and also to the Lord Chancellor; to the former yesterday between services, to the latter to-day. I fear they will both set me down for an egregious fool[3]."

If Sedgwick was serious in thinking that the Chancellor would ridicule him for saying "No," the answer which he was not long in receiving must have caused him considerable pleasure.

My dear Sir,

    I read both your letters to the Chancellor yesterday. He more than once interrupted me to express his warm approbation. When I had done, I asked him what I was to say to you. "Say to him," the Chancellor answered, "all that is kind and respectful on my part"; and he then proceeded in very forcible terms to eulogise your disinterestedness. He also descanted upon your claims, and trusted something would turn up that would enable him to prove that they had not been overlooked. I have seldom heard him more warm in his commendation, and I only regret I cannot tell you all he said. I hope however that I have recollected enough to satisfy you that you have been dealing with a person capable of appreciating merit.

        Yours very truly,
        D. LE MARCHANT.

[1] *Life of Sir C. Lyell*, i. 374.   [2] The Lord Chancellor's principal secretary.
[3] This letter bears neither date nor postmark.

Soon afterwards Sedgwick had an interview with Lord
Brougham, and it was probably in consequence of what was
then said that he wrote with even more than his usual
vivacity to Ainger :

LONDON, *March* 17, 1832.

"What strange things are in the womb of Time!
Who would have thought that the Lord Chancellor would
ever offer me a living worth perhaps £1000 a year, and that
I should refuse it? But this very event has come to pass
within the last week. You may well think, as some of my
other friends have done, that I am raving mad. No matter.
The Chancellor thinks me in my sober senses, and has
promised me one of the very first stalls which is vacant—
which I can hold both with my Fellowship and Professorship,
and make my hammer ring more merrily than ever against
the rocks....

On Monday I return to Cambridge, where I have some
geological papers to finish, and about the first week in May
I hope to be off for North Wales, when I shall, I hope, be
following my vocation for five months. On my return I shall
begin to write a book with which I have been pregnant for
seven or eight years."

It was to be expected that the high spirits which
prompted this letter would not be maintained for long at the
same level. On the very next day a reaction had set in,
as Lyell tells us. His estimate of Sedgwick is severe, but
events proved him to have been right.

"Sedgwick asked me to walk home with him. I found a gloom
upon him, unusual and marked. I most carefully avoided all
allusion to the rejected living, but now, when the first excitement
of the declining the boon is over, and that others have expressed
their wonder at it, and that he finds himself left alone with his glory,
he is dejected. He told me, Thursday last, that he wished before
he left Cambridge to *do something*. 'Now if I take a living instead
of going to Wales, I abandon my Professorship, and cannot get out
the volume on the primary rocks with Conybeare,' etc. Then he
hinted that in a year, when this is done, he may retire on some

1832.
Æt. 47. living, and marry. But I know Sedgwick well enough to feel sure
that the work won't be done in a year, nor perhaps in two; and then
a living, etc. won't be just ready, and he is growing older.   He has
not the application necessary to make his splendid abilities tell in a
work.   Besides, every one leads him astray.   A man should have
some severity of character, and be able to refuse invitations, etc.
The fact is that to become great in science a man must be nearly as
devoted as a lawyer, and must have more than mere talent[1]."

On reviewing the whole case—with the help of what we
know of Sedgwick in the closing years of his life—we are
inclined to agree with Lyell, and to decide that he was wrong
in refusing Farleigh.   No doubt he would have regretted
Cambridge at first; but he had a happy capacity for accepting new surroundings and new occupations.   In a few months
the rector of Farleigh would have been as much at home
there as the Prebendary of Norwich became afterwards in
the Cathedral Close.   Nor would he have found the duties of a
parish clergyman—especially as those duties were understood
fifty years since—incompatible with the pursuit of geology.
His best geological friends, the brothers Conybeare, were
both beneficed clergymen, and we believe that they did not
neglect either parish or science.   Lastly, Sedgwick made a
fatal mistake, when he cut himself off, irrevocably, from
marriage; and that he deliberately chose a bachelor life is
evident from what he says about his readiness to accept a
stall which he could hold with his Professorship and Fellowship.   We are not aware that he ever owned to regrets for
Farleigh; but in the loneliness which is inseparable from old
age within the precincts of a college he not seldom dwelt
upon what might have been, had he been blessed with a wife
and children.

Sedgwick must have needed some distraction after the
excitement of such a decision as has just been recorded; and
he found it in the entertainment of the celebrated Mrs

---

[1] *Life of Sir C. Lyell*, i. 375.   The above extract is from Lyell's diary for
Tuesday, 20 March.   The passage immediately preceding the extract speaks of a
party at Murchison's on the previous Sunday, i.e. 18 March.

Somerville and her husband, whom he persuaded not only to spend a week with him, but to occupy rooms in College.

To Dr Somerville.

Tuesday morning,
[*3 April,* 1832].

My dear Sir,

Your letter delighted me, and I am sure you have decided wisely not to rusticate at the Observatory[1].

The time you have fixed is the best of all possible times, and I hope you will write as soon as possible to finally fix the hour of your arrival. I have a plan in my eye which I think quite excellent. Mr Sheepshanks' rooms, on my staircase[2], are now empty, and I believe he does not return into residence next week. In that case we will mount a regular matrimonial four-posted bed, and try to domesticate you and Mrs Somerville within the College walls. This experiment was tried and approved of by Mr and Mrs Murchison. The rooms in question are very good; have a dressing-room with a fireplace attached to them, and a small mathematical library in which Mrs S. may disport herself when she is tired of duller subjects. The day you arrive I can either give you a quiet dinner, or ask a few friends to meet you. I mention this alternative because Mrs Somerville may perhaps anticipate fatigue, and not wish to meet a party the first evening. Only express your wishes on this matter, and they shall be law. I shall write by this post to Sheepshanks, and if by any mischance I should be disappointed in my present plan I will secure rooms for you opposite Trin. Coll. Give my

---

[1] Then occupied by Geo. Biddell Airy, M.A., Plumian Professor of Astronomy 1828—1836. He and his accomplished wife were intimate friends of Dr and Mrs Somerville.

[2] At this time Sedgwick occupied the rooms in the Great Court, on the upper floor of the building between King Edward's Gate and the Master's Lodge, which he held till his death. He succeeded Professor Clark, who accepted a college living in 1825. Sedgwick's rooms as an undergraduate were between the Chapel and the Great Gate, as mentioned above (p. 78). We do not know where he lived between his degree, when he would probably change, and 1825.

1832.   kindest greetings to Mrs Somerville and your family, and
Æt. 47. believe me most truly yours,

A. SEDGWICK.

Dr Somerville replied that he and his wife had "no
habits in hours, food, or in any other circumstances. Dispose
of us as you list; we are ready to feed in seclusion, *petit
comité*, or in any party you like to form on Monday, meaning
not to be fatigued by the journey, which rather recruits
Mrs Somerville." To this communication Sedgwick promptly
replied :

To Dr Somerville.

TRIN. COLL. Thursday evening,
[5 *April*, 1832].

My dear Sir,

Your letter delighted me. I have ordered dinner on
Monday at half-past six, and shall have a small party to
welcome you and Mrs Somerville. In order that we may not
have to fight for you we have been entering on the best
arrangements we can think of. On Tuesday you will I hope
dine with Peacock—on Wednesday with Whewell—on Thursday at the Observatory. For Friday Dr Clark, our Professor
of Anatomy, puts in a claim. For the other days of your
visit we shall (D.V.), find ample employment. A four-posted
bed (a thing utterly out of our regular monastic system) will
rear its head for you and Madame in the chamber immediately
under my own; and your handmaid may safely rest her bones
in a small inner chamber. Should Sheepshanks return we
can stuff him into a lumber-room of the Observatory; but of
this there is no fear, as I have written to him on the subject,
and he has no immediate intention of returning. You will of
course drive to the great gate of Trinity College, and my
servant will be in waiting at the Porter's Lodge to shew you
the way to your academic residence. We have no Canons at
Trin. Coll. others (*sic*) we would fire a salute on your entry.

We will however give you the warmest greeting we can. 1832.
Meanwhile give my best regards to Mrs S. and believe me Æt. 47.
most truly yours, A. SEDGWICK.

The visit lasted for a week, and though Sedgwick unfortunately was "sadly out of sorts" during part of the
time, he evidently enjoyed himself thoroughly. Whewell,
whose health and spirits were always in good order, was quite
as enthusiastic in his commendation of the lady. Before she
came to Cambridge, he had probably known her only as the
authoress of *The Mechanism of the Heavens*, and was somewhat
surprised to find her not only accomplished in music, drawing,
and various languages, but "a very feminine, gentle, lively
person, with no kind of pretence to superiority in her
manners or conversation[1]." In consequence, she soon became a great favourite with the ladies of Cambridge, and a
tradition of her grace and affability long survived among
them. When the week was over Sedgwick and Whewell
escorted their friends to Audley End, where they enjoyed
their society for four days more, and then, as Sedgwick put it,
"they moved to London, and we returned to our respective
dens in College." A few days afterwards Mrs Somerville
expressed her satisfaction by letter:

CHELSEA, *April* 25, 1832.
My dear Sir,

  Fruitless as the endeavour would be to express how
highly we have been gratified with the delightful week spent within
the hospitable walls of Trinity College, I still feel it to be due
to you, and all our kind friends, to assure you in language as
rigorously true as ever was conveyed by x, y, that our reception has
made an impression upon us not to be forgotten. Our anticipations
were sanguine, but they were surpassed by the reality, and you will
only do us justice by believing that the attentions so liberally
bestowed are duly appreciated. That my studies should merit the
notice of such men as adorn your University, it would have been
presumption to expect; their approbation therefore, so handsomely
given, is the more gratifying. The two acts of our little drama, the
first at Cambridge, and the second at Audley End, form a very
agreeable episode in our life.

[1] Whewell's *Life*, by Mrs Stair Douglas, p. 142.

1832.
Æt. 47. I trust you have completely recovered from the cold which I fear you caught while kindly devoting your time to me, and that we shall soon have the pleasure of seeing you in town. I beg you will let me know of your arrival before you have made engagements among the numerous friends who are so desirous of your society, for I can assure you there is no one who will value the privilege of being included among them more than yours very sincerely,

MARY SOMERVILLE.

Sedgwick started for Wales towards the end of May, as he had proposed; but, anxious as he was to complete his work there, he allowed himself a week's holiday in June at Oxford, where the British Association held its second meeting. He had rather sneered at the notion of such a gathering when it was first started, and had protested that he would not leave Wales for either York or Oxford. Murchison, however, made him break his resolution in favour of the latter city, and Buckland clenched the matter by a humorous invitation which nobody could well have refused. "I exhort you," he wrote, "by all your love for Professorial Unity and the eternal fitness of things, to locate yourself in a fraternal habitat within my domicile during the orgies of the week beginning on the 3rd of June[1]," and then went on to tell him of the arrangements that Mrs Buckland had made for his comfort, and the friends whom he would probably meet. Still he went unwillingly, and, not many days before the meeting assembled, wrote to Murchison :

CAERNARVON, *June* 5, 1832.

...I shall be glad to make myself of use [at Oxford], but in the bustle of the meeting, and among friends, philo- sophical reporters, blue-stockings, and big-wigs, I shall not find much time. If I say anything it must be *extrumpery*, and I suppose about Snowdonia, which I now know something about. It is, however, a terrible hard crust for sucking geo- logists to mumble, and as for the ladies (God bless 'em !) it will I fear turn their stomachs. I am, in short, willing to be

[1] From Rev. W. Buckland, 19 April, 1832.

of use, but I have not good cards in my hand ; and if other
people are there who are better prepared (and I defy them to
be worse) I shall be very glad to have an excuse for sparing
my breath....Yours to the top end of his hammer,

1832.
Æt. 47.

A. SEDGWICK.

Sedgwick's reluctance vanished in the congenial society
he found at Oxford. The Report of the meeting shews that
he not only took an active part in the business of the geological
section, but accepted without a murmur the office of President
for the following year, when the Association was to meet at
Cambridge, saying in a public speech "that it would be at all
times and in all situations one of his greatest pleasures to
contribute his assistance to the British Association[1]." Six
weeks afterwards, writing calmly and deliberately to a friend,
he admitted that the meeting had "gone off admirably." As
soon as the meeting was over Murchison hurried back to his
work in Wales, and Sedgwick, after a short excursion with
Whewell "to Stratford on Avon and other not distant
places[2]," followed his example. He was evidently anxious
that their separate investigations should become part of a
common whole, and therefore kept his friend informed of his
whereabouts and his plans, while, with indomitable energy,
he followed the strata over hill and dale, principally on foot,
through a wide extent of rugged country. The rough pen-
and-ink sections which illustrate the first of the next two
letters are of especial value as exhibiting the views which
he held when it was written.

BARMOUTH, *July* 23, 1832.

Dear Murchison,

On Saturday the 30th of June, I had from Rodney's
pillar a glimpse of you and the Colonel on the summit of
Moel-y-Golchfa. I landed that evening at Llanymynach, and

---

[1] *Second Report of the British Association*, p. 102.
[2] Whewell's *Life*, p. 141.

spent next day with my friend Evans in a proper clerical
manner.   I did, however, after church, go up on the hill north
of the town, and enjoyed what I think one of the very finest
views I ever beheld.   I was certain, at the first glance, that
there was a great deal of work before me in Montgomeryshire
which I had very little expected.   The porphyritic system of
Snowdonia is there told over again, as I made out from the
look of the country, and the information I obtained from
Evans.   The next day, July 2, very early, we started for
Llanrhaidr, where we breakfasted; then visited the celebrated
Pistill Rhaidr, which is caused by great ribs of porphyry
passing through the greywacké, which it binds together and
saves from degradation.   The water tumbles over the greywacké and porphyry through a perpendicular height of about
230 feet.   It is a gorgeous fall for a geologist, though the
artists think it formal and unpicturesque.   We then scaled
the Berwyns, and went along the top as far as the highest
point (Gaderferwyn), and descended by a tributary valley to
Llanrhaidr.   Thence I drove to Llangynog late in the evening.
The Berwyns for many miles N. and S. dip to the W. or
W.N.W.; there must therefore be an anticlinal line in Montgomeryshire ranging somewhere N.E. and S.W., and probably
passing near Llanfyllin.   E. of that town the beds again roll
over to the S.E. so as to bring in the newer rocks between the
Vernwy and the Severn which form the base of the system in
which you are working.   Before the end of the summer I shall
endeavour to make traverses on the line of Llanidloes, New-
town, (perhaps Pool and Montgomery), Llanfair, Llanfyllin,
and Llangollen; and, if I have time, I shall then make a long
run towards the south, so as to make one or two long
traverses in South Wales.   In this way our work will link
together.

Next day, July 3, I crossed the Berwyns to Bala.   Through
the whole ascent, and nearly to the base on the W. side of the
chain, the dip is about W. by N., working gradually to N.W.
in the prolongation of the chain towards Corwen.   On the

western side of the chain an anticlinal line strikes through the region about N.N.E. and W.S.W., in consequence of which some bands of black shelly limestone I found at the top of the Berwyns are brought out again with an opposite dip, viz. E.S.E. These bands of black limestone are absolutely identical with the transition lime which separates the greywacké of Westmoreland from the great system of greenslate and porphyry of the central mountains of Cumberland. They form a very grand base-line, which I have now traced from Glyn-Diffws (five miles N.W. of Corwen) to Dinas Mowddy, a distance, as the crow flies, of about 30 miles.

1832.
Æt. 47.

From Bala I examined the range of the limestone, and made excursions to the Arenig chain, extending my rambles almost to the great Bangor road. The whole region forms the side of a great saddle, dipping about E.S.E., much interrupted by vast unstratified masses of porphyry, which are, however, more or less tabular, and range with the strata, without altering their dip. A little to the W. of Penmachno the great Merioneth anticlinal strikes in, ranging N.N.W. and E.S.E. (i.e. geometrically parallel to the four Caernarvon anticlinals, and also to the beds of shell limestone above mentioned). This line I have traced into the sea near Barmouth, and I have examined all the country on the W. side of it. In short, I have toiled like a slave, and have made myself ill, so that I am now almost confined to the house. Tomorrow I hope to be in working condition again. There are no porphyries on the W. side of the great Merioneth saddle between Festiniog, Harlech, and Barmouth, but an enormous fault, which cuts slap through Caernarvonshire from Llanllyfni to Tremadoc, strikes in three miles N. of Harlech, and may be traced into the sea two miles from Barmouth. It is a very grand geological phenomenon, connected, as I believe, with the vast eruptions of syenite in the S. parts of Caernarvonshire. The country S. of the Barmouth river I hardly know anything about, having for the two last days been laid up by a very feverish cold and sore throat. This is very provoking, as the weather is glorious, and the porphyritic peaks are glittering in the sun as if in mockery of my infirmity.   I have, however, once scaled Cader Idris.   It forms a portion of the eastern side of the great Merionethshire saddle, and differs in no essential respect from other parts of the county.   The crater on its southern side is nothing more than a deep pool of water in soft calcareous greywacké!!

I did not intend to make you pay for a double letter, but I have already scrawled over the last page before I once thought of the direction.   You must see, by what I have written, that I have had a good harvest; its reaping is, however, most laborious, as the tracing the geological parallels compels me to climb almost every mountain.   When I have worked down through the Cader Idris region to Machynlleth, my labours will become more light, as the hills are much lower, and there are great uninteresting tracts I shall not be compelled to look at.   I now understand what formerly I was puzzled with; the transition or primary system of North Wales, though enormously thick, is not one tenth part the thickness one might at first imagine.   In consequence of the anticlinal lines, the system of rocks three or four miles S.E. of Bangor reappears in the centre of Merionethshire; and the shell limestone E. of Bala lake reappears, if I mistake not, near Meifod in Montgomeryshire.   Pray write to me at Machynlleth, and tell me what you think of this notion, and how far it agrees with what you have seen.   The dark calcareous beds near Meifod &c., &c., I have not yet seen; so I am arguing on an imperfect case.   I will fill up this sheet with one or two sections for your amusement.

Most truly yours,

A. SEDGWICK.

MACHYNLLETH, *August* 10.

To the same.

"I am just arrived here, and have got your letters, which I have read, or rather tried to read, with much interest. You must see at once it is impossible we should meet.   I am

EXPLANATION OF SECTIONS.

SECTION. No. 1.

1. *Porphyry of the Menai.*
2. *Slate and Porphyry.*
3. *Porphyry.*
4. *Great zone of roofing slate and coarse grey-wacké.*
5. *Porphyry of Mynydd Mawr.*
6-10. *Slate and Porphyry blended together in endless confusion, but not destroying the traces of strati-fication: in consequence of the 5 anticlinals the same stratified masses are repeated again and again, so that at the Merioneth anticlinal (x, y) we are on the parallel of No. 4.*

SECTION. No. 2.

a = *Bands of black shell limestone.*
d' = *Do. at the top of the Berwyns.*
d'' = *Supposed reappearance of the same beds of lime-stone on the East side of Montgomeryshire.*

SECTION. No. 3.

a = *The great fault above described.*
b = *The great Merioneth anticlinal.*
c = *Black shell limestone near Llany Mowddy.*

SECTION 1.

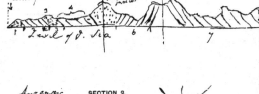

SECTION 2.

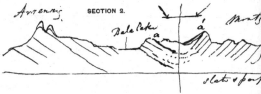

SECTION 3.

Pont Aber-glaslyn   Cnicht   Moelwen   Ffrwddmawr   Arrenrig   Bala lak

9   1 0   (X)   (y)   11

a''

Fowddy   valley above Dinas Mowddy.

E. S. E.

c

SECTION 1. *In this section you will see that I have neglected all proportions; but it will convey a true notion of the great geological facts.*

SECTION 2. *The next section is in part ideal. It begins at the right hand of the preceding, and continues it over the Berwyns.*

SECTION 3. *Section from the coast thro' Rhinog Fawr and Arran Fowddy to the valley above Dinas Mowddy.*

off for Towyn, and hope in two days to finish Merionethshire.
I shall then work into Cardiganshire as far as Plinlimmon and
the Devil's Bridge. For this I will hypothetically allow a
week. Then I double back, and make traverses in Montgomeryshire, partly to work out the anticlinals, partly to lock
my work into yours. I don't quite twig your sections; indeed
I can't read 'em, but I see you have done excellent work.

I find, as a general rule, that the moment I get into a low
country the strata begin to roll and reel about, but while they
are in elevated ridges they keep their strike, and dip beautifully, the changes being produced by parallel anticlinals. I
hope before this sun is down to trace the great Merionethshire
anticlinal into the sea, north of Towyn. I shall then have
followed it over hill and dale for more than 50 miles."

To Ainger, as usual, he wrote in more general terms.
After describing himself as "burnt as brown as a pack-
saddle, and a little thin from excessive fatigue," he proceeds:

"I have been rambling in various parts of North Wales,
for days, and almost for weeks, together, as much secluded as
if I had been in the centre of New Holland. Now and then
I stumbled on a struggling Cantab, with whom occasionally I
also contrived to spend the evening. These were, however,
rare occasions. North Wales is Cumberland over again,
only on a rather larger scale, and expanded over a wider
surface. The valleys of North Wales are many of them
glorious; but they want the beautiful lakes of your county.
After all, the Lake Mountains for my money. The Welsh
are a kind-hearted, but rather dull set of people; just made to
be beaten by the Saxons. It is, however, wrong to judge of
a people whose language one does not speak. I like to talk
to country people, and to see their humours, but from this I
am shut out among these children of Caractacus. This it is
which has made my solitude doubly solitary. As soon as the
weather changes, for it is now detestable, I shall look again
towards the south, and endeavour to effect a series of long

traverses in South Wales; but in what direction, you know at this time pretty nearly as much as I do myself. In short, I shall hoist sail, and sail before the wind. Before my final return I hope to spend a week with my friend Conybeare in Glamorganshire; this will, however, depend on the cholera, which is raging not far from him, and may frighten me from my present purpose[1]."

Sedgwick's share in the foundation of the Cambridge Philosophical Society has been already related. His sanguine anticipations of success had been more than realised in the thirteen years which had elapsed since 1819. The members were numerous; the meetings well attended; the papers valuable, and of varied interest. Moreover the Museum, begun in the very first days of the Society's existence, had assumed respectable proportions, through the exertions of Professor Henslow, Mr Leonard Jenyns, and their friends; and the reading-room—which was not restricted to Fellows of the Society—had become a place of popular resort. This rapid development might of itself have warranted a new departure in the position of the Society, even if an accidental circumstance had not rendered immediate action necessary. The meetings were held, for the first few months, "in the Museum of the Botanic Garden," and afterwards in rooms in Sidney Street, facing Jesus Lane. These were commodious, and well-suited to the purposes of the Society; but their tenure of them was limited, and early in 1832 it became known that the owner declined to extend it. Thereupon, at a special General Meeting held 7 April, 1832, it was decided, on the motion of Mr Peacock, to be "expedient that the Society should possess a house of their own, built expressly to suit the objects of the institution;" and, as a preliminary step, that a charter of incorporation should be obtained[2].

It happened that these important measures, amounting

[1] To Rev. W. Ainger, dated "Llansilin, near Oswestry, 29 August, 1832."
[2] We believe that the Cambridge Philosophical Society was the first Society, out of London, that obtained the distinction of a charter.

almost to a second foundation of the Society, were adopted during Sedgwick's tenure of the office of President, to which he had been elected in May, 1831 ; and further, his name appears alone upon the charter. As President, he was, for the moment, the head of the Society ; and the adoption of a single petitioner, in lieu of several, diminished the fees exacted by the Stamp Office. But, whatever may have been the reason, it must be esteemed a fortunate circumstance that he, who took so deep an interest in the Society, should be for ever associated with its permanent establishment. It is almost needless to add that he was himself especially delighted at being placed in such a position. Writing to Murchison, 7 November, 1832, he says:

"Yesterday after lecture I presided at a public meeting held by our Society for the purpose of accepting a Charter. We afterwards adjourned to an Inn and had a blow-out. Finally three or four of my friends came to my rooms and kept me up till two this morning—for which I do not now much thank them."

We can readily imagine Sedgwick's enthusiasm on this occasion, probably the first of those annual celebrations in which he bore so jovial a part ; and among the reminiscences with which he used to amuse the company, none recurred so frequently as the story of the charter, when, as he used to say, " I was the Society."

Two other matters, of very diverse nature, occupied much of Sedgwick's time and thought during the Michaelmas Term of 1832. The first was a movement—as we should now call it—to call forth " some expression of national gratitude to the memory of Sir Walter Scott." Sedgwick became a member of a committee appointed for that purpose, and did his best to rouse the enthusiasm of his Cambridge friends—but without success. The scheme proposed, to purchase Abbotsford, and to secure it to Sir Walter's children, did not commend itself to the common-sense of those to whom he tried to recommend it. It was in fact, as he himself said—" a strange round-about

1832.
Æt. 47.

1832.   way of shewing respect[1]"—and before long he found that a
Æt. 47.   few very modest subscriptions, given probably more out of
regard to himself than from any other motive, would represent
the generosity of Cambridge.

This was succeeded, after a brief interval, by a contested
election for the University, in which Sedgwick, Thirlwall, and
a few other ardent spirits made an energetic but unsuccessful
effort to persuade the constituency that Mr John William
Lubbock, Vice-President of the Royal Society, as "a man
distinguished for his literary and scientific attainments," would
be a more suitable representative for a learned body than a
mere politician. Sedgwick became chairman of the Cambridge
Committee, and an active canvass was set on foot. But the
"vehemence of some of his Whig friends[2]," and the lukewarm-
ness of others, who agreed to vote but declined to canvass,
boded ill for the success of the attempt. After a ten days'
struggle Mr Lubbock withdrew, and Mr Goulburn, with the
Right Hon. Charles Manners Sutton, late Speaker of the
House of Commons, were duly elected. The following letter
from Dr Samuel Butler reached Sedgwick soon afterwards:

SHREWSBURY, December 10, 1832.

My dear Professor,

It is better to retire in time than to give a great many
friends an expensive and hopeless journey at such a season of the
year, and when many of them have their votes and interests pre-
occupied elsewhere.   I had written this morning to bespeak chaises
all along the road, especially for the night hours, which I was very
anxious about.   By to-morrow's post I can rescind the order.

I am almost sorry that Lubbock offered.   I am no cynic; I hate
to *snarl* and not to *bite*; a grin and a growl does nothing but make
one laughed at.   Another thing which I have observed in the course
of this canvas was, that people did not like to be called upon for so
long and expensive journies with apparently so little prospect of
success.   It is an unwise waste of strength and interest.   I should
not be surprised if some of our well-wishers took their names off the
boards to avoid such frequent solicitations, especially when they find
themselves on the losing side.   I say this because I hope that on
some future occasion time and man may be well chosen, and that we

---

[1] To R. I. Murchison, 7 November, 1832.
[2] Whewell's *Life*, p. 149.

shall act with good hope of success, and not without at least a fair 1832.
chance of it. Nothing should have kept me, or any persons whom Æt. 47.
I could influence, from the poll on this occasion, but physical
impossibility; but wavering or luke-warm persons will not be equally
zealous, and the oftener they are called upon without any prospect
of success, the less inclined will they be to serve us. Therefore, I
say, look for *influential candidates*, and *magna nomina*.

To all that can be said of Mr Lubbock's merits I most willingly
subscribe. But look at the array against him. All the force of Eton
from attachment to the Speaker; all his family connections; and all
the Speaker's parliamentary friends. He was a well-chosen opponent,
whom no private person, however high in character for talents or
virtues, could hope to conquer, and I have no doubt he would have
headed the poll. We could not conquer even Goulburn, who has
a powerful party among the Saints, and whose interest is increased
by the talents of his son. Now look at the last election. Many
voted for Lord Palmerston from old connections, who would not
have voted for us now. The high connections of Mr Cavendish
were all on the alert, and stood him in more stead than his own high
merits, and made many take the trouble of coming to serve him,
especially among the aristocracy, who would not stir a step for any
individual of private family.

All this I say, that you may ponder it, and either keep to yourself, or communicate to the confidential friends of our party, as you
may judge best. But I repeat my exhortation that you will carefully
look out for highly connected, and, if you can, highly popular, as well
as highly gifted, candidates. Let us give them no possible advantage
that we can help. There is time to look about before the next
election. Let those of our friends who can be *depended on*, be
prepared in time, and be secret.

My dear Professor, I have one grand article of faith: that no
speech of four hours in either House ever did good to the cause it
presumed to advocate; and that no sermon should be above half an
hour long if the preacher means to carry with him the good will and
attention of his congregation. Now I am afraid my preachment to
you is thirty one and a half minutes. God bless you. We shall
meet triumphant yet. Truly yours, S. BUTLER.

Before the year ended, it fell to Sedgwick's turn to preach
the sermon at the annual Commemoration of Benefactors in
Trinity College, on Monday, 17 December. It seems at first
sight impossible that such a sermon, preached every year in
the same place on the same subject, could ever be treated with
any marked originality; and in fact, allusions to the founders,
benefactors, distinguished members of the college, and those
who have passed away since the last occasion, with the obvious

1832. lessons to be drawn from such occurrences, form the staple of
Æt. 47. these discourses, which vary only with the rhetorical power of
the writers. Sedgwick, however, struck out a new line, and
produced a work which not only made a sensation at the
time, as is proved by the fact that it ran through four editions
in two years, but which still possesses an historical interest,
not merely as evidence of the breadth of his own studies and
speculations beyond the range of his particular science, but as
a protest against the metaphysical and ethical doctrines then
commonly accepted. Unfortunately it is now presented to us
in a form so different from that in which it was originally
delivered, that it is impossible to realise the effect produced
on those who heard it. The subject, *The Studies of the
University*, had probably occupied Sedgwick's thoughts
for a long while; but, with more than usual procras-
tination, he did not put pen to paper until the Thursday
preceding the day of delivery[1]. In consequence, when asked
to publish, he found himself in a considerable difficulty. As
he says in the preface : " having animadverted with much
freedom on some parts of the Cambridge course of reading,
[the author] felt himself compelled, before he dared to give
what he had written to the public, to enter at more length on
a justification of his opinions. On this account, his remarks
on the classical, metaphysical, and moral studies of the
University were cast over again, and expanded to at least
three times their original length." In fact, only one-third of
the work remains as it was originally written; and this, it
must be reluctantly confessed, is not in Sedgwick's happiest
vein. The style is heavy and laboured, and there are none of
those eloquent passages which made his speeches so animated
and so delightful. The matter, therefore, rather than the
manner, must have caused those who heard it to wish to read

---

[1] To R. I. Murchison, 8 December, 1832. " I cannot be up in Town on
Monday the 17th,...when I have to preach the Anniversary Sermon. How I shall
get through I don't know, as I have not yet written it, and till next Thursday
night shall not have much time to think about Divinity."

it quietly at home. This wish, expressed to Whewell as
Senior Tutor, reached Sedgwick in the following letter:

From Mr Whewell.

TRINITY COLLEGE, *December* 23, 1832.

My dear Sedgwick,

When you had scribbled down the last sentence of your
sermon after the bell had stopt, and had succeeded by a sort of
miracle in reading your pothooks without spectacles, omitting however half the sentences, and a quarter of the syllables of those which
remained, I daresay you thought you had done marvellously well,
and had completed, or more properly had ended, your task. In this,
however, you were mistaken, as I hope soon to make you acknowledge.

The rising generation, who cannot err, inasmuch as they will
discourse most wise and true sentences when you and I are laid in
the alluvial soil, declare that their intellectual culture requires that
you should print and publish your sermon. I will give you a list on
the other side of the names of the persons who have joined in
expressing this wish. I undertook very willingly to communicate
this their desire to your reverence, inasmuch as I thought your
sermon full of notions, as the Americans speak, which it will be very
useful and beneficial to put in their heads; or rather to call them
out, for a great number of those good thoughts are already ensconced
in the excellent noddles of our youngsters, like flies in a bookcase in
winter, and require only the sunshine of your seniorial countenance
to call them into life and volatility. I do not know anything which
will more tend to fix in their minds all the good they get here than
to have such feelings as you expressed, at the same time the gravest
and the most animating which belong to our position, stamped, upon
a solemn and official occasion, as the common property of them and
us. And I also think it of consequence that when they on their side
proffer their sympathy in such reflexions, we, on ours, that is, in the
present case, your dignified self, should not be backward in meeting
them, by giving to all parties the means of returning to, and dwelling
upon, these reflexions.

Such is my thinking about this matter, and therefore I have
undertaken to urge their request; and I hope you will be able to
extract from some abysmal recess your manuscript, and to place it
before the astonished eyes of the compositor. It is probable that he
will look, as Dante says the ghosts looked when they peered at him,
like an old cobbler threading his needle, but never mind that. The
fronts of compositors were made to be corrugated by good sentences
written in most vile hands; so let him fulfil his destiny without loss
of time....

Yours ever,

W. WHEWELL.

Petitioners for the printing of Professor Sedgwick's sermon :

Ld Lindsay	Monteith	Cator	R. Morgan
Blakesley	Forsyth	Childs	Whiston
Spedding	Merivale	English	Seager
Kemble	Campbell	Phelps	Potts
Garden	Brookfield	Good	Scratchley
Tennant	Wright	E. Morgan	Newton.
Alford			

To this cordial request Sedgwick no doubt gave an equally cordial assent, and prepared to print without delay. But ill-health (as usual), and other occupations, intervened, and November, 1833, came, before the *Discourse*, as it was then termed, saw the light. It will, however, be best to complete our account of it while noticing its first delivery; though, on some grounds, we should have preferred to wait until we had reached the publication of the last edition in 1850. By that time the modest octavo of one hundred and nine pages had swelled into a ponderous tome, in which four hundred and forty-two pages of preface, and two hundred and twenty-eight of appendix, include between them ninety-four pages of discourse : "a grain of wheat between two millstones," as Sedgwick himself admitted. In conversation he has been heard to describe the publication as "the wasp," because it had so small a body, and so large a head and tail. But, whether he spoke seriously or in jest about it, it was the pet child of his brain; and, as is the way with indulgent parents, he perhaps found virtues in it which others, especially at this distance of time, may fail to discover.

A detailed criticism of the *Discourse* would lead us into discussions unsuitable for a biography. We will therefore content ourselves with a very brief sketch of the subject-matter. It must be remembered at the outset that the studies of Cambridge are approached from the moral rather than from the intellectual side; that the author is speaking as a clergyman from the pulpit, not as a philosopher from the desk. To him Cambridge was a place not merely of sound learning, but also of Christian education. This obvious limitation of his survey has sometimes been lost sight of, and in consequence an erroneous, not to say an unjust, conception has been formed of his work. The studies of the University are reviewed under a threefold division : (1) the laws of nature; (2) ancient literature and language; (3) ethics and metaphysics. Under the first of these divisions natural science is considered in the light of the results to which a reverent study of it ought to lead ; and it is pointed out that its various branches, Astronomy, Anatomy, Geology[1], minister to natural religion, and "teach us to see the finger of God in all things animate and inanimate." In other words, this part of the discourse may be described as a rapid but forcible exposition of the argument from design. Under the second head ancient languages and ancient history are briefly considered. Sedgwick was no scholar, though fond of classical reading, and he does not seem to understand the necessity of thoroughness in the study of ancient literature. When he deplores the time wasted in "straining after an accuracy beyond our reach," he loses sight of the fact that without

1832.
Æt. 47.

---

[1] In the portion of the Discourse devoted to Geology the following passage occurs : "By the discoveries of a new science (the very name of which has been but a few years engrafted on our language), we learn that the manifestations of God's power on the earth have not been limited to the few thousand years of man's existence. The geologist tells us, by the clearest interpretation of the phenomena which his labours have brought to light, that our globe has been subject to vast physical revolutions. He counts his time not by celestial cycles, but by an index he has found in the solid framework of the globe itself. He sees a long succession of monuments, each of which may have required a thousand ages for its elaboration. He arranges them in chronological order ; observes on them the marks of skill and wisdom, and finds within them the tombs of the ancient inhabitants of the earth. He finds strange and unlooked-for changes in the forms and fashions of organic life during each of the long periods he thus contemplates. He traces these changes backwards and through each successive era, till he reaches a time when the monuments lose all symmetry, and the types of organic life are no longer seen. He has then entered on the dark age of nature's history ; and he closes the old chapter of her records." These remarks, and others of a similar kind, so shocked the Rev. Henry Cole, "late of Clare Hall," that he attempted to refute them at length in : *Popular geology subversive of divine revelation! A letter to the Rev. Adam Sedgwick...being a scriptural refutation of the geological positions and doctrines promulgated in his lately published Commencement Sermon.* 8vo. Lond. 1834.

accuracy no work, whether in language or science, can be satisfactorily carried out. At the same time he cordially approves the cultivation of Greek and Latin authors with certain limitations; pleads for the further study of their philosophical and ethical works; and in justification of this points out that "the argument for the being of a God, derived from final causes, is as well stated in the conversations of Socrates, as in the Natural Theology of Paley."

The third part of the discourse contains a severe criticism of Locke, and of "the utilitarian theory of Morals" as expounded by Paley. In this Sedgwick anticipated the views which Whewell subsequently set forth in his *Philosophy of the Inductive Sciences* (1840), and in his various writings on Moral Philosophy. In the letter dedicating the former work to Sedgwick he says: "the same spirit which dictated your vigorous protest against some of the errors which I also attempt to expose, would have led you, if your thoughts had been more free, to take a leading share in that Reform of Philosophy, which all who are alive to such errors, must see to be now indispensable." It is not improbable, having regard to the intimacy between Whewell and Sedgwick at that time, that the views set forth in the *Discourse* may be the result of conversations between them. Moreover there was a strong taste for metaphysical speculation in Trinity College in those days; and Thirlwall, Hare, and their friends, who were Sedgwick's friends as well, would "tire the sun with talking" on these subjects. In heading a reaction against Locke and Paley, Sedgwick merely gave expression to opinions deliberately arrived at by the most thoughtful men with whom he was associated. One passage from his criticism on Locke, though it has been often quoted, is so graphic, and illustrates so well his position, that it may find a place here:

If the mind be without innate knowledge, is it also to be considered as without innate feelings and capacities—a piece of blank paper, the mere passive recipient of impressions from without? The whole history of man shows this hypothesis to be an outrage on his

moral nature. Naked he comes from his mother's womb; endowed
with limbs and senses indeed, well fitted to the material world, yet
powerless from want of use; and as for knowledge, his soul is one
unvaried blank: yet has this blank been already touched by a
celestial hand, and when plunged in the colours which surround it,
it takes not its tinge from accident but design, and comes forth
covered with a glorious pattern.

1833.
Æt. 48.

Interest in the *Discourse* was not confined to Cambridge.
The *Quarterly Review,* though it did not devote an article to
it, called it "the most remarkable pamphlet since Burke's
Reflections;" and John Stuart Mill said all that could be said
against it in an elaborate article which he afterwards reprinted
in his *Dissertations and Discussions*[1].

After this digression we will return to Sedgwick's own
proceedings. He had been more than usually unwell during
the Michaelmas Term, and therefore, as soon as the Commemoration was over, went down to Wensleydale in Yorkshire with Mr Lodge, and thence proceeded to Dent, which
he found a good deal altered. "The architectural beauties
of this metropolitan city," he wrote, "are sadly on the wane,
which makes me rejoice in the icon I have at Cambridge of
Dent in its glory. In rambling about among the scenes of
my childhood I sometimes fancy I am a young man again.
The delusion however passes away when I look about me, and
see the young fry playing about the old parsonage the very
antics I was myself playing forty years since[2]." During his
visit he gave himself a complete rest, and when he returned
to Cambridge at the end of January he was able to announce
that "my Christmas in the North has given me most ram-
pagious health, but has put me most dreadfully in arrears[3]."

Among these arrears was the delivery of a short course of

---

[1] *Dissertations and Discussions,* by John Stuart Mill, 8vo. Lond. 1859,
i. 95. There is a curious article in *The Phrenological Journal* for September
1834, in which Sedgwick is reproved for not alluding to Phrenology in his
Discourse. The writer is known to have been the celebrated phrenologist,
Mr George Combe.

[2] To Rev. W. Ainger, 15 January, 1833.

[3] To R. I. Murchison, 5 February, 1833.

lectures, to make up for some which he had been compelled
to omit through ill-health in the Michaelmas Term of 1832.
Then came the revision of proofs of papers for the Geological
Society's *Transactions ;* a correspondence with Murchison on
the speech which he would have to deliver as President in
February, and which Sedgwick evidently revised and corrected[1]; and lastly, arrangements for the visit of the British
Association to Cambridge in June.   But the year was barely
three months old when a fresh attack of illness—evidently a
severe one—put an end to all work, and very nearly rendered
the President elect incapable of meeting the Association.

Early in June he tells Murchison : " I returned home on
Thursday from Walton-on-the-Naze very much recovered.
Indeed I consider myself now fairly off the sick list.   Perhaps
I ought more properly to say that I am restored to my senses ;
for during the last seven or eight weeks I have been under a
strange mental obscuration, unable to do a stitch of work
requiring thought or attention[2]."

But, when the Association met, on Monday, 24 June, his
ailments were forgotten in the excitement of welcoming an
overflowing assemblage of men of science, not merely from
England, but from various parts of Europe.   The importance
of the Association was becoming recognised, and the meeting
was in every way memorable ; not merely from the numbers
gathered together, but from the importance of the work done.
And, whatever was going forward, whether a general meeting
in the Senate House, or the more private business of the
geological section, or a dinner in the Hall of Trinity College,

---

[1] After the serious discussion of various important topics, the following amusing
passage occurs in a letter dated 5 February : " In regard to the animating paragraph you talk of, you ask for what is almost impossible. You cannot take a
flying leap in cold blood, you must lead up to it by the animation of the chase.
Anything I could write would be flat as ditch-water, and certainly would be out of
tone and keeping.   Put your own notions in black and white, and before they are
printed or spoken I will look them over and purge them if I think they want it,
and you will endure the treatment."

[2] To R. I. Murchison, 4 June, 1833.

Sedgwick was the animating spirit, delighting everybody by his geniality, or thrilling them by his unpremeditated eloquence. Had he prepared his speeches, he said, " the intensity of present feelings, would, like a burning sun, have extinguished the twilight of a remembered sentiment." Probably on no occasion in his life did he speak so often and so effectively. Dr Chalmers, who was present, spoke of "the power and beauty" of part of his farewell address to the Association; and is reported to have said in conversation afterwards that he had never met with natural eloquence so great as that of Sedgwick. But this excellence, so delightful to his contemporaries, prevents us from enjoying more than a very faint reflexion of his brilliancy. The reporters could not follow him; and he was himself too indifferent to fame to correct their travesties of what he really said as fully as we could wish[1]. The official *Report* is of course a better authority than the *Cambridge Chronicle*, but even there we have to content ourselves, for the most part, with dry bones. One passage from his opening address, in which he announced a grant from the Civil List to the great chemist Dalton, will bear quotation, not merely from his personal interest in Dalton[2], but because it has the true ring of authenticity:

> There is a philosopher sitting among us whose hair is blanched by time, but possessing an intellect still in its healthiest vigour; a man whose whole life has been devoted to the cause of truth; my venerable friend Dr Dalton. Without any powerful apparatus for making philosophical experiments, with an apparatus, indeed, which many might think almost contemptible, and with very limited external means for employing his great natural powers, he has gone straight forward in his distinguished course, and obtained for himself in those branches of knowledge which he has cultivated a name not perhaps equalled by that of any other living philosopher in the world. From the hour he came from his mother's womb the God of nature laid His hand upon him, and ordained him for the ministration of high

1833.
Æt. 48.

---

[1] Sedgwick, writing to Murchison 7 July, 1833, says: "I have now been working four days at it [the Report], till my head is almost in as much confusion as the short-hand notes. You never read such a chaos. Our reporter was not up to the work, and what he has done is almost worse than nothing."

[2] See above, p. 248.

1833.
Æt. 48. philosophy. But his natural talents, great as they are, and his almost intuitive skill in tracing the relations of material phenomena, would have been of comparatively little value to himself and to society, had there not been superadded to them a beautiful moral simplicity and singleness of heart, which made him go on steadily in the way he saw before him, without turning to the right hand or to the left, and taught him to do homage to no authority before that of truth. Fixing his eye on the most extensive views of science, he has been not only a successful experimenter, but a philosopher of the highest order; his experiments have never had an insulated character, but have been always made as contributions towards some important end, as among the steps to some lofty generalisation. And with a most happy prescience of the points to which the rays of scattered observations were converging, he has more than once seen light while other eyes all was yet in darkness; out of seeming confusion has elicited order; and has thus reached the high distinction of being one of the greatest legislators of chemical science[1].

The rest of the year 1833 was very uneventful. Sedgwick's exertions to entertain the philosophers brought on a fit of the gout, and when that had passed away the reports of the meeting had to be corrected and made ready for press. This business despatched, the season was too far advanced for a campaign in Wales; so, after a week at Leamington, he attacked Charnwood Forest, accompanied by Whewell and Airy, and "made out its structure in considerable detail[2]." An amusing note, written from Leicestershire to Mrs Murchison, indicates his difficulties there, and his subsequent intentions:

MONT SORREL, *August 11th*, 1833.

My dear Mrs Murchison,

You offer me most provoking temptations, but it is quite in vain for me to try to meet you and Mr Murchison. It will be the end of the month before I reach Cumberland, where I want to make a few sections on the spot, especially some on the coast near Whitehaven; and by the time they are over it will be high time for me to face about for

[1] *Report of the Third Meeting of the British Association*, p. x.
[2] Salter's *Catalogue*, p. xviii; Whewell's *Life*, p. 155. Charnwood Forest was one of the places of which Mr Conybeare had suggested the investigation. See above, p. 325.

Cambridge. The last motive you mention would operate rather as a repulsive than as an attractive force, for the lady you talk of is, as I have been told, a most formidable and cruel tyrant, who has slain her tens of thousands without pity. I should not like to be offered up as a burnt offering before the shrine of any woman sprung from Eve, only to be told that I was suffering in good company, and without being permitted to take a single cup from the living fountain of hope.

1833.
Æt. 48.

But in the name of wonder what have I to do with love and hope and such flimsy matter? I am wedded to the rocks, and Mount Sorrel (does not the word set your teeth on edge?) is my present mistress. By the way she is a little coy and hard-hearted, and refuses to tell me her pedigree, and to introduce me to her old relations. But I am going with Professors Whewell and Airy to knock at one of her back doors tomorrow morning, and perhaps we may make an entry, and establish ourselves in one of her larders. Should that be the case we shall dish her up to some tune. But talking of love has made me run into figurative language, which you know is quite out of place in geology; so leaving all figures let me again tell you how much I thank you, and how miserable I am that I cannot join your party, and how delighted I shall be to meet you all again in London. Yours, my dear Mrs Murchison, to the bottom of Lyell's *Hypogene* rocks,

A. SEDGWICK.

At Whitehaven Sedgwick met with a kindred spirit in Mr Williamson Peile, manager of Lord Lonsdale's collieries, and he frequently revisited Whitehaven in subsequent years. Mr Peile's duties had naturally led him to study geology from its practical side; but with Sedgwick's help his knowledge was largely developed, and in 1835 a joint communication was made to the Geological Society: *On the range of the Carboniferous Limestone flanking the primary Cumbrian*

*Mountains, and on the Coalfields of the N. W. coast of Cumber-
land.*

On returning to Cambridge in October, Sedgwick was engaged in writing the appendix to his *Discourse*, which, as mentioned above, was published in November. "Just as the last term was waning to its end" he wrote in February, 1834, "my sermon broke its shell; and in consequence, no doubt, of the long incubation, turned out to be six times the orthodox stature. Whether it deserves to be sent to the flames, for being of these most heretical and monstrous dimensions, you will best judge for yourself when you see it[1]." Later in the year, when the third edition had appeared, he could announce: "It has been very well received, and will, I hope, be the means of doing some good to our young men[2]."

While Sedgwick's *Discourse* was passing through the press, a slanderous pamphleteer named Beverley was preparing an elaborate attack on the University; and, by a curious coincidence, the two works appeared in the course of the same month. The pamphlet and its author were equally worthless, and the whole question, though it created a prodigious excitement at the time, might well be allowed to rest in the oblivion to which it has long since been consigned, had not Sedgwick devoted much time and energy to exposing and refuting the malicious calumniator. On this account the matter cannot be passed over; but our account of it shall be as brief as possible.

In October, 1816, Robert Mackenzie Beverley, a native of the town of the same name in Yorkshire, was admitted a pensioner of Trinity College. He did not proceed to the degree of Bachelor of Laws until 1821, and he continued to reside in Cambridge for some months afterwards. If his own account of himself could be believed, he was a virtuous, hard-reading, student; but, on the other hand, whenever he has a

[1] To Rev. Charles Ingle, 16 February, 1834.
[2] To Bishop Monk, 1 November, 1834.

particularly disreputable story to tell, the experience of one
of his own friends is given as the authority for it.   He could
not therefore have known the best set either in his own
college, or in the University at large.   He seems to have
been chiefly remarkable for personal vanity, which shewed
itself in the set of his cap, and the carriage of his gown ; and
for an effeminate delight in dress.   "For chains and chitter-
lings, for curls and cosmetics, for rings and ringlets, no man
was like him.   He was indeed a finished and a fragrant
fop—a very curious coxcomb.[1]"

   In those days he professed to belong to the Church of
England, but after taking his degree he became a dissenter,
and, with the ardour of a convert, set himself to the task of
vilifying, and, if possible, of pulling down, the body of which
he had once been a member.   He began with *A Letter to his
Grace the Archbishop of York, on the present corrupt state of
the Church of England.*   This was written in September, 1830,
and published early in 1831.   Coarse and vulgar as the production is, it was evidently suited to the taste of those for
whom it was intended, for before the end of the year it had
reached a sixth edition.   Beverley was delighted at his unex
pected success.   "Though it becomes not me to say so," he
writes, "yet it cannot be concealed that my 'Letter to the
Archbishop of York' has produced a practicable breach in the
walls of the Establishment."   He therefore lost no time in
publishing *The Tombs of the Prophets, a Lay Sermon on the
Corruptions of the Church of Christ.*   Both these publications
were indited with the avowed object of effecting "a total
separation of the Church from the State, and a speedy con-
fiscation of that which is falsely called Church Property."
The success of the sermon was fully equal to that of the
letter.   It was largely sold, and honoured by more than one

[1] Sedgwick's *Four Letters to the Editors of the Leeds Mercury in reply to R. M. Beverley, Esq.* 8vo. Camb. 1836, p. 55.   See also *Remarks upon Mr Beverley's Letter to the Duke of Gloucester.*   By a Member of Trinity College. 8vo. Camb. 1833, p. 27.

*1833.
Æt. 48.*

1833.
Æt. 48.

reply. Beverley, fond as ever of display, and perhaps not insensible to the pleasure of making money by the sale of pamphlets which must have been easily written, began to give himself the airs of a Luther. His exposure of the corruptions of the Church had awakened the nation; his efforts must next be directed against the Universities from which the Church draws "its mischievous strength." As even he, with all his presumption, could not affect a knowledge of Oxford, he confined his operations to Cambridge, and in November, 1833, brought out *A Letter to his Royal Highness the Duke of Gloucester, Chancellor, on the present corrupt state of the University of Cambridge.*

With affected candour, Beverley begs to be allowed to instruct the "illustrious Prince" whom he is addressing on certain important matters. "It is not to be supposed," he says, "that you can be acquainted with the *arcana* of that mother and nurse of arts and wickedness." He then passes in review the morals, the religion, and the learning, of the University. All the scandalous stories which he had heard while in residence, or with which his correspondents[1] had supplied him, are gathered together. His ignorance is only equalled by his falsehood and his malignity. Silly tales, such as no one but a freshman would credit for an instant, are gravely set down as undisputed facts. Exceptional

[1] The way in which his evidence was collected is shewn by the following notice "To Correspondents," at the end of his *Reply to Professor Sedgwick's Letter:* "I take the opportunity of thanking my correspondents whose letters are not yet answered. Two letters received the first week in December may be of service.

"One correspondent, however, should remember that it is impossible to rely on any *anonymous* information. As the Revelations of Verax might, if properly authenticated, be useful, it is the more to be lamented that he withholds his name and address. He may with confidence venture his name, which will never be disclosed.

"The testimony from Emmanuel College is not forgotten. All communications must be directed to the care of the Publisher, and *the postage must be paid;* for want of attending to this established rule, some letters and notes have been refused admission." Well might Sedgwick term him "the hucksterer of scandal, the advertising broker of impurity." *Four Letters,* p. 37.

instances of folly and depravity are assumed to be the rule. 1833.
Rioting, drunkenness, gambling, immorality, extravagance, Æt. 48.
are stated to be universal; the Fellows are as bad as the
undergraduates; religion is a farce; even learning is a thing
of the past. Lastly—for Beverley's real object in writing his
Letter is artfully concealed until near the end—the Dissenters
are excluded from a place "which should not be styled a University, but a Particularity," by an iniquitous system of tests.
But, before it can be made fit for the education of their sons,
Reform must have reached the root of the whole mischief.
The only practicable course is "to confiscate all the University property, to declare it lapsed to the Crown, and to
remodel it *de novo*."

This farrago of blunders and misrepresentations had an
immense circulation. Three editions appeared before the end
of 1833, and the author's friends among the dissenters, who
had read his previous works with satisfaction, probably
accepted his accusations against the University as true.
Those who knew better were not slow in replying. Ten
pamphlets, most of them written by indignant undergraduates, appeared as rapidly as the editions of the libel.
Some of these take the letter to pieces, paragraph by paragraph, and point out that the picture there drawn of Cambridge is a gross caricature; others hold the author up to
ridicule in satiric verse. One of the former says, with much
truth:

"You have wilfully and deliberately belied the Undergraduates
of Cambridge. You have taken particular exceptions and built
generalities upon them. You have gloated over the recollections of
your own College intemperance till the foul corruption has quickened
into life, and your imagination, drawing its stores from the scenes
of debauchery in which you once revelled, has presented as the
general portrait of Cambridge what forms the rare and disgraceful
exception[1]."

---

[1] *A Letter to R. M. Beverley, Esq., from an Undergraduate of the University
of Cambridge.* 8vo. Cambridge and London 1833, p. 6. The writer is known to
have been William Forsyth, B.A. 1834, afterwards Fellow of Trinity College.

Another, parodying Canning's *Knife-Grinder*, thus apostrophises Beverley:

> Silly Lie-grinder, what have you been doing?
> Great is the scrape your pen has got you into,
> Deep are our threats: your head has got a crack in't,
> So have your brains, Sir.

> Tell me, Lie-grinder, how came you to write such
> Lies against Cambridge? Was it out of spite, or
> Was it in hopes of sharing in the Uni-
> versity plunder?[1]

These well-intentioned answers no doubt afforded pleasant reading to members of the University; but it might be doubted whether any of them would reach the north of England, or Beverley's native town, where he had now become, as it is said, a popular dissenting minister. Sedgwick therefore wisely determined to carry the war into the enemy's camp, and finding that *The Leeds Mercury* had quoted some passages from the *Letter to the Duke of Gloucester* with approbation, addressed a first letter to the editors 7 January, 1834, in which he pointed out, with indignant severity, that Beverley knew "no more of the University of Cambridge than might be learnt of the glorious history of this country from the records of its jail-deliveries or the calendar of Newgate." At the same time, while he disposed satisfactorily of Beverley's general charges, by quoting his own long experience, and the special knowledge gained when he was proctor, he damaged his case not a little by asserting roundly that Cambridge had been free from the vice of gambling for the past six years. This statement drew the following remarks from Mr Conybeare:

> "I cannot think that you have handled Beverley one whit too roughly; his uniform scurrility admitted no other kind of answer, and might have deceived that large portion of the public unacquainted with our Universities had not the fellow been gibbetted in his true colours. And with regard to him you have certainly proved

---

[1] *An Anglo-Sapphic Ode, dedicated (with French leave), to Robert Mackintosh (sic) Beverley, Esq., entitled The Friend of Veracity versus the Lie Grinder. Not by a Can-ning, but a Can-tab.* 8vo. Cambridge, 1833.

yourself a most correct herald, and blazoned him party per pale cox-   1834.
comb and knave, both proper.   Æt. 49.

" I have heard only one serious objection to what you have said,
namely as to your hypothesis of the non-gambling of the last six
years, and this you have qualified in your second letter.   But I was
very sorry to find before that all the *younger* Cambridge men thought
you certainly at fault on this scent[1]."

Sedgwick's first letter was of the nature of a preface ; what
was to come next is indicated in the following passage :

" I deny not the existence of immorality in the University, but I
do deny the charges of Mr Beverley; and I assert that they are
preposterously exaggerated, or impudently untrue.   Before long I
will appeal to specific facts ; and out of what he has written I will
again and again convict him of shameful and deliberate falsehood.
Out of his own mouth I will utterly damage and destroy his
testimony, and turn him out of the courts of honour and common
sense, as a witness not to be believed.   If I fail in doing this, I
deserve to be pointed at, and to become a byword among my equals,
and to be held up to scorn by honest men."

This programme was strictly adhered to.   In three
further letters which appeared in *The Leeds Mercury* during
the first half of 1834[2], the wretched Beverley has full and
signal justice dealt out to him.   He is hunted from one position after another, and shewn, by the citation of irrefragable
testimony, to have been guilty of direct falsehoods, falsehoods
of implication, and falsehoods of ignorance.   So severe is the
castigation, that, had he been less criminal, one would almost
pity him.   But, if the *Letter to the Duke of Gloucester* be read
in conjunction with Sedgwick's masterly reply, it will be
acknowledged that his scornful mockery of his antagonist's
pretended religion, his exposure of his ignorance, and his
indignant denunciation of his profligacy while at Cambridge,
are more than justified[3].   When the work was done, Sedgwick described his feelings while doing it to Bishop Monk:

[1] From Rev. W. D. Conybeare, 10 May, 1834.   It should, however, be stated
that Mr Forsyth, in his *Letter* quoted above (p. 413), maintains that "the extent
of gambling that goes on here is incredibly small."

[2] The second letter appeared in *The Leeds Mercury* for 8 February 1834 ; the
third is dated 15 May ; the fourth, 2 June.

[3] Beverley attempted to answer his redoubtable antagonist in the columns of
The Leeds Mercury, but without success.

" Our opponent I believe I have effectually silenced ; and many years, must, I think, elapse before any party will dare to bring forward Beverley as an implement of mischief. I had a most revolting task to perform, such as no man can go through without dirtying his own fingers. If you saw my letters, I hope you remembered that I was not writing for gentlemen or scholars, but for the instruction of a multitude of bitter blackguards in the shape of Yorkshire dissenters[1]."

The following letter is specially interesting as shewing that all dissenters were not prepared to agree with their self-constituted champion.

From Mr T. M. Ball.

61 COLEMAN ST., LONDON,
10 *February*, 1834.

Sir,
        I am a dissenter. In common with thousands of all creeds I read Beverley's Letter to the Duke of Gloucester. Of Cambridge and its noble University I know but little, but by common report. The picture drawn of its condition by the writer of that Letter was indeed horrible, but I for one could not and would [not] believe all he had written; it bore evidently the stamp of malice, and hatred, and every unchristian feeling. You may suppose then that it was with much pleasure I read, and I did every word of your excellent, eloquent, and convincing reply copied into *The Times* from a Leeds paper. I have also this morning read another in the same Journal, and I write now for the purpose of expressing my hopes that your promised Letters will appear, not in a country paper, but as pamphlets, for I should, for one, wish to possess them, and you may be sure that I am not the only person who feels this desire. Although a dissenter I am no enemy of the Church, no dishonest longer to grasp what is her's by right and law. Trusting you will excuse this intrusion, which only a love of truth, and strong admiration of your admirable replies prompts me to venture thus addressing you, and claiming your attention, and ardently hoping it is your intention to do as I have expressed my hope,

I am, with great respect,
Your obedient servant,
T. M. BALL.

Early in January, 1834, while staying at Milton Park, Sedgwick met with a severe accident, by which his right arm

[1] To Bishop Monk, 1 November, 1834.

was disabled for several months. "The day after I arrived at
Milton," he says, "I started with a party of ten for Croyland
Abbey, and in passing carelessly under one of the branching
trees, whether by the swerving of my horse, or by incautiously
raising my head too soon, it was caught among the extreme
branches, and I was pulled off my horse[1]." The extent of the
injury was unsuspected at the time, and the patient was
treated for a severe sprain. On his return to Cambridge he
sent for the celebrated surgeon Mr Okes, "who saw the whole
extent of the mischief in an instant, and pointed out the
existence of a great *transverse fault*, throwing down the
metacarpal bones in such a way as to bring one of them to
the end of the radius, and thrust the thumb below the palm
of the hand." The bones were soon put into their right
places, while Sedgwick "howled loud enough to shake all
the windows in the Great Court;" but his recovery was slow,
and for several months any work that entailed legible writing
had to be done by one of his friends. Even the letters
against Beverley were dictated to either Romilly or Kemble[2].
His efforts at lefthanded penmanship did not go beyond a
letter to a relative or an intimate friend, nor could it be said
of him, as of a celebrated Puritan divine,

> "though of thy right hand bereft,
> Right well thou writest with the hand that's left."

Such a condition was ill-suited to a man of his bodily and
mental activity, and he likened himself, no doubt most truthfully, to "a chained bull-dog."

Before long, in despite of his maimed condition, and the
advice of doctors to avoid excitement, he became the central
figure in an agitation which threw the University into confusion for more than six months, having for its object the
abolition of tests on proceeding to degrees. For the moment
he and his friends were unsuccessful, and thirty-seven years

---

[1] This and the following extracts describing the accident, are from a letter to
R. I. Murchison, 8 February, 1834.

[2] John Mitchell Kemble, of Trinity College, B.A. 1830.

elapsed before tests were completely swept away. In the interval, whenever an occasion presented itself, Sedgwick shewed unflagging interest in the cause, and one of the last occasions on which he spoke in public was a meeting at St John's College Lodge, to assist the movement which resulted in the Test Act of 1871.

The movement of 1834, in which Sedgwick bore so prominent a part, began with a petition, drawn up under the following circumstances. In December, 1833, Professor Pryme had offered Graces to the Senate suggesting the appointment of a Syndicate to consider the abolition or modification of subscription on proceeding to a degree. These were rejected by the Caput. In February, 1834, Dr Cornwallis Hewett, Downing Professor of Medicine, offered a similar Grace, with special reference to the faculty of medicine. This also was rejected by the Caput, on the veto of the Vice Chancellor, Dr King, President of Queens' College. Finally, 12 March, 1834, the Senate petitioned to be heard by counsel in respect of the charter of the London University[1]. Thereupon several members of the Senate met at Professor Hewett's rooms, Sedgwick was called to the chair, and it was resolved to draw up a petition to both Houses of Parliament—not as coming from the body at large, but as expressing the opinion of certain individuals, who, from the tactics of their opponents, had no other mode of recording their opinions[2]. After expressing their attachment to the Church, and the University, and their conviction that "no system of civil or ecclesiastical polity was ever so devised by the wisdom of man as not to require, from time to time, some modification, from the change of external

[1] It is difficult to understand why this demand should have given so much offence, but Sedgwick himself enumerates it among the reasons for the action of the petitioners in his letter to *The Times*, dated 8 April, 1834. The University had merely prayed to be heard by Counsel in support of the insertion of a clause in the Charter, "declaring that nothing in the terms of the Charter should be construed as giving a right to confer any Academical distinctions designated by the same titles, or accompanied with the same privileges, as the degrees now conferred by the Universities of Oxford and Cambridge."

[2] Sedgwick's Letter to *The Times*, *ut supra*.

circumstances, or the progress of opinion," the petitioners make 1834.
the following statement : Æt. 49.

" In conformity with these sentiments, they would further suggest
to your honourable house that no corporate body like the University
of Cambridge can exist in a free country in honour or in safety unless
its benefits be communicated to all classes as widely as is compatible
with the Christian principles of its foundation.

Among the changes which they think might be at once adopted
with advantage and safety, they would suggest the expediency of
abrogating by legislative enactment every religious test exacted from
members of the University before they proceed to degrees, whether
of bachelor, master, or doctor, in Arts, Law, and Physic. In praying
for the abolition of these restrictions, they rejoice in being able to
assure your honourable house that they are only asking for a restitu-
tion of their ancient academic laws and laudable customs. These
restrictions were imposed on the University in the reign of King
James I, most of them in a manner informal and unprecedented,
and grievously against the wishes of many of the then members of
the Senate, during times of bitter party animosities, and during the
prevalence of dogmas, both in Church and State, which are at variance with the present spirit of English Law, and with the true
principles of Christian toleration."

As it was thought desirable to get the petition presented
before the Easter recess, time was precious. Accordingly, it
was not circulated publicly, but lay for signature at the rooms
of Mr Thomas Musgrave[1] in Trinity College, from Friday
14 March, to Monday 17 March, while those interested in its
success solicited support by private canvass. It received the
signatures of sixty-two resident members of the Senate.
Among them were two Masters of Colleges, Dr Davy of
Gonville and Caius, and Dr Lamb of Corpus Christi ; nine
Professors, Hewett, Lee, Cumming, Clark, Babbage, Sedgwick,
Airy, Musgrave, Henslow ; several Tutors of Colleges, and
distinguished Masters of Arts. Some of these were either
conservatives, or very moderate liberals. It was presented to
the House of Lords (21 March) by Earl Grey, at Sedgwick's
personal instance ; and to the House of Commons (24 March)

---

[1] Fellow of Trinity College, B.A. 1810. He was Lord Almoner's Reader in
Arabic from 1821—1837, when he was made Dean of Bristol. He became Bishop
of Hereford a few months afterwards, and Archbishop of York in 1847.

by Mr Spring Rice, member for the town of Cambridge. By
both houses it was received with respect, and became the
subject of animated debate.

As might have been expected, it was succeeded, after about
ten days, by a *Declaration*, signed by 101 residents. "We do
not admit," said this laconic document, "that 'the abolition
of' the existing 'restrictions' would be, as alleged, 'a restitu-
tion' of the 'ancient laws and laudable customs' of the University: neither do we acknowledge that any of 'these restrictions
were imposed in a manner informal and unprecedented'."
As these words directly controverted the statements of the
petition, Sedgwick, as "chairman of a party of the resident
members of the Senate who agreed to the words of the
petition lately presented in parliament," addressed a long
letter to *The Times* (8 April) in vindication of himself and his
friends, which may be taken as an official statement of their
position. It deals chiefly with the historical question, and it
is only towards the end that he gives a short account of the
motives by which the petitioners had been actuated, and the
circumstances under which the document had been drawn
up.

By this time the excitement in the University had become
very great. As the number of resident members of the Senate
did not exceed one hundred and eighty, sixty-two of whom
had signed the *Petition*, and one hundred and one the *Declaration*, nearly every resident was directly interested in the
question. The promoters of the *Declaration*, elated at their
success, gave notice of a Grace at the next congregation
(16 April), to affix the University seal to a petition to
both Houses of Parliament praying for the maintenance
of existing tests. Non-residents came up in considerable
numbers, but only to find that Dr Hewett had availed
himself of his right of veto as a member of the Caput, and
thrown out the Grace. This manœuvre, however, could
scarcely be called successful, for the petition was immediately deposited in the hall of Queens' College, and before

long received two hundred and eighty signatures. On the following day it was taken to London by the Vice Chancellor, and within a week presented to the House of Lords by the Chancellor, and to the House of Commons by Mr Goulburn.

Professor Hewett's action was eloquently defended by Sedgwick in a letter to *The Cambridge Chronicle* (16 April), the publication of which, taken in conjunction with his previous letter to *The Times*, and his *Seventeen Reasons for adopting the prayer of the Petition signed by sixty-two Resident Members of the Senate*, involved him in further controversy, notably with a correspondent of *The Cambridge Chronicle* who signed himself *A Member of the Senate*, a designation which concealed his old antagonist, Dr French. From these ephemeral publications we will pass on to a letter written to Bishop Blomfield, as containing a dispassionate statement of the whole question from the point of view of himself and the petitioners[1].

TRINITY COLLEGE, *April* 27, 1834.

My Lord,

I have this moment, under your Lordship's frank, received a copy of your speech delivered in the House of Peers on April 21st[2], and sit down at my breakfast table to reply to one or two paragraphs in which you seem to misapprehend the wishes of the sixty-two petitioners who first moved the question. In my present condition I am compelled to write with my left hand, and have consequently a mechanical difficulty in expressing myself. I must be as plain and short as I can...

Your Lordship's speech seems constructed on the supposition that Dissenters, under the contemplated Act, would have

---

[1] This letter is printed from a copy taken by the Rev. J. Romilly, and preserved by him in the Registry of the University. A few paragraphs, not specially important, have been omitted.

[2] Hansard's *Parliamentary Debates*, Ser. 3. xxii. 994. The occasion was the presentation of the petition signed 16 April against any removal of tests.

1834.
Æt. 49.

1834.
Æt. 49.

the right of admission.   What Mr Wood's[1] intentions were I
know not—I wish heartily the getting up of the Bill had not
been with a Dissenter—but our intentions were to give no
such right, and I have in two letters written some time since,
pressed this very strongly on Lord Grey....We wish no man
to be forced on the University; and if Mr Wood adopts the
suggestions sent up last night and agreed to at my rooms,
the Bill will not touch the rights of the admitting officers in
the several colleges.   A man is not to come up as a Dissenter;
he is not to be considered as such by any official college act;
he must conform to discipline, and we give him a degree
without exacting subscription.   A moderate, well-informed
Dissenter will come up under such a system (this is not
conjecture but fact) and he will take a degree.   A bigot—a
man who would haggle about organs and surplices—will and
must keep away, and we do not want him.   A right to a
degree without signing a test does not do away with the
necessity of discipline, or of conforming to college rules; nor
does it give (as far as our wishes are concerned) any right of
admission which is not sanctioned by the voluntary acts of
the admitting officers.   If Dissenters were to come up as such,
and allowed to force themselves on the several colleges, I
should then agree with every syllable in the speech I have
before me.   But we look to no such result; and if it come at
all it will come as a future consequence of the exclusive policy
which is now maintained.   The Universities cannot maintain
their old position and continue Universities.   This your
Lordship seems in part to admit, as you contemplate the
lopping off of the Medical Faculty.   I may be mistaken, but
I cannot bear the thoughts of this, and I think the policy that
suggests the possibility of it perfectly suicidal.

[1] Mr G. W. Wood obtained leave (17 April) to bring in a Bill to grant to His
Majesty's subjects generally the rights of admission to the English Universities,
and of equal eligibility to degrees therein, notwithstanding their diversities in
religious opinion—degrees in Divinity alone excepted.   It passed the House of
Commons by large majorities at its different stages, but was rejected by the House
of Lords.   Hansard, *ut supra*, 902; xxiv. 492, 632, 1087; xxv. 815.

You say, my Lord, that when Dissenters—but be it
remembered only after having been admitted and having
kept terms like other men—have a legal right to academic
honours they will not long consent to be subject to college
rules relating to chapel, lectures, etc. I am compelled to
say that I do not see the force of this. A Dissenter knows
our organization when he comes up, and if the advantages of
our education have induced him to conform during years
past, *a fortiori* he will be willing to conform when he can
thereby have also the advantage of a degree. This appears
to me perfect demonstration. Dissenters may have some
foolish expectations from the operation of the intended Bill,
but we cannot help this. Again, I affirm with perfect confidence that the operation of the Bill implies no change
whatsoever in the college lectures. We have had amongst us
during the last twenty years Roman Catholics, Methodists,
Presbyterians, Quakers, Congregationalists of every shade,
and they all attended lectures, and never, I believe, made a
single objection to a lecture given in College. Let me appeal
to your own experience, and to that of your friends the
Bishops of Gloucester and Lincoln....

I feel so confident in the truth of what I am now saying
that I have not in my heart been able to acquit of a charge
of insincerity some of those who have accused the sixty-two
petitioners of attempting to destroy root and branch the
system of religious education in this University. Were our
wishes carried into effect I verily believe we should be compelled to compromise nothing on which a good and charitable
churchman ought to make a stand. No man is now forced to
attend the sacrament. The attendance at chapel would,
I think, be better after the proposed change than it is at
present, and the public Professors of Theology would probably
be called to renewed and more effective exertions in behalf of
the doctrines of our Church. In parochial instruction it is, I
doubt not, impossible to blend together men of different
persuasions. But may it not probably be far otherwise, or

1834. rather, I ought to say, has it not been far otherwise in a
Æt. 49. system of academic instruction? The two cases are so dissimilar, that, with all deference, we cannot, I think, argue
from one to the other.

*   *   *   *   *   *

You say that the proposed Bill would be *an infringement
of our privileges*. In which respect? To give degrees in the
faculties is our privilege: the tests with which these degrees
are clogged are no privileges, and we want to wipe them out
as worse than nothing. This is our prayer. We may be
right, or we may be wrong, but we want *to preserve our privileges*. And those who resolve to keep these tests at all
hazards know that they are supporting that which, if upheld,
will lead to *an infraction of our privileges*, and virtually cut
off from us the medical faculty. Your Lordship does not
know the enormous injury this amputation would inflict on
us. Our Professor of Physic gives an admirable course
of lectures, and his pupils attend the Hospital. The
Professor of Anatomy gives a new and extended course.
We have built a Museum and purchased specimens and
anatomical models at the expense of thousands. We have
an extended course of Chemical and Botanical lectures with
reference to the Anatomical Class. Are all these things
to vanish away? Yes! say those who oppose the sixty-two
petitioners: Perish Science and live the Tests! We will
not allow even so much as a Syndicate of inquiry! And
yet the same persons would enter on a negotiation with
Sir Henry Halford which would virtually swamp our whole
medical faculty, as well as the lectures which have risen in
consequence of it!

Since your time, my Lord, Cambridge has improved in
vitality. We have a chartered Philosophical Society which
has produced five large volumes of *Transactions* rivalling in
original matter the first scientific memoirs in Europe. We
have a noble Observatory in full action, and in honourable
correspondence with all the other public Observatories in the

world. And who first started and set afloat these noble monuments of Cambridge zeal and learning? Some of those who took a leading part among the sixty-two Petitioners. And yet they are to be set down as innovators, lovers of movement, and disturbers of the consecrated institutions of their country! So far they are men of movement that when they see everything about them stirring they know they cannot remain immovable without being left in helpless solitude. They believe that the scientific character of Cambridge is not only its honour but its *security*. As a great learned and scientific University giving degrees in all the learned faculties—incorporated as a lay body, and only regarded as such in the eye of the law of England—...Cambridge may stand firmly....But if she once be considered as a mere school for the Church Establishment her endowments will be thought out of all reasonable dimensions, and before many years are over we may see our noble edifices beginning to crumble about our ears.

When the Act of Uniformity was in force, with all its terrific train of penalties, and all who refused the injunctions of the Test and Corporation Acts were excluded from offices of trust and honour, our exclusive system was in harmony with the law. Now it is not so. And we wish to put ourselves right with the actual constitution of our country, so that we may still be the nurseries and fountains not merely of the Church but also of the Commonwealth.

*   *   *   *

Pray excuse this formidable visitation; accept my best thanks for your speech; and believe me, my Lord,

Your most faithful servant

A. SEDGWICK.

Before long Sedgwick's attention was engaged by another matter arising directly out of the same agitation. The strength of public feeling on the question of tests had manifested itself in a number of pamphlets, of which Dr Turton's[1] *Thoughts on the Admission of Persons, without regard to their Religious Opinions, to certain degrees in the Universities of England,* was perhaps the ablest, and certainly the most widely read. It was promptly answered by Connop Thirlwall, at that time an Assistant Tutor of Trinity College. He was one of Sedgwick's intimate friends, and had cordially co-operated with him in the matter of the petition, though there is no evidence that he had taken any very active part in promoting it. In the course of his reply to Dr Turton he was led to inquire whether colleges might be held to be schools of religious instruction; and, having answered this question in the negative, went out of his way to denounce the existing system of compulsory attendance at chapel as a positive evil. Within a week after the appearance of this pamphlet, the Master, Dr Wordsworth, called upon the author to resign his office, and Thirlwall, almost without hesitation, complied. An exercise of authority so despotic, and so unprecedented, added to the unpopularity of the Master with many of the Fellows, was received with a loud outburst of indignation. Sedgwick, who happened to be in London, was promptly informed by Whewell of what had taken place. After briefly recording the facts, and his own disapproval of the Master's conduct, he went on to say:

"What will happen next I have no guess, for I have talked with none of Thirlwall's friends about the case, but I much fear they may attempt some violent and rash measure; and what I wish to beg of you is that you will be our good genius, and moderate instead of sharing in, our violence....You have more influence in the College than any other person, and have perhaps the power of preventing our present misfortunes being followed by any fatal consequences[2]...."

Sedgwick hastened back to Cambridge, and did all that could be done, under the circumstances, in conjunction with

---

[1] Thomas Turton, D.D., Regius Professor of Divinity from 1827—1845, when he was made Bishop of Ely.

[2] From Rev. W. Whewell, 27 May, 1834.

Musgrave, Sheepshanks, Romilly, and Peacock[1]. But, as Thirlwall had resigned, their efforts could effect nothing except a dignified submission to the inevitable. From our point of view the matter is chiefly interesting, as shewing the position which Sedgwick had attained in College, and which he kept throughout the rest of his long life.

The following letters furnish an appropriate conclusion to the busy episodes in Sedgwick's life which have just been narrated.

From Mr Robert Southey.

KESWICK, 10 *February*, 1834.

My dear Sir,

I am much obliged to you for your discourse, and for the pleasant letter that accompanied it. It is indeed most gratifying to see you employing your sledge-hammer against the Utilitarians; and counteracting the mischief which has been done by Locke and Paley. Heavy as the hammer strikes, your name and character carry with them equal weight; and I do not think any other person could at this time have done so much good.

This too I can truly say, that in these dark times, nothing has cheered me so much as the part which you have thus taken.

Believe me, my dear Sir,

Yours with sincere respect and regard,

ROBERT SOUTHEY.

From Mr William Wordsworth.

RYDAL MOUNT,
May 14th, 1834.

" My dear Sir,

I am much indebted to you for a Copy of your discourse on the Studies of the University; and which has been read to me twice. It is written with your usual animation, and I hope will in the course of time prove of beneficial effect, if the Universities are to continue to exist; which from some late proceedings in your own, I am disposed to doubt.

In every part of your Discourse I was interested, but was most nearly touched by your observations on Paley's Moral Philosophy, which,

---

[1] Mr Romilly records in his diary, 29 May, "Sedgwick and Sheepshanks arrived from town to-day to look into Thirlwall's case. Sedgwick and Musgrave drew up a Paper, and took it to the Master: 'We the undersigned resident Seniors request you to call a Seniority to inquire into the proceedings which led to Mr Thirlwall's resignation of the Tuition.' Signed by Sedgwick, Musgrave, Peacock, Romilly, Sheepshanks."

1834.
Æt. 49.

tho' like all his works a Book of unrivalled merit in certain points, is deplorably wanting in essentials. In fact there is no such thing as Morals as a Science, or even as Philosophy, if Paley's system be right. You and I, I remember, talked upon the subject when I had last the pleasure of seeing you—so that I need not say more than that I heartily concur with you in what your Discourse contains upon it.

Thank you for the drubbing you have given that odious Slanderer, Beverley. I was sorry to learn from those letters that you had had so severe an accident. You have the best wishes of all this family for your entire and speedy recovery.

Should I be silent upon the part you have taken as the Public Leader of the 62 or 63 Petitioners, I should not be treating you with sincerity, or in the spirit of that friendship which exists between us. This is not the place for me to discuss the subject, and tho' I feel that my opinion, as an opinion merely, may not be entitled to much respect, as your personal friend I cannot hold back the declaration of my conviction that the Petitioners are misguided men,—that part of them, at least, who have signed this Petition with a hope, that by so doing, they are contributing to the Support of the Institutions of the Country, the Church included.

Farewell! God bless you, and be assured that whatever course you pursue either in public or private life, there is no likelihood that I shall ever have occasion to doubt that *you* act from pure and conscientious motives. At the same time allow me to say, that I have no dread of being accused of presumption by you, for not having bowed to the scientific names which stand so conspicuous upon this ill-omened Instrument.

"Ever faithfully yours,

"WM. WORDSWORTH."

The history of the Woodwardian Museum during the four years comprised in the present chapter must now be briefly noticed. The unfitness of the room for the purposes of a Museum, occupies, as heretofore, a considerable space in the reports of the Inspectors. In 1830 they extend their observations to the Professor's college rooms, where they find not only cabinets belonging to the University, but "ten or twelve packing-cases from Germany and other parts of the continent, as yet unopened, for which the Professor cannot find room either in the Museum or in his chambers." In 1833 these defects are remedied to some slight extent by the acquisition of the two rooms at the west end of the Divinity School, (now the Music Room and the Newspaper Room of the

Library) hitherto used by the Registrary ; but Sedgwick " is compelled to state that this addition is by no means adequate for the reception of the present collection, much less for its proper exhibition, or for such augmentations as the present state of geological science requires." Meanwhile, thanks to Sedgwick's own exertions, important additions were being made in nearly every year. We read of casts of *Plesiosaurus* and *Ichthyosaurus* presented by Viscount Cole and Mr Chantrey (1832) ; of a collection lately the property of Dr E. D. Clarke, removed from the cellars in the Botanic Garden after his death, and of specimens collected in North and South Wales by the Professor (1833); and lastly, of a bust of Cuvier presented by his widow (1834).

Sedgwick's plans for the summer had been settled early in the year. " I propose," he wrote in April, " to spend the early part of the Long Vacation in Wales, thence to find my way by steam to Glasgow, to mount up to and batter the Grampians, to descend by the west coast of Ayrshire, and then to thread my way among the hills of the Lammermuir chain, so as to end my work in September in time for the meeting of the British Association at Edinburgh, where I shall have to resign my office of President of the Philosophical ἁμαξόβιοι[1]."

This programme, in its main outlines at least, was faithfully carried out. Early in June he started for Wales with Murchison. On this visit he began work on the south side of the Principality, instead of on the north, for the purpose of examining, under his friend's guidance, the ground he had already gone over, and thus, as he said afterwards, learning the alphabet of the Silurian tongue[2]. The excursion was specially gratifying to Murchison, who parted from Sedgwick under the firm conviction that he was fully convinced of the accuracy of his determination of the sequence of the rocks examined. " Although I think and hope," he wrote to

[1] To Rev. W. Ainger, 2 April, 1834.
[2] Salter's *Catalogue*, p. xix.

Whewell, "that he endeavoured to pick every hole he could in my arrangement, he has confirmed all my views, some of which, from the difficulties which environed me, I was very nervous about until I had such a backer[1]." After six weeks spent in "marches and countermarches in Hereford, Brecon, Caermarthen, Montgomery, and Salop," the friends parted at Ludlow (10 July). Sedgwick hastened back to his old ground in North Wales, where he probably spent most of the time until claimed by the Association.

Sedgwick has unfortunately left no account of his proceedings at Edinburgh, where he was the guest of Dr Alison, father of the historian. We know from the *Report* of the Association that he resigned the office of President in an eloquent, though not specially noteworthy speech, but that he did not himself contribute anything to the geological section. Before leaving Edinburgh he had the pleasure of making the acquaintance of Professor Agassiz, who had just come to Great Britain for the first time. A life-long friendship between the two geologists was the result of this interview.

After the meeting Sedgwick and Murchison, according to previous arrangement, started together for the south. From this point the story can be told in his own words:

TRIN. COLL., November 15, 1834.

My dear Mrs Alison,

The day I left you I had a delightful drive along the banks of the Esk, which contrast so finely with the country on both sides of them. Straight furrows and rotation crops may gladden the farmer's face, but they had few charms for my companion and myself. We halted, however, at the cliffs of Dunbar (they are very curious and if you have not seen them pray look at them the next time you pass that way), and we were overjoyed at the sight of the noble glen of Dunglass.

---

[1] Murchison to Whewell, 18 July, 1834, quoted in Geikie's *Life of Murchison*, i. 223.

Perhaps your brother[1] has told you of our expedition to   1834.
St Abb's Head.   Nothing could turn out better—the geology   Æt. 49.
most instructive—the scenery grand and varied—the sea as
smooth as glass—the cliffs sublime, and every headland re-
flecting the lights of a glowing sun.   We were all excited to
the highest pitch; sometimes speculating on the strange
frolics dame Nature had loved to play thousands of years
before strathspeys were thought of; then talking of Walter
Scott, the Master of Ravenswood, and the Middle Ages;
and ever and anon, as conversation seemed to flag, plying
Sir John Hall's bottles, not so much from the love of
wine, as in the hope of seeing the embers of imagination
blaze out afresh.   The day after our sea-trip Murchison
went helter-skelter after the foxhounds, and had the good
luck to come back with a fox's brush, and an unbroken
neck.

On the Saturday we were again under way, and continued
together to Newcastle, where we parted; but I had the good
fortune to pick up another friend and knight of the hammer,
with whom I struck up a league offensive and defensive, and
we forthwith commenced an action of assault and battery
against the ribs and shoulders of the mountains which range
in a lofty unbroken chain from Stainmoor through Cross Fell
to the frontier of Scotland.   I will not torment you with any
narration of our battles and victories; sufficient to say that
we parted at Carlisle—that I found my way to another
friend's[2] house near Whitehaven—that I joined some young
people in a pedestrian tour to the Lakes, and contrived to
reap a rich harvest of joy from the exuberant spirits of
my youthful companions—that I talked a day and a half
with Wordsworth, who is the best talker I have the happiness
of knowing, and who talked in his best fashion—that I found
my way to my native valley, which will bear looking at after

[1] Duncan Farquharson Gregory, afterwards Fellow of Trinity College, B.A.
1838, M.A. 1841.
[2] Probably Mr Williamson Peile.   See above, p. 409.

1834.
Æt. 49.
the fairest prospects of the Lake region—that my friends
were all well, my nephews and nieces springing like mush-
rooms, and the old vicarage house about as noisy as it used
to be when I was myself a child.

But my sheet is ending and it is high time for me to end
with it. Give my kindest remembrances to everybody at
Woodville and Heriot Row, and to your brothers and sister,
and believe me, Dear Mrs Alison,

Your obliged and affectionate friend,

A. SEDGWICK.

By the end of October Sedgwick was back in Cambridge,
restored to good health and spirits by his summer in the open
air, busy with his lectures, and with preparations for a visit
from Professor Agassiz early in November. In order to lay
before the " famous foreign fishmonger," the fare he specially
fancied, Sedgwick had been for some time in correspondence with his friend Gwatkin, through whose good offices his
own meagre table was to be garnished with " a dish of fish"
from Barrow in Leicestershire. " A Yorkshireman hates to
buy a pig a poke," he added " but I am sure I may trust my
old fellow-hammerer to make a good bargain for me[1]." The
fish in question did justice to Gwatkin's discrimination, and
were duly added to the Woodwardian collection ; but, at the
last moment, to Sedgwick's great annoyance, Agassiz was
prevented from coming.

Soon afterwards an event occurred which gave Sedgwick
a new position, and, for a part of each year at least,
diverted the current of his thoughts into a new channel.
In the middle of November Lord Melbourne's ministry broke
up, and just as Lord Chancellor Brougham was lamenting
that Sedgwick and Thirlwall were the only clergymen who
had deserved well of the liberal party for whom he had been

[1] To Rev. R. Gwatkin, 3 October, 1834. Mr Gwatkin was then vicar of
Barrow on Soar, Mount Sorrel, Leicestershire.

unable to provide[1], came the news that a stall at Norwich, and a rectory in Yorkshire, were vacated by the death of the gentleman who had held them both. Brougham gave the stall to Sedgwick, and the rectory to Thirlwall. The pre- ferment, as Sedgwick said, "was saved as from the fire," for it was only presented to him formally on the day before Brougham gave up the Great Seal. Sedgwick accepted without a moment's hesitation, and, as the next letter tells us, hurried through the formalities of induction with equal rapidity.

1834.
Æt. 49.

CLOSE, NORWICH,
Dec. 15, 1834.

My dear Ainger,

My poor brain is turned topsy-turvy, and my memory has fled so far from me that I cannot tell to whom I have written, and to whom I have not, since I became Prebendary of Norwich. Here however I am, in my own Residence, as good a Prebendary as you can see on a winter's day, though still without a shovel hat. My friends in College have been putting about a shilling sub- scription to buy me a gorgeous shovel hat. I shall receive it with due gratitude, and hang it on a peg to be looked at, but, as to putting it on my nob, that is another question. I doubt not you have heard of my appointment. Perhaps I informed you of it myself. If I did, I have forgotten it. One of the last acts of the ex-chancellor was to put the great seal to my presentation. The very day I received the notification of this act, I heard from a friend at Norwich who told me that the Chapter had heard of Lord Brougham's intention, and that, if I received the presentation in time, it would be very agreeable to them that I should take my pre- decessor's turn, which commenced on the 1st of December. By so doing I should secure all the domestic arrangements of the Chapter. There are six Prebendaries, and each resides

---

[1] Lord Houghton in *The Fortnightly Review*, February, 1878. At the end of 1833 Lord Brougham had induced Sedgwick to revise the MS. of his *Discourse on Natural Theology.*

two months. Now December and January just suit me, as
the greater part of these two months falls in our Christmas
vacation.   Partly therefore on account of my Brethren of
the Chapter, and partly on my own private account, I lent
a willing ear to this suggestion—went up to Town without
an hour's delay—procured my presentation from the Chancellor's office—took an early coach to Norwich—arrived on
Saturday, November 29, in time for the Dean's breakfast—
took the oaths and signed the books—presented my deed
with the great seal affixed to the Dean after the First Lesson
in the morning service—was formally installed in the presence of the congregation—read in on Sunday the 30th—
and commenced Residence in my own house on Monday, the
1st inst. Is not this doing business? My servant arrived
after a day or two, Lady Jane Wodehouse (the wife of
one of my Brother Prebendaries) provided me two excellent
maid servants.   I have taken my predecessor's furniture and
wines at a valuation, and am gradually settling down into
my proper place. Our Residence while it lasts is severe.
We are not permitted to be away from our houses for a
single night. Attending service regularly, and preaching
generally once each Sunday, are duties which are looked for.
We have also to give certain dinners of ceremony to the
officers of the Cathedral.   Giving and receiving dinners con-
stitutes a formidable service in a city like this.

What my stall may do for me in the end I cannot say,
but I am quite sure that for the first year it will make me
poorer than I have been since I knew how to spell my
own name. My fees and furniture will run me into debt
to the tune of six hundred pounds at the very least. I
wish some good Christian would just now give me a thousand pounds, it would just make a poor body comfortable.
If my life be spared the stall will I doubt not turn out a
very comfortable thing. I hope I may count upon its producing me nearly £600 a year. This, together with my
Senior Fellowship and Professorship, must surely enable me

soon to lift my head above water. My clerical employment here is a good thing, and I mean not to flinch from
it. The preaching I spoke of is not compulsory; but has
been commenced of late years by some of the new comers.
Pray write to me soon.

Yours ever,

A. SEDGWICK.

An amusing incident respecting Sedgwick's first visit to
Norwich, deserves to be recorded. He called on Dean Pellew,
as the above letter shews, in time for an early breakfast; and,
on being shewn into the drawing-room, found there his
daughter Minna, aged three, playing at bricks. Sedgwick
at once went down on his knees, and assisted her to build
a tower of Babel, in which occupation, the Dean, to his great
amusement, found his new Canon busily engaged. Sedgwick
never forgot either the child, or the incident, but maintained
a close friendship with her until his death, writing to her
frequently, and generally sending her a present on her birth-
day. These letters invariably contained either some allusion
to the "early lessons in architecture which you gave me," or
some such passage as the following : "Perhaps you think me
wrong in calling you my *oldest friend*" he says in 1850; "at
any rate yours is the oldest friendship which I formed in
Norwich; and it has never been interrupted since I began to
build castles with you on your carpet on the 29th November.
1834[1]."

The following letter, though written two months later,
completes the history of Sedgwick's first experience of his
new dignity.

[1] Among the congratulations, humorous and serious, which were showered
thick upon Sedgwick, may be quoted an epigram, by C. V. Le Grice. Lord
Brougham is supposed to have presented the Stall with the following couplet:

"Dear Adam, if, as I believe,
You'll one day wish to have an Eve,
Then on the Eve of such event,
At Norwich snug I've pitched your tent."

SUNDAY EVENING, 11 P. M. 8 *February*, 1835.

My dear Ingle[1],

After chapel I went to drink tea with Prymc, our
City Member, and have had a very long talk with him; and
on returning to my den I just looked in upon Thirlwall, who
starts for Yorkshire tomorrow, at an hour when I shall probably be recumbent, and between a pair of sheets. He tells
me that he shall halt at York. I am therefore seated at the
desk scribbling a page, and perhaps two, which he promises
to convey for me as far as York, and perchance to your door.
You don't know him, it seems; but it is clear that you ought
to know him; and I hope this act of great benevolence on his
part will move your bowels, and make you friends as long as
you both last.

I only returned to College yesterday. My Cathedral
Residence ended indeed on Saturday the 31st January; but it
took me five days to pay bills, pack up odds and ends, and
unhook myself from a hundred little engagements. I, however, moved off the stocks on Thursday night—halted a day
with Dr Bayne at Bury, arrived in Cambridge yesterday,
and here I am to-day (Sunday). Your letter delighted me;
not because of the grease and butter which covered its first
page, but because it convinced me that you were enjoying
your oldest and best flow of animal spirits, and that you
could receive some pleasure from the recollection of old and
good days when Charles Ingle was a burly school-boy begging
salt at a sizar's door. My residence at Norwich forms a
strange episode in my history. Now that I am once again
in my old haunts, I can hardly believe that I have not been
dreaming. While there, I was in the position of Vice Dean.
In the absence of the Dean I was the official representative of
the dignity of the Chapter—called upon to practice a series of
formal hospitalities in a queer, old-fashioned, in-and-out, ugly,
old, house. Several times I was afraid of being on my beam

---

[1] The Rev. Charles Ingle was then vicar of Osbaldwick, near York.

ends ; but by some special providence I was saved from ship-
wreck, and am at last safe in port. Everybody was kind and
hospitable ; indeed I have been almost killed with kindness;
and all the good old Tory inhabitants of the rookery seemed
mightily anxious to see how such a monster as a Whig
Prebendary would behave at meals; and you may depend
upon it they have all been much built up with the sight. I
did, however, contrive to bring together more heretics and
schismatics within my walls than ever had been seen before
in a Prebendal house since the foundation of the Cathedral.
Independents and Highchurchmen were seen licking out of
the same fleshpots, and Quakers crossed my threshold without
fear and trembling.   By the way some of the Quakers are my
delight.   J. J. Gurney is an excellent and learned man,—
brother of Mrs Fry and Mrs F. Buxton,—reads Hebrew, and
spouts the Greek Fathers by the hour together.   I don't
believe there is a better man living.   Friend Amelia[1] you
know well.   I like her much; but I never dared to rumple
her cap in the way you mention.   I have also been much
given to preaching, holding forth twice, and sometimes thrice,
on a Sunday.   But, if I begin to preach now, Thirlwall will be
asleep.   So good night and God bless you.

Yours always

A. SEDGWICK.

[1] Mrs Opie.

CHAPTER X.

(1835—1840.)

CAMBRIDGE OCCUPATIONS. ELECTION AT DENT. PRESENTATION
AT COURT. BRITISH ASSOCIATION AT DUBLIN. SKELETON
OF IRISH ELK. VISIT OF AGASSIZ TO CAMBRIDGE (1835).
LECTURES AND SOCIETY AT NORWICH. GEOLOGICAL TOUR
IN DEVONSHIRE WITH MURCHISON. DEATH OF MR SIMEON
(1836). ILL HEALTH. PAPER ON GEOLOGY OF DEVONSHIRE.
CRITICISM OF BABBAGE. DEATH OF BISHOP BATHURST.
FOUNDATION OF COWGILL CHAPEL. GEOLOGY IN DEVON-
SHIRE. BRITISH ASSOCIATION AT LIVERPOOL. INUNDATION
OF THE WORKINGTON COLLIERY (1837). EXPLORATIONS AT
BARTLOW. DEVONIAN PAPER. QUEEN'S CORONATION.
BRITISH ASSOCIATION AT NEWCASTLE. OPEN-AIR LECTURE
(1838). THE SILURIAN SYSTEM PUBLISHED. FOREIGN TOUR
WITH MURCHISON (1839). ILL HEALTH. CHELTENHAM.
PAPER TO GEOLOGICAL SOCIETY (1840).

SEDGWICK was evidently much gratified with his first term
of residence at Norwich. He was conscious of having won a
dignified piece of preferment by his own merits, without
interest or favour; he was pleased with his new friends, and
did not, at first, find his duties irksome. Before long, it
must be confessed, as the novelty of the situation wore off,
he became less enthusiastic, and for a while was listless and
ill at ease.

This can surprise no one who reflects for a moment on
what his previous life had been. When he became a Preben-

dary of Norwich Cathedral he was fortynine years of age,
and had resided in Trinity College for just thirty years. His
pursuits, his habits, his affections, were all bound up with the
interests of the College and the University. His intimate
friends, with the exception of Ainger and Murchison, were all
working with the same objects in view. The duties entailed by
his position as a Fellow and a Professor were not onerous,
even in term-time; and, had he been a better economist of
time, he might have devoted almost as many hours as he
pleased in each day to his own pursuits. But at Norwich
he could not call a moment his own. He had to lead an
essentially public life; to submit to incessant interruptions;
to be at the beck and call of anybody who chose to ring his
bell. Even the services in the Cathedral—so different from
those to which he had been accustomed in the college chapel—
were exceedingly distasteful to him. "These long services," he
writes, "cut my time to shreds, and destroy the spirit of
labour. We have the shadow of Catholicism without a grain
of its substance, for not one of the Chapter thinks himself
better for these heartless formalities, or nearer heaven. A
cold empty Cathedral, and a set of unwilling hirelings singing
prayers for an hour together. The bell tells me I must be
off....I am just returned, after a full hour and a half of shiver-
ing. And what the congregation? One single old woman in
addition to the officials. As soon as my fingers are warm I
have to go a mile to the County Hospital to read morning
prayers; for this month I am chaplain. On my return I
shall have time barely for a short walk (or ride if the horrid
weather take up) and then another long Cathedral service
from which I shall come home dog-tired and unfit for work[1]."

These expressions, and many others that might be quoted
from letters written in 1836 and 1837, must not be taken too
literally. Still it can hardly be doubted that for some time
he was out of his element at Norwich. Gradually, however,
like a tree that has been transplanted into genial soil, he

[1] To R. I. Murchison, 25 January, 1837.

became thoroughly happy in his new surroundings; he regarded his old-fashioned house in the Close, especially after
it became filled with his nephew's children—as a second
home; and probably no member of the Chapter performed
his duties, whether public or private, with so much regularity,
heartiness, and success. In a subsequent chapter we shall
throw together some reminiscences of his life in Norwich
which have been collected by those who knew him and loved
him well. These, however, will be more intelligible after
some of the principal persons with whom he was there
associated have appeared in the general narrative of his life.

Sedgwick was back in Cambridge by the end of January,
1835. "Since my return," he writes, "I have been almost
driven off my feet: lectures, college business, arrears of
correspondence, disagreeable domestic news involving me
in the botheration of lawyers' consultations, etc., etc. and
more than all together the oppressive consciousness of having
more work before me than I have any chance to get through:
all these causes have driven me out of my senses[1]." Of the
above-mentioned lectures one, towards the end of the course,
was delivered in the field, a mode of instruction which under
Sedgwick's guidance became exceedingly popular, even with
those who cared nothing for geology. On this occasion a
cavalcade of seventy horsemen started from Cambridge, and
rode across the fens. Before the day was over Sedgwick had
given five distinct lectures; the last, on fen-drainage, from the
top of Ely Cathedral[2].

In the course of this term the Senate determined upon
a step which must have given Sedgwick great satisfaction,
inasmuch as it held out a prospect, at last, of providing
decent accommodation for the geological collections. A
scheme for providing a new Library, with Museums and
Lecture-Rooms beneath it, after being under consideration for
six years, had been abandoned for want of funds. A Syndicate

---

[1] To R. I. Murchison, 17 February, 1835.
[2] *Diary* of Rev. J. Romilly, 9 April, 1835.

was now appointed to do what ought to have been done in the first instance, namely, to solicit subscriptions. Of this Syndicate, which included all the Professors, Sedgwick was a member in virtue of his office. It is hardly likely that he would busy himself very actively in work which would have been singularly uncongenial to a man of his temperament; but his correspondence shews that he successfully solicited a few of his friends at a distance, while he himself contributed the substantial donation of one hundred guineas.

Notwithstanding the distraction of these diverse occupations, he found time to prepare one of his most important papers, *On the Structure of large Mineral Masses*, and to read it to the Geological Society (11 March), before he was called to Dent by an election for the West Riding of Yorkshire. The contest ended in the return of the liberal candidates, Viscount Morpeth and Sir G. Strickland, to Sedgwick's great satisfaction and amusement. To Dr Ainger, whose son, then at Sedbergh School, was of the same politics as his father, he wrote triumphantly: "Well! have we not worked the Tory noodles of the West Riding? Why in the name of wonder have they disturbed our fraction of the county? I turned mob-orator, and had unbounded success, so that all the music, fun, and noise was on our side; and in keeping my stiff-necked Dalesmen from drinking and fighting I really think I did something little short of a miracle. Before I came they had (though in a good cause) shewn a little over-zeal; and your son, who came as flag-bearer of the blues in Fawcett's carriage, went away, I fear, in a rather dirty envelope[1]." To

---

[1] To Rev. W. Ainger, 18 May, 1835. It was on this occasion that Sedgwick was accused of having delivered "a political harangue from the pulpit." The truth was that he had told the congregation that giving a vote was a solemn duty, to be discharged "as unto God and not as unto man;" and that above all they must avoid the sin of intemperance. His brother John writes to a friend (27 November, 1835): "I am happy to say his exhortations seemed to produce the effect of raising the standard of morality amongst the people; for it is a striking fact that not a drunken man was to be seen during the two days' contest, amongst more than 1000 people gathered together, with about two hopeless exceptions."

Canon Wodehouse he explained his political opinions in more
sober language :

"Our party came in at a canter ; and why were the Tories
so foolish as to start the race ?   If the country is to be saved,
it must be by the union of such men as Morpeth and Wortley.
Contests such as I have witnessed put off indefinitely any
reasonable hopes of a broad and firm coalition of good men.
And who is to gain in the mean time ? The tories ? Certainly
not.   The radical party gain a cog at every movement of
the state machine.   Tory domination, in any sense of the
words, is gone for ever.   Yet your party can't see that ; and
think it a goodly triumph if they can ruin a whig in any
corner of the land.   Do you think it possible that the men
who were joined with Sir R. Peel in the late Ministry could
go on with him in the measures he contemplated without
utterly ruining their characters ?   I think not : and this at
least we know, that some of them started with a direct violation of pledges that they had given on the hustings.   I don't
think the Whigs a strong party ; yet I hope they will be strong
enough to get through both Houses a good drastic measure
respecting the Irish Church.   Till that is on some resting
place no Ministry can stand six months.   After two or three
tumbles, we may perhaps live to see a coalition : but it may
come too late[1].

Early in June he was called to Yorkshire again by the
sudden demise of Miss Sill, an old lady of fortune, who had
made him her executor, and as it turned out, one of her
residuary legatees.   "For the first time in my life," he wrote,
"I am to be well paid for my work ; " and in fact, when the
accounts were made up, he found himself the fortunate
possessor of the thousand pounds which a short time before
he had wished some good Christian would give him.   Before
leaving Cambridge he had made up his mind to attend
a Levee and a Drawing Room, as in duty bound, but, with
characteristic carelessness, he had neglected to take the steps

---

[1] To Canon Wodehouse, 25 May, 1835.

prescribed by etiquette. In this dilemma he made a diverting appeal to Murchison : " A card, or notification, is to be left with some person at some place, to convey some information about my courtly intentions, and without these things I cannot be received. Will you then do this unknown operation for me ? As I am coming up partly on purpose, it would be folly to fail in mere forms[1]." The next letter shews that he did not journey in vain :

1835.
Æt. 50.

TRIN. COLL., *June* 30, 1835.

My dear Ainger,

Now be it known to you that I reached London on Tuesday, that I kissed hands at the Levee on Wednesday, and that I exhibited my handsome face at the Drawing Room on Thursday. In short I am now a finished courtier.

We are already beginning to hear the cry of the Installation[2]. Four and twenty years since I enjoyed the festivity intensely. My capacity for certain noisy robust enjoyments is certainly less now than it was then; but on the whole I am full as happy now as I was then ; at least so I think, and I ought to know best. Yet Shakespeare says "past and to come seem best, things present worst," does he not ? and have not poetical generalities, like other generalities, their exceptions ? Let me then remain an exception on the right side. My love to your household.

Yours ever,

A. SEDGWICK.

Sedgwick had determined to do no field-geology this year until after the meeting of the British Association at Dublin in August, and therefore remained at Cambridge until it was time to start. On reaching Liverpool with the intention of

[1] To R. I. Murchison, 17 June, 1835. He was presented by the Lord Bishop of London.
[2] H.R.H. the Duke of Gloucester, Chancellor of the University, had died 30 November, 1834 ; and John Jeffreys, Marquess Camden, had been elected in his room without a contest, 12 December, 1834.

1835.
Æt. 50.

crossing by the ordinary night steamer, he found that a special steamer, the *William Penn*, had been placed at the disposal of the Association by her owners, and was to start early on the following morning, Sunday, 9 August. He therefore agreed to wait, and to officiate as chaplain. As they proceeded down the Mersey, along a new channel discovered and laid down a short time before by Captain Denham, it was suggested that his infant son, who happened to be on board with his mother, should be baptized by the chaplain before morning service. This was accordingly done; and, says an eye-witness, "by one of those strange and fortuitous accidents which often lend an air of romance to the realities of life, it happened that just as the service began, the vessel arrived close by a newly invented iron boat, bearing an apparatus and a bell, which rings constantly as the boat is rocked by the waves, and warns mariners of their position when fogs are so thick that they cannot discern guides of any other kind; it now fairly rung the inmates of the *William Penn* into church, and Annesley Turner Denham, aged three months, was made a Christian almost within its sound." When prayers were over Sedgwick preached on a text from one of the Psalms. The peculiarity of the occasion gave a spur to his eloquence; and the assembled passengers, together with the crew, listened to him with rapt attention while he enforced the true end of all scientific and philosophical pursuits; and from the least as well as the greatest discoveries of man, traced the whole to a Being of infinite power, wisdom, and goodness. A Roman Catholic gentleman, who described the scene to Miss Edgeworth, had tears in his eyes as he spoke of the effect the sermon had produced[1].

Sedgwick was Vice-President of the Geological Section, and, conjointly with Murchison, read a paper *On the Silurian and Cambrian Systems, exhibiting the order in which the older*

[1] From Miss Edgeworth, 23 November, 1836. The rest of the account is taken from *The Literary Gazette* for 1835, p. 513, and from a letter written by Sedgwick to his brother John, 21 August, 1835.

Sedimentary Strata succeed each other in England and 1835.
Wales. The other incidents of the meeting, and Sedgwick's Æt. 50.
proceedings afterwards, are related in the following letters;
but he omits to record that while he was clambering along a
steep slope near the Giant's Causeway he lost his head, and
nearly fell into the sea[1].

To Rev. John Sedgwick.

FLORENCE COURT, ENNISKILLEN,
August 21, 1835.

"We had a glorious passage, and such a reception
as has eclipsed the remembrance of all former meetings. All
the public bodies vied with each other in hospitalities. There
were no drawbacks; the week was uninterruptedly fine, and
every face seemed to be suffused with happiness. More than
all this, all the philosophical sections were most actively and
successfully employed in the work for which they were called
together.

Tuesday I left Dublin and went as far as Cavan, through a
most wretched and beggarly country. I had no notion of the
external misery of this strange people before I saw it with
my own eyes; and it contrasted painfully with the splendour
I had witnessed during the preceding week. Wednesday
brought me to Lord Enniskillen's, where I am spending the
remainder of the week[2]. It is a noble domain, surrounded
with mountains, and from the windows we have a view of
three magnificent lakes; but when you quit the confines of the
park filth and misery are again seen on the wayside, though
certainly in a less offensive form than in the county of
Meath."

---

[1] Geikie's *Life of Murchison*, i. 231.

[2] It was on this occasion that Lord Enniskillen "had the satisfaction of seeing
Murchison and some other guest *glorious*, and Sedgwick *comfortable*." (Lyell to
Sedgwick, 25 October, 1835, Lyell's *Life*, i. 457). Lyell proceeds: "Depend
upon it the building of the Museum [by Viscount Cole] and subsidies for what the
old Lord once condemned as 'damned nonsense' will go on with good spirit,
after his finding that the hammer-bearers are such a jolly set."

1835.
Æt. 50.

To Mr Lyell.

DENT NEAR KENDAL, *September* 20, 1835.

I received your letter[1] in Dublin; but as for writing, I had not a moment's time; at least, during the ten days I was there. The hot weather, close packing, and perpetual festivities, not to mention the serious labours of the sections and general meetings, were almost the death of me; so that during the week after, which I spent with Lord Cole at his father's seat, I was almost confined to the house by English and Irish cholera.

Let me tell you that I read certain extracts from your letter in one of our sectional meetings, and that the questions you started were discussed, not, however, with much power, as there was no one present, except Phillips, who had a sufficiently specific knowledge of the subject. He doubts Des Hayes' conclusion, to say the least of it; and makes fight on the crag species. I was not present when the discussions took place last spring at the Geological Society; but I do not believe a word about the *different epochs* of the crag. It is all of one epoch, and, geologically speaking, not a long one, at least so I think.

I was much amused at your discussions on elevation craters etc. etc. with the geological conclave at Paris. I don't care one fig about the question, and am disposed to think that more fuss is made about it than it deserves. This may, however, only arise from my ignorance of the phenomena exhibited by modern volcanos. I suspect the truth is between the two parties. All the protruded masses of igneous rock (granite, porphyry etc.) constantly produce that collocation of stratified masses which is presented by the so-called craters of elevation. Why should not the local elevatory forces do over again what they have formerly done? I have of course read your paper[2] published in the *Philosophical Transactions*,

[1] The letter, printed in Lyell's *Life*, i. 450, gives an account of some shells from a bed in the Suffolk Crag, supposed of older date than the upper Crag.

[2] *On the Proofs of a gradual Rising of the Land in certain parts of Sweden:* Phil. Trans. 1835, pp. 1—38.

and was by no means surprised at the fact of the travelled blocks of the north of Europe belonging to a very recent period. This is what I should have expected. Your ice theory will, I think, only let you slip into the water, and give you a good ducking. Erratic blocks are diffused in latitudes where there are no icebergs, and never were. How do you get your icebergs to shove the Shap Granite over Stainmoor to the Yorkshire coast; or the Wastdale Granite across Morecambe Bay, over the plains of Cheshire, to the Derbyshire hills, and the outskirts of the Welsh mountains? I think I have ascertained this summer that the greatest part of the erratic blocks in the south of England have passed over marl beds full of recent marine species. But of this by the way.

1835.
Æt. 50.

Agassiz joined us at Dublin, and read a long paper at our section. But what think you? Instead of teaching us what we wanted to know, and giving us of the overflowing of his abundant ichthyological wealth, he read a long stupid hypothetical dissertation on geology, drawn from the depths of his ignorance. And among other marvels he told us that each formation (e.g. the lias and the chalk) was formed at one moment by a catastrophe, and that the fossils were by such catastrophes brought from some unknown region, and deposited where we find them. When he sat down I brought him up again by some specific questions about his ichthyological system, and then he both instructed and amused us. I hope we shall before long be able to get this moonshine out of his head, or at least prevent him from publishing it. His great work is going on admirably well, and we voted another hundred pounds in promotion of it. I think it is by far the most important work now on hand in the geological world.

Griffith[1] exhibited a geological map of the whole of Ireland, coloured from his own observations. It is a thousand pities he did not publish it fourteen years since. As far as I

[1] Mr (afterwards Sir Richard) Griffith. *Report of British Association*, 1835, p. 56.

1835.
Æt. 50.

have examined the demarcations, they appear to be very well laid down—much more correctly I think than in our friend Greenough's first edition[1]. The description and discussion of this map took two mornings. The only new zoological fact brought to light was the existence of fossil fish in the New Red Sandstone of the north of Ireland. Murchison threw off one morning on his Silurian System; and I followed on the lower division of the Cambrian rocks. Phillips read two elaborate papers on the distribution of the *Astacidæ* and Belemnites in our secondary strata. That on the *Astacidæ* was, as far as I was able to judge, admirable. The paper on Belemnites was not perhaps quite so well worked out. There were several papers on local details connected with Irish Geology—some of them good of the kind. We had one ignorant and impudent paper by a Mr Williams on the coal deposits of North Devon[2], which he refers, like De la Beche, to the greywacké. Murchison is gone to have a brush at it, and I suspect strongly will succeed in turning it out of the older system, and putting it where it ought to be. I am the more inclined to this belief now, as he has proved De la Beche to be wrong in a part of Pembrokeshire where he has coloured a part of the undoubted coal-measures as greywacké.

Since I left Florence Court I have had (along with Mr and Mrs Murchison, Lord Cole, Colonel Montgomery, and Mr Griffith) a delightful ramble along the coast of Antrim; and on my return to this place halted a few days with Sir Philip Egerton[3]. The weather has been glorious though terrifically hot; but now it is both wet and cold, and will continue so among the mountains till these equinoctial blasts pass over.

---

[1] *Geological Map of England and Wales*, with a Memoir. By G. B. Greenough. Lond. 1819.

[2] *On certain Fossil Plants from the opposite Shores of the Bristol Channel.* By the Rev. David Williams. *Brit. Ass. Report* 1835, (Sections) p. 63.

[3] Sir Philip de Malpas Grey-Egerton, Bart. F.R.S., a distinguished geologist, resided at Oulton Park, Tarporley, Cheshire. He died 5 April, 1881.

In this year Sedgwick succeeded in obtaining for his
Museum an almost complete skeleton of the extinct Elk
(*Cervus megaceros*) commonly called the Irish Elk. In the
autumn of 1834 he had heard, through Lord Cole, of a fine
skeleton which had been discovered in a bog near Enniscorthy. It had passed into the possession of Dr Macartney,
a medical man of that town, but it was understood that he
might be persuaded to part with it. Sedgwick, who was then
at Edinburgh, lost no time in ascertaining what help he
might expect from the Woodwardian trustees, and then wrote
to Lord Cole :

"I have heard from our Vice-Chancellor about the great
Irish fossil, and he wishes me to say that he will help me to
the tune of £100, provided it be not the skeleton of an Irish
Bull. With this encouragement, I wish you to consider me
as a customer, but every farthing beyond £100 will probably
have to come out of my own pocket, including expenses and
carriage. Therefore pray have mercy on a poor body, and
don't screw him to death. If I can do no way else (for by
the powers I don't think I can raise the wind myself) I must
raise a subscription among my friends to lift me out of the
bog. I hope your correspondent will think £110 a good
offer. As a good Yorkshireman, I hate to buy a pig in a
poke, but I trust to the faithfulness of the account given by a
person whom you consider of a respectable character, and
incapable of imposing on us by a false description[1]."

In the course of 1835 several letters passed between
Sedgwick and Dr Macartney, whose reluctance to part with
his prize increased as the time for packing it up drew near,
and it was finally settled that the price to be paid should be
£140, inclusive of the cost of packing and carriage to Dublin.
In justification of this large sum Dr Macartney urged that the
skeleton was an exceptionally fine one, and that as the bones
had been found close together there was "reason to conclude

---

[1] To Viscount Cole, 12 September, 1834.

they all belong to the same individual[1]." As the Wood-wardian Museum of that day was too small to contain so large a specimen, the bones were confided to the care of Professor Clark. Sedgwick was still absent when the cases arrived at Cambridge, and on his return was horrified to find the animal tailless. On this defect he wrote " a doleful letter of inquiry"[2] to Lord Cole:

FLORENCE COURT [*erased*].
TRIN. COLL., CAMBRIDGE,
October 23rd, 1835.

Dear Lord Cole,

What a scatterbrain I must be to begin my date with Florence Court! Expunge the two first words, and you will find me where I really am, in Trinity College, Cambridge. But why am I troubling your Lordship? I will tell you. I reached Cambridge yesterday, and went to our Anatomical Professor, to whose care the Enniscorthy Bog beast was committed. It has turned out most beautifully—horns, head, legs, and body, but, horrible to tell, *the tail is wanting!* The straw in the packing-cases was examined with the utmost care, and not one single caudal vertebra was found skulking in it. This has put me at my wits end. The very sight of my stump-tailed beast has given me a sympathetic sciatica— a horrible tic in the regions of the rump which wrings groans from me enough to melt the heart of a flint! Perhaps Dr Macartney can administer a sedative. In plain English then, had the beast a tail? if so, where is the tail now? When the second question is answered, how am I to get the said tail? These are grave questions, and I should be greatly obliged to you to put them gravely to the Doctor. His honour is concerned in seeing that my beast is not funda-mentally at fault. If I knew his address I might try to put salt on his tail myself, but he told me he was about to quit

---

[1] From Dr Macartney, 6 May, 1835. The Woodwardian trustees paid £130; the rest was probably paid by Sedgwick.

[2] To Sir Philip Egerton, 23 October, 1835.

his old den, and I wish not to send my salt to a wrong
market. Do therefore lend me a helping hand in my hour of
need. My lectures begin on Monday, so for the next month
I shall be fixed down in this place, and shall hope to hear
from you, and that you will have the humanity to pity my
case, and give me my cue (*queue*).

Present my best remembrances to your father, whose hospitality and kindness I must ever remember with gratitude.

Ever most truly yours,

A. SEDGWICK.

In answer to this letter Lord Cole undertook to do what
he could ; and, as the skeleton can now boast of a proper tail,
it may be concluded that either the requisite number of
vertebræ were found in Dr Macartney's possession, or extracted from one of the numerous bogs in Ireland where
these gigantic stags are still plentiful.

Early in the Michaelmas term Sedgwick had the pleasure,
so long deferred, of receiving Agassiz in Cambridge. Soon
after his return he had written : "I am once more settled
in the University, and have buckled on my harness; and
my fish are all gaping in expectation of seeing their great
law-giver." At last, 2 November, the famous ichthyologist
arrived. No account of the visit has been preserved, but it is
interesting to know that the collection of fossil fishes which
Sedgwick's energy had got together in the Woodwardian
Museum was laid under contribution for the *Recherches sur
les Poissons fossiles*, on which Agassiz was then engaged.
That such was the case is evident from a subsequent letter,
undated, in which Sedgwick enquires : "When are my fossils
to be drawn ? My fish are gaping for the artist, and scolding
him for his long neglect of their beauteous faces."

In December Sedgwick left Cambridge for his second
term of residence at Norwich. Some months previously the
authorities of the Museum had asked him to deliver a course
of geological lectures during the winter, and he had acceded

to their wishes, without due consideration of the difficulties of
the task. " I wish very much you would come to my first
lecture this evening," he writes to Canon Wodehouse, "if you
can do so conveniently.   I shall want your advice on more
points than one ; about the length, topics, points, etc. etc. of
my pattern lecture.   I don't know my audience, and therefore
I want to feel my way. Geology introduces some tender
topics which require delicate handling.   I must speak truth,
but by all means avoid offence if I can.   Above all I must
try not to make my lecture a bore, which may be done easily
in two ways—by firing over their heads, or by running out to
an unwarrantable length."   We shall recur to these lectures
in a subsequent chapter ; for the present it is sufficient to say
that the dangers enumerated in the above letter were avoided,
and that the interest shewn by everybody, and especially
by the ladies of Norwich, was so great that at the end
of the course he could report: " I have had a merry
time of it at Norwich. Among other amusements I gave a
course of geological lectures to a class of three or four
hundred.   Half the stockings in Norwich are turned blue
in consequence[1]."   These sentences were written after the
event, when distance was lending enchantment to the view
of a successful achievement.   But, while the course was still
proceeding, Sedgwick was in a very different frame of mind.
His anxiety may be measured by the ill-humour of the
following passages from a letter to Murchison :

NORWICH, *January* 15, 1836.

" I am looking forward to my return to my old den and
old habits with some anxiety, as I am almost worked off my
legs in this place, and tired of the life I am leading, but it
is not a life of mere eating and drinking, as you seem to
fancy.   Each Sunday I have to attend three services, and to
preach twice.   My sermons cost me some trouble, and towards
the end of the week I have to rise at six in order to find two

[1] To Rev. W. Ainger, 19 February, 1836.

or three hours I may call my own. Besides, I am giving a short course of lectures which cost me much trouble. I state these things as an explanation why I have not looked over your MS. I have not done *your work*, solely because I have not had time; and for the same reason I have not done my *own work*. I don't think I am a selfish person. Last year I did nothing for myself, but I was working (a considerable part of it) very hard for other people. If you knew the time I had given up to my friends I believe you would be disposed to set me down for a fool. With such convictions on my mind I felt both hurt and angry at some expressions in your letter. You say if I won't do your wishes 'you must pocket the affront'. Now you have no right to talk of being affronted because I have not time to do all you wish. If I am to see this or any other MS. on such conditions, I must formally and positively decline the task. You sent me work enough for three or four entire days. Had they been at my disposal I should have given them to the task with very great pleasure; but they have not, and so there is an end of the matter. I will look over the preface as soon as I can and return the bundle."

The hardworked Canon had, however, some compensations. About ten days after the above letter was written Mr Romilly spent a week with him, and his diary affords us some charming glimpses of Sedgwick in his lighter moments, when he threw lectures and sermons to the winds, and surrendered himself unreservedly to the company in which he took especial delight—that of ladies and young people. We have already heard of his introduction to the Dean's daughter, his "oldest friend in Norwich," and now, thanks to Mr Romilly, we make the acquaintance of Miss Clarke, who, down to nearly the close of Sedgwick's life, held a foremost place among his lady friends. In after years, as Mrs Guthrie, she was widely known for her beauty, her wit, and her active benevolence; but in 1836, when she was barely twenty-one,

(margin: 1836. Æt. 51.)

with her naturally high spirits unchecked by sorrow or respect for convention, she was evidently the spoilt child of Norwich in general and of Sedgwick in particular. Her sister had married Captain Willoughby Moore of the Enniskillen Dragoons, and when the regiment was quartered at Norwich, Sedgwick, as Canon in residence, called upon the officers. It is conceivable that in Miss Clarke's eyes the Woodwardian Professor might have been more amusing than the Bishop or the Dean ; and that he in his turn found more congenial relaxation in her natural gaiety than in the more artificial courtesies of the matrons of the Cathedral precincts. Whatever their common ground for friendship may have been in the first instance, they were on a footing of close intimacy when Mr Romilly arrived. His first visit was to " Mrs Moore and her charming sister Miss Caroline Clarke (Sedgwick calls her *Cara*, or *Carissima*) who is a most fascinating creature. She does not look above eighteen, but she comes of age next February." Presently it was time to go out riding, then a matter of almost daily occurrence, and some officers joined them. " Sedgwick " we read, " was in tearing spirits. By their noise and laughter they attracted a great crowd, who shouted on seeing them charge down the street." During the rest of the visit Miss Clarke's name constantly recurs : in the evening she draws caricatures of Sedgwick's figure on horseback ; next morning she comes to breakfast with him dressed as a Quakeress, and plays her part admirably, to the "infinite delight" of her host and his friend ; on another occasion she dons a gown and cassock, and personates, first Sedgwick, and then an old shuffling canon. " It was vastly comical," says the narrator. Lastly, he describes a gathering in Sedgwick's house, with Lady Jane Wodehouse and her children, who were dressed up as a Turk and his hareem by Miss Clarke, while Sedgwick exhibited a magic lantern.

Amid these diversions Sedgwick's residence came to an end, and he soon after returned to his Professorial duties at

Cambridge. But his lectures had hardly begun before he was   1836.
compelled to adjourn them *sine die*, and hurry down to Dent   Æt. 51.
to discharge his duties as executor[1].

DENT, *February 24th*, 1836.

My dear Murchison,

      I have been here ever since Saturday week; up to
the ears in papers, and muddling my poor brain in accounts,
some of them of thirty years standing. At first I was in
despair; but by working very hard, ten or eleven hours a day,
I begin to see light, and to comprehend the nature of the task
that has fallen on me. My poor co-executor is on his death-
bed, unable to put pen to paper, and may die any hour. I
am lingering from day to day, waiting the event which will
enable me to transact some very important business by my
single signature. My position is most uncomfortable, as my
lectures ought at this very time to be going on in Cambridge.
Could you send me a single line? How are you getting on
with your book? How did the Anniversary go off? Was
our President [Lyell] very eloquent? etc. etc.

      My adventures during the last stage before I reached Dent
were laughable enough. I took a post-chaise from Kirkby
Lonsdale and was deluded to attempt the high, rugged,
mountain-road; but it was so dark and misty, accompanied
with sleet and wind, that the driver twice got off the road,
and the post-chaise was once on its side, though easily
righted. This I did not like, so I mounted by the side of my
Jehu, and before ten minutes were over we again missed the
road, and were within an ace of rolling neck and crop into a
hollow made by one of the mountain-streams. With some
difficulty we got the horses again into the old wheel-tracks,
for there was hardly the appearance of a formed road. I
then took out one of the lamps, and walked for seven miles
with a hairy cap on my head, and a boa about my neck,
all bespattered with sleet and snow, and looking like an old
grizzly watchman. The driver followed my light, and I led

[1] See above, p. 442.

him safely to the top of the pass which overhangs my native valley. All's well that ends well. My friends gave me a hearty welcome and a blazing fire. With them I have remained ever since, with the exception of two walks to Sedbergh, for the purpose of talking with a solicitor, and bringing home a cart-load of papers. I never thought this calamity would have befallen me. I was intended for the sleeping partner; all I meant to do was to come down to sign papers and receive money.

Yours ever,
A. SEDGWICK.

Early in March Sedgwick was back in Cambridge, and announced the continuation of his lectures; but before many days were over the gaieties of Norwich were resumed in Trinity College. Here again Mr Romilly is our guide:

Wednesday, 9 *March,* 1836. At two o'clock arrived from Norwich Captain Moore and the fascinating Miss Clarke. I was not present at the arrival, but came a few minutes after to Sedgwick's rooms to lunch. I found the brilliant "young person," as she sometimes calls herself, sparkling with joy. After no long time Whewell came in, and we proceeded to the Senate House and King's Chapel. We went to the top of the chapel, and between the roofs. At six we dined with Sedgwick; a small party of six—Sedgwick, Cara, Moore, Whewell, Lodge, and I. It was very delightful.

Thursday, 10 *March.* Breakfasted with Whewell to meet Sedgwick and his guests. Then to our Library, and afterwards to Sedgwick's Museum. At six we dined with Sedgwick to meet a large party. When Sedgwick took the ladies into the drawing-room after dinner they locked the door upon him, and would not let him out till he had sung them a song. He sang an odd one: *An Alderman lived in the City;* omitting however the most objectionable verse. Heard afterwards that *Carissima,* while with the other ladies, won all their admiration by her charming conversation. She sang them two songs.

Friday, 11 *March.* Breakfasted with the Marchesa[1] to meet Sedgwick and his guests; Worsley[2] and Whewell also there. We

---

[1] The Marquis Spineto was at that time Teacher of the Italian Language to the University. He and his wife, a Scotch lady of good family, were extremely popular at Cambridge fifty years ago.

[2] Thomas Worsley, Trin. Coll. B.A., 1820. In 1824 he became Fellow of Downing, and in 1836 Master, an office which he held till his death in 1885. He was intimate with many of the Fellows of Trinity, but especially with Whewell.

then lionized Jesus Chapel, the Mesman[1] and Fitzwilliam[2] galleries.   1836.
After this there was a grand cavalcade consisting of Miss C., Moore,   Æt. 51.
Sedgwick, Whewell and Worsley.   On Parker's Piece a dog barked at
Miss C.'s horse, and flew up at her habit which was waving in the
wind.   Moore dismounted, and lashed the poor dog severely, at
which our tender-hearted fascination cried bitterly.   The party dined
with me.

    Saturday, 12 *March*.   Breakfast with Sedgwick and his two
guests—a *partie quarrée*.   Today Sedgwick had to lecture, so Miss
C. spent an hour in the Public Library, and then paid a second visit
to the Fitzwilliam Museum to admire Claude's *Liber Veritatis* under
the exhibition of Worsley.   Sedgwick joined us on our road to
Worsley's, where we all lunched.

    Sunday, 13 *March*.   We breakfasted with Whewell.   Sedgwick,
Miss C., and Moore went to church at Simeon's, and heard a good
and characteristic sermon from him.   Dined with Sedgwick at four.
Whewell was the fifth person.   We went to Chapel, and heard that
beautiful anthem *Plead thou my cause*, from one of Mozart's
Masses.   After the service Sedgwick had desired Walmisley to play
the Dead March in *Saul*, the Minuet in *Ariadne*, etc.   We, of
course, stayed these out.   After tea Whewell exhibited some books
of the architecture of the Alhambra, and of some churches in
Germany, and gave us a capital lecture on Saracenic and Gothic
architecture.   In spite of this, however, the evening went off languidly,
as we knew to-morrow morning came the parting.

    Monday, 14 *March*.   Breakfast with Sedgwick—*partie carrée*.
Our fair visitor was out of spirits, and we most sorry to lose her.
She and Moore went away to London by the Telegraph.   The
departure of so fascinating a creature has left a sad blank in our
existence.

The next three months were taken up with work which
needs no special comment—the examination for scholarships
in Trinity College, geological business in London, Chapter
business at Norwich, and a hasty visit to Dent for the marriage
of his sister Margaret.

Towards the end of June Sedgwick and Murchison broke
ground together in a new field of geological exploration.   For
several seasons in succession they had been at work contem-

---

    [1] A collection of pictures formed by Daniel Mesman, Esq.   They came into
the possession of the University in 1834, the donor's brother, the Rev. Charles
Mesman, having renounced his life-interest in them.   They were hung in the
large room of the Pitt Press, now the Registry of the University, from 1834 to
1848, when they were removed to the Fitzwilliam Museum.

    [2] The Fitzwilliam Collections then occupied the old Perse School in Free
School Lane.

poraneously but independently among the older rocks, and though Whewell, when President of the Geological Society, had spoken of their labours as "on all accounts to be considered as a joint undertaking," they had, in fact, started from wholly different points, and employed different methods. Nor is there any evidence that they themselves considered that they were working with a common end in view. Now, however, they combined their forces, and brought them to bear on a disputed question outside the special fields in which each had been engaged. Their labours ended in the establishment of the Devonian System, but, when they began, they had no such ambitious views. They proposed to themselves to settle the age of the carboniferous deposits of central Devon, known as the Culm-Measures. These rocks had been referred to the grey-wacké by Mr De la Beche in 1834, and again by Mr Williams at the Dublin meeting of the British Association in 1835[1]. Murchison therefore suggested to Sedgwick, or possibly Sedgwick suggested to Murchison, that they should take this work in hand conjointly, and get the question settled. The matter had probably been often discussed in conversation; and soon after Sedgwick's return to Cambridge from Norwich in February 1836 he drew out a geological itinerary for the next summer:

TRIN. COLL. 6 *February*, 1836.

"I am anxious to talk with you on many things.... Whewell tells me you have ratted to the iceberg theory. I give you joy of your conversion; but the sooner you turn back the better; or rather, scamper round the hypothetical ring and come again to the grand stand, and you will be where you were before.

"I want to talk about a plan for next summer. What do you say to an early start (middle of May?)? First: Quantocks and Horner's country, with plenty of calcareous beds and organics. Secondly: South Devon and a few points in

---

[1] See above, p. 448. Murchison appears not to have gone to Devonshire in 1835 as Sedgwick expected. Geikie's *Life of Murchison*, i. 231. The above account is borrowed from Mr Geikie's work, p. 248.

Cornwall, working the Plymouth beds well by the way. In 1836.
Cornwall I could put you into cover, and we could floor all Æt. 51.
the cross points in a week or two, and we might meet De la
Beche and Boase at a point or two. Thirdly: return by
North Devon, see the Ilfracombe calcareous beds and ... greywacké coalfield ...[1]. Fourthly: cross and meet Griffith in
the south of Ireland, and touch up a Silurian point or two.
Fifthly: make a long scud and go to Scotland about the
Mull of Galloway, and run up the west coast examining the
coalfield and Red Sandstone, and so working up to the Old
Red through the part we left unfinished in 1827. All this we
might finish, and find good matter for one or two general
papers. I think this a good plan. What say you to it?"

It is almost needless to say that this very comprehensive
scheme was not carried out. The two friends started in June,
and approached their work by way of Somersetshire, Ilfracombe, and Barnstaple[2]. What had been accomplished up to
the middle of July is pleasantly touched upon in the following
letter to Canon Wodehouse.

PLYMOUTH, *July* 20, 1836.
My dear Wodehouse,

To catch a letter from a geologist when under way
is about as easy as to catch a spark from the tail of a rocket.
Our lights are, however, quenched by a pelting shower, and
here we are in shelter, with a tea urn hissing on the table,
and some broiled fish within nostril scent. How long they
will remain in their present position before they pass down
the Gulf Stream remains to be seen. To leave, however,
such sensual, epicurean, topics let me tell you that I have
been zigzagging through the N. W. coast of Somerset; pound-
ing the sides of N. Devon, and burrowing through Dartmoor
to this place. Murchison, whom you know by name, is my
fellow-labourer. As you care less than nothing about things

---

[1] A piece of the letter has been unfortunately cut out, for the sake of the signature.

[2] Geikie's *Life of Murchison*, i. 250.

1836.
Æt. 51.

under the earth I will not torment you with any account of our excursions to the earth's centre (a right merry region, for there you know the operations of gravity are suspended), or our traverse to the antipodes.

Suffice it then to say that on passing through Ilfracombe I tapped at the Post Office window and found a letter from the Dean of Norwich, which told me many things, and among the rest informed me that you were out of health. Now, my dear Brother, though it be quite true that geologists get very crusty on the outside, don't suppose that their bowels are ever petrified. I was very sorry to hear the news of your illness, and I avail myself of the first moment of leisure (and it is but a moment) to tell you so, and to beg of you to write to me— above all to tell me you are quite well again. Any tit-bit of Norwich news will be delightful, and any fireside news will be most welcome. How is little Boppity? How is William? How Mrs Bunch? and how all the rest in the ascending order, till we mount to Charles the First? Lady Jane is, I hope, quite well. What a charming thing to have such a kind nurse! If I had a wife I would sham ill now and then in order that she might make a pet of me.

It is raining worse than ever, but I must finish this sheet in order that I may finish my breakfast, and go with my impatient friend to a Museum, the keeper of which is waiting for us. My kindest remembrances to Lady Jane, and my love to all your children.

Your affectionate friend

A. SEDGWICK.

No further information respecting their route or their proceedings has been preserved, but when the British Association met at Bristol in August they were able to present to the Geological Section *A classification of the old Slate Rocks of the North of Devonshire, and on the true position of the Culm deposits in the central portion of that County.* In this communication they shewed that the Culm-Measures lay at

the top of the rocks afterwards named by them Devonian,
and belonged to the carboniferous system. The work they had accomplished so far made them desirous of undertaking a more extended research into the structure of Devonshire; and Murchison being called away by the state of his mother's health, Sedgwick set out for Devonshire alone, much exhausted by the week's work at Bristol, and eager for fresh air[1].

On his way from Bristol to Exeter he stayed a night at Broomfield with Mr Andrew Crosse, who had made a communication to the Association "on the effect of long-continued galvanic action of low intensity in forming crystals and other substances analogous to natural minerals[2]." Sedgwick writes of this visit: "Spent the morning in Mr Crosse's laboratory; splendid experiments; but in two respects disappointed—neither the variation in diurnal intensity, nor the effect of light on the galvanic process of making crystals, are I think proven. For the variation of intensity Crosse depends on his sensations—a bad mode of measuring, as the effect may depend on a variation of nervous susceptibility. I hope some better test may make the fact out; but it is not yet proven[3]."

After leaving Exeter he was joined by Mr Austen[4], a young geologist whom he describes as "a clever, good, independent, workman;" and in his company made further investigations into the extent and relations of the Culm-Measures. This necessary work despatched, Mr Austen returned home, and Sedgwick went alone into Cornwall, chiefly examining the coast-line. The letters in which he informed Murchison of what he was doing are too technical for reproduction here; and the results will be best noticed when we come to speak of their scientific value. A single letter to Whewell, whom he had met last at Bristol, gives a graphic *resumé* of his personal discomforts.

---

[1] To Canon Wodehouse, 29 August, 1836.
[2] *Report of British Association*, 1836, p. 47.
[3] To R. I. Murchison, September, 1836.
[4] Robert Alfred Cloyne Austen, afterwards Godwin-Austen.

CAMELFORD, *October 7th*, 1836.

"Here I am on my way homewards; but I have still much work to do, and I have very little time to effect it. What is worse, I have the prospect of broken weather. Could I engage ten fine days I could then leave the country with clean hands....Since we parted I have been very busy. One day with that lightning-monger Crosse; a day among the whetstone pits; many days on the east side of Dartmoor, where I traced the Culm-Measures as far as Newton Bushell. Then I passed along the south coast of Devon, retouching points Murchison and I had touched before. From Devon I went to the coast of Cornwall, and paced my way through the wild and rugged cliffs in the old fashion, with horse and panniers....The rain disagrees with me, and I have twice been laid up, and forced to call in a doctor. I traced organic remains almost all the way from Plymouth to the Lizard. They abound on the north coast, and such raised beaches! they are enough to make your mouth water. The sea has changed its relative level in some places by 30 or 40 feet, and old waterworn strands may be seen up in the cliff as plain as the nose on your face. To-morrow I take a peep at the point on the coast (about five miles from this place) where the Culm-Measures abut against the Cornish slates. Having made good the demarcation in this county I hope to move on to Devonshire, and track my way along the northern skirts of Dartmoor."

On the day before Sedgwick wrote this letter, Whewell had put to him the important question, Should he, or should he not, agree to become President of the Geological Society? The question and the answer are as follows:

LONDON, *October* 6, 1836.

My dear Sedgwick,

I have just seen Murchison, who has given me such notice of your whereabouts as he thinks will enable me to hit you at a long shot. He tells me that you go on setting the strata of Devonshire to rights, which is good; but he tells me that you have had bad health, which I grieve much to hear. Pray take care of yourself, and avoid hard work, bad weather, and Dr Sedgwick's physic. Consider that the two former cannot be expected to agree with a rosy prebendary, as they did with a lean geologist, and that the doctor in question knows too much of other things to make his prescriptions good for anything.

    I want very much to have your advice and opinion on a point on which it will have great weight with everybody, and with me will be decisive. Your leading geologists think they see strong reasons for not taking Buckland for their president next year, and are at a loss whom to pitch upon. Now it appears to me that if, in this embarass- ment, they are tempted to turn their eyes to a person who is not a geologist, and who cannot be supposed, by any one acquainted with the subject, to have any detailed geological knowledge, they will lose that eminence and respect which has hitherto belonged to the office in the geological world of Europe. And, if they were likely to do this, I think that no private regards would prevent your giving your opinion against such a step. The case in short is, that Murchison and Lyell have proposed the office to me ; and the extreme pleasure which I cannot but feel at finding it possible that such a proposition should have entered anybody's head, is dashed by my own conscious- ness of my want of the qualifications which I think are requisite. I am not going to trouble you with disqualifying speeches, nor need I tell you, I think, what I think of the Society and the office. I hold it to be a dignity much superior to that of President of the Royal Society. And I want to fortify myself, by your authority, that it demands a person more professedly and notoriously versed in geology than I am. I should be very much obliged to you to write to me about this as soon as you can; and if you think, as it appears to me that you must think, that I am not the proper person for your generalissimo, you will of course make no scruple of saying so, and I will then put an end to all further thoughts of this scheme. You need not fear hurting my vanity by telling me that I ought not to have the presidentship ; for I am quite sufficiently gratified by the opinion which the proposal implies in such men as Murchison and Lyell to need no humouring in the mode of telling me what I know to be true. So there is my case for you, and now let me have your advice.

Always yours truly
W. WHEWELL.

LAUNCESTON, *October 12th*, 1836.

My dear Whewell,

    Here I am, and when I shall be able to budge I know not, for the elements are in most dreadful disorder. Thank God! I am now well, though I have had one or two severe attacks brought on by fatigue and bad weather. By

(margin note: 1836. Æt. 51.)

1836.
Æt. 51. working hard between showers and storms I have done a great deal of good work, and fixed many geological land-marks, which will, I hope, stand all weathers. Were the sky to clear I could rub off my score in about three days; and it would vex me to the very marrow of my bones to quit this country without settling the boundary of the Culm basin. But to business, for it is rather late, and I am very much tired, having been twice half-drowned since I started in the morning.

Don't hesitate one instant, Mr President elect. Did you not pick geological rubbish out of my eyes in 1820? Have you not figured in the mines of Cornwall? Have you not drawn the granite veins in a way that would do honour to Michael Angelo[1]? Have you not been Professor of Mineralogy? Have you not given the only philosophical view of that science that exists in our language[2]? Have you not written the best review of Lyell's system[3] that has appeared in our language? etc. etc.? You are just the man we want. Again, some great questions connected with the theory of electric currents; the existence of joints in given directions in crystalline rocks (e.g. the granite of Cornwall and Devon, where also the major axes of the felspar crystals are arranged on given lines, and the master-joints are nearly Mag. N.); the mechanics of the theory of elevation; and many etceteras, want your hand to guide them, and your back to bear them. So you must for your own honour and our good stand in the gap for us. In short you are made for the office, and the saddle must be guided to your back. I rejoice much that Lyell and Murchison have turned their thoughts to you. Again, I think it will do us honour to link ourselves with men of exact science, and I dream't last night that I saw you riding into the Geological

---

[1] See above, pp. 330, 331.

[2] *An Essay on Mineralogical Classification and Nomenclature.* By W. Whewell, 8vo. Camb. 1828. He was Professor of Mineralogy 1828—32.

[3] Whewell reviewed the first volume of Lyell's *Principles of Geology* in *The British Critic*, No. 17, publ. in January, 1831; and the second volume in *The Quarterly Review*, No. 93, publ. in March, 1832. Todhunter's *William Whewell*, i. 51, 60.

1836.
Æt. 51.

Society on the back of a great tidal wave as high as the top of Somerset House.

Tell Murchison that in the quarries a mile and a half south of this place there are fossils without number, that the Culm-Measures make a great plunge to the south (I have this day during a most pelting storm traced them to the foot of Brent Tor), and abut against Dartmoor, getting well frizzled at their edges. I can find plenty of Culm beds, but unfortunately not much thicker than this paper—just enough to soil the fingers, but not enough to warm the tip of your toe. By the way I saw this morning a *quaternary formation of ink*. In a yellow pyritous soil, containing sulphate of iron, there was a copious black precipitate, wherever the roots of the oak-tree passed through the soil, and you might have fancied that some one had upset one of Day and Martin's blacking-pots. What shall we call this new geological deposit? Melanoune? no, that will not do. What's the Greek for ink? My best regards to my Cambridge friends. I hope to be trundling in among you about the time I mentioned.

Yours ever

A. SEDGWICK.

When the observations made hastily in Devonshire in June and July were communicated to the British Association, the authors announced that they proposed to lay a more complete paper before the Geological Society. This had now to be prepared; but, as might have been expected, the difficulties which had occurred in 1827, and to which collaboration must be always exposed, were repeated, but in an aggravated form. Sedgwick had now more calls upon his time. At the beginning of this particular term he wrote: "My lectures begin the day after to-morrow, and I have three papers on the stocks. My hands were never so full before. I wish the days were forty-eight hours long, or that I could find some patent mode of getting through the term without eating or sleeping[1]." In the next place,

[1] To S. Woodward, 26 October, 1836.

1836.
Æt. 51.

he could never count upon being well enough to work, especially when under pressure. "Could I command my stomach, I could clear off my work easily enough; but the moment I try to employ my pen, all the inner man gets out of order, and then, as for brain, my head might just as well be filled with brick-dust[1]." His position too, as prebendary of Norwich, had given him a new set of duties, and exposed him to be called away from Cambridge when absence might be specially inconvenient. For instance, when Murchison was crying out for substantial help towards his *Description of a raised Beach in Barnstaple Bay* (read to the Geological Society 14 December) Sedgwick was executing "a shuttle-cock movement" between Cambridge and Norwich, and it was only at the eleventh hour that he could seize a brief interval in which to scribble a peroration, which his friend managed to tack on to his own work. This paper, however, dealt with only a single special point, and Sedgwick's short-comings in the way of collaboration were comparatively unimportant. What happened when the general essay on the structure of Devonshire was announced was far more serious, as will be narrated directly. Meanwhile the following letters, in which the serious and the jocose are blended in Sedgwick's happiest manner, will fitly conclude our account of the year 1836.

To Canon Wodehouse.

TRIN. COLL., *November* 4, 1836.

My dear Wodehouse,

Many thanks for your kind letter, which I should have answered sooner had I not been called up to Town on unexpected business. I am only just returned; and, as I have much lecture-work on my hands, I will try to come point blank to business. Your *orders* appear admirable, and I must of course abide by them. Touching the dining-room I will implicitly follow your advice, especially as you say you

---

[1] To Rev. W. Ainger, 20 November, 1836.

have consulted the *learned Barnesius*[1]. I should have called her *Barnesia;* but I dare say you were right, as, in spite of her petticoats, she is certainly a being of right masculine understanding. In the wake of such a learned upholsteress one cannot go far wrong. Again, what you and she *condemn* I will *execute.* As for the *spots* on the *paper,* I care naught about them. The only fear I have is, that if they remain, Barnesia may, like Lady Macbeth, become a sleep-walker. What a horrible thing it would be to see her every night about two o'clock sliding into my bed-room, with a lamp in her left hand and a mop in her right, and muttering through her chattering teeth: "Out damned spot! out I say! one, two; why then 'tis time to do't. Fie Mr Prebendary! a Parson and be afear'd! what need we fear who knows it, when none can call our paper to account? Wash your hands; put on your night-gown, look not so pale, I tell you yet again your paper must be cleansed"!!! *Repeated* scenes like this would be bad discipline for a nervous man like myself. So, having given you such formal and intelligible directions, I may be content to leave the question of paper currency with your wisdom. I can easily bring a *Megatherium* or *Dinotherium* to cover the spots.

I have had a letter from the Dean, who seems to wish us to begin our residence as if our members were now reduced to four, i.e. each of us to take three months. With my present engagements here such an arrangement would be impossible, at least for this year. It would, I believe, compel me to resign my Professorship—a move I am by no means inclined to take. Pray excuse the trouble I am compelled to give you. My kind remembrances to Lady Jane, and to all your children, whom I hope to find at home well and right merry during the Christmas vacation. This fickle weather puts my old crazy machinery a little out of order. Always, my dear Brother, Most truly yours

A. SEDGWICK.

1836.
Æt. 51.

---

[1] Mrs Barnes, Sedgwick's housekeeper. She had been engaged for him by Lady Jane Wodehouse.

30—2

To Rev. W. D. Conybeare.

TRIN. COLL. *December* 5, 1836.

My dear Conybeare,

I am trying to steal a few minutes during the intervals of swallowing my breakfast. And why in such an hour? Because my servant is occupied in packing up my outward garniture, not to mention a thousand odds and ends necessary to my comfort during a three months' incubation in the cathedral rookery at Norwich. Because (driving hard, as they say in the north) I am looking out for sections, and other materials, for my concluding lecture at 12 o'clock. Because, after the said lecture, I have to return to the audit; after the audit to eat a great audit-dinner; after the dinner to put myself into a fly, and by aid of its wings to pass Newmarket Heath in time for the night-coach to Norwich. Any one of these reasons would be enough to put me on the move, but all together they make me stir about like Whewell's anemometer[1] during the late gales. How deep then must be my regard for my friends at Axminster to think of them and write to them in such a whirlwind! But it is with memory as with a muddy ditch, and with thought as with air-bubbles rising therefrom. You must stir the water to make them rise rapidly, and poke the mud to have them in plenty.

Since my return I have often intended to send you my best greetings, but good intentions are often our worst stumbling blocks. A thousand times in my life I have broken my shins, and soiled my manners, over them. But I have been very busy with my lectures, and not in very good health. One entire week I was almost laid up with dyspepsia. The dull round of my mill-horse life was broken in upon last week by a trip to London to the anniversary of the Royal Society. Had I time I would tell you all about it, but I have not. During the past term Simeon's death and funeral have been

---

[1] Mr Whewell described an anemometer invented by himself at the meeting of the British Association held at Dublin in 1835; and more fully at that held at Bristol in 1836. *Report of the Association*, 1835, p. 29; 1836, p. 39.

by far the most exciting events that have passed amongst us. 1836.
The greatest part of the University—graduates, and under- Æt. 51.
graduates—assembled to do his memory honour, and while
the procession moved down the magnificent chapel to the
grave, and while its vaulted roof was reverberating the almost
supernatural notes of Handel's *Dead March*, I do not think
there was one person (including many hundred spectators)
who was not for a while almost carried away by a powerful
emotion[1]. William[2] and your friends here are all well.
Whewell is printing his great book[3], which will be out early in
the spring, perhaps before I return from Norwich. Among
other events, I may mention that last week a lioness had four
cubs at Cambridge, and a sensible undergraduate suggested
four names for the said cubs which have been adopted,
and they are accordingly called, Whewell, Peacock, Sedgwick,
and Simeon, to the no small amusement of some people who
don't look deep for pleasure. Should " Sedgwick " ever be
brought in a cage to Axminster, tell Mary to poke him gently.

Buckland's book[4] I have of course read carefully, and must
read again. The natural history is excellent, though of course
not equal. But some of the best parts of it ought to have
appeared in separate memoirs, and not in a book of Natural
Theology, which has to do with laws, and not with minute
details. The part on belemnites pleased me most, it is quite
perfect. Trilobites very good, and views in part very original.
Concamerated shells in general excellent, but he refines too
much, and some of the illustrations (e.g. from gothic architecture) are false. The siphon theory[5] puzzles me ; it is at least

[1] The Rev. Charles Simeon, Fellow of King's College, died on Sunday,
13 November, 1836; and was buried in the College Chapel on the following
Saturday.

[2] William John Conybeare, then an undergraduate of Trinity College, B.A.
1837, and afterwards Fellow.

[3] *History of the Inductive Sciences*, 3 vols., 8vo. Camb. 1837.

[4] *Geology and Mineralogy considered with reference to Natural Theology* : the
Sixth Bridgewater Treatise, 2 vols. 8vo. Lond. 1836.

[5] Proofs of design derived from the structure of the siphuncle in *Nautilus
pompilius*. Vol. i. pp. 314—332.

ingenious. The account of the *Megatherium* disappoints me. Some remarkable peculiarities are overlooked. The account of the teeth, however, is in his best manner. Pentland tells me that he is quite wrong in putting it in armour; that the coat of mail belongs to another beast. Let them fight it out.

I wish I had not filled up the first page with mere padding, as I want to tell you more about this book. The moral and theological part is, I think, a great failure. In showing unity of design he is good; but the argument is broken up too much into fragments. He ought to have had one grand sweeping chapter on that head instead of 500 corollaries; several direct arguments fail of their aim. They ought to have appeared at the end, by way of removing objections. We can establish wisdom and benevolent design from the happy tendency of laws; so far the argument is direct. But in the working of the laws we find much particular and individual evil. If we argue *directly* we lead to the conclusion that God wants benevolence, or wants power. But if we argue *indirectly*, we show (and this Buckland has done, though he has not put the argument well, and has overstrained it), that much good comes out of apparent ill, and that the evil on the surface of things may be resolved into something connected with our own imperfection and shortsightedness. I have not room or time to explain further. The concluding chapter is not good. Instead of a grand outpouring of his own soul, pregnant with high thoughts and bursting for their delivery, he gives us a mere mosaic, without coherence, and a discontinuous string of quotations. I cannot bear such writing on such a grand occasion. By the way the chapter on Mineralogy is decidedly bad and jejune. Had he read Whewell's philosophical *Report*[1] he must have seen where Mineralogy was among the sciences, and what argument it might supply. Instead of that he has been dabbling among mere chips and fragments. You must not tell Buckland what I am now writing, for he would not

---

[1] *On the recent progress and present state of Mineralogy.* Brit. Ass. Report, 1831—32, pp. 322—365.

forgive me. As a book of Palæontology it is very good, in
many parts admirable; as a book of Natural Theology it is
not good. It exhibits a want of power, and a want of skill.
The descriptive style is very good, but the moral and didactic
parts are sometimes mouthy and turgid. When I read the end
of one of his chapters where he compares the footsteps of a
tortoise on the Corn Cockle sandstone to the footsteps of a
conqueror, I thought the whole passage in such bad taste that
I almost threw the book down in a passion[1]. I must, however,
have done; and once more do not tell Buckland one word of
what I am writing, I am certain he would not bear it. With
all his faults he is most kindhearted, and a most valuable
friend; and in his own way he is perhaps the cleverest man I
have had the happiness of knowing.

Ever yours,
A. SEDGWICK.

To Rev. Charles Ingle.

NORWICH, *December* 13, 1836.

My dear Ingle,

Another year has rolled away, and I am once
again fixed in the monotonous round of a Cathedral residence.
One of my brother Prebendaries died since you and I parted,
and our respective terms of servitude now extend over one

---

[1] "The Historian or the Antiquary may have traversed the fields of ancient or
of modern battles; and may have pursued the line of march of triumphant Con-
querors, whose armies trampled down the most mighty kingdoms of the world.
The winds and storms have utterly obliterated the ephemeral impressions of their
course. Not a track remains of a single foot, or a single hoof, of all the countless
millions of men and beasts whose progress spread desolation over the earth. But
the Reptiles that crawled upon the half-finished surface of our infant planet, have
left memorials of their passage, enduring and indelible. No history has recorded
their creation or destruction; their very bones are found no more among the fossil
relics of a former world. Centuries, and thousands of years, may have rolled away
between the time in which these footsteps were impressed by Tortoises upon the
sands of their native Scotland, and the hour when they are again laid bare and
exposed to our curious and admiring eyes. Yet we behold them, stamped upon
the rock, distinct as the track of the passing animal upon the recent snow; as if to
show that thousands of years are but as nothing amidst Eternity, and, as it were,
in mockery of the fleeting perishable course of the mightiest Potentates among
mankind." Buckland, *ut supra*, i. 262.

fifth part of the year. The consequence has been, that my turn commenced this morning instead of the 1st of December, and will end on the 23rd of February instead of the last day of January. You now know the limits within which you may be sure to find me, as far as anything is sure in this world of changes; for within the limits I have given you I am compelled by our Chapter Rules (so far like the Rules of King's Bench) to sleep in my own house every night. Pray then come and see me, and cast a light over the deep shades of our cloisters. Some of the stars that shone last winter are away; but some are left; with kind hearts too, and ready to give you a hearty welcome. You owe me a visit; and, you graceless dog, you have never written to me, as you promised, since your continental tour. Come then, and make us all laugh, as you made Lyttelton, when you spent your last day in Trinity; or, make us laugh a little less, as he tells me his ribs were sore for a month after he met you, and that his intercostals have not forgotten it to this day.

As for myself I spent a very laborious, and I hope not an unprofitable, summer, in rambling through the cliffs of Devonshire, Cornwall, and Pembrokeshire. Eleven packing-cases (heavier far than the gates of Gaza) have been the fruits of this toil; and from them are to germinate other fruits in the shape of Memoirs for the Geological and Cambridge Philosophical Societies. But these are dull things to write about, especially to one, like yourself, who is almost stone-blind to the beauties that lie concealed beneath Dame Nature's outer garments. But the mouth speaketh of the heart's fulness, you know; so I must needs talk of things inanimate and underground. To-morrow is the anniversary of the Norwich Museum; which finishes, like other good things, with a good dinner, intended to cement friendship, and to set tongues, as well as teeth, in active movement. At this symposium I am to preside; and before my hinder parts are permitted to reach a position of stable equilibrium on the cushion of an armchair, I am to be presented with a splendid silver inkstand, as a

token of thanks for lectures given, and, I suspect also, as a
retaining-fee for lectures to be given hereafter. You remember
the scene got up last winter after my concluding exhibition,
and the formidable mass of melted butter that was spilt on my
head by a little dapper gentleman, one of the progeny of
Æsculapius. After having stood that, I am prepared to stand
everything ; but I am resolved to be as informal in my reply
as is compatible with the rules of courtesy ; and in telling
them that I feel most kindly towards them and heartily wish
them well, I shall truly only tell them what I both wish and
feel.

Your friends in Cambridge are all well, and several of them
are I hope coming this winter to see me. Peacock is the new
Lowndean Professor in the place of Lax ; and promises to do
his duty in a less lax manner than his predecessor. He is in
great force, and the stars in his astronomical tail shew most
gloriously. The Cathedral bell has just tolled midnight, so I
must to bed. Good night then, my dear old friend ; write to
me by return of post, and tell me you are coming, and believe
me,

Affectionately yours
A. SEDGWICK.

P.S. So our friend Charles[1] is now the Venerable Archdeacon of Craven—a craven Christian soldier in name only,
and not in deed.

The paper on the raised beach at Barnstaple had contained
a sufficient quantity of geological heresy to arouse some slight
opposition from the orthodox ; and Murchison began to be
exceedingly nervous as to what would happen when the still
more heretical views respecting the Culm-Measures were laid
before the Society. The correctness of Mr De la Beche's
inferences respecting the structure of Devonshire would have
to be impugned ; he had, as he told Sedgwick, "toiled day
after day for months, in the district, examining every hole and

1836.
Æt. 51.

---

[1] Rev. Charles Musgrave, whose intimacy with Sedgwick has been frequently
mentioned.

cranny in it[1]," and was not likely to submit to the exposure of a serious error without making a vigorous attempt to justify himself; his friends were numerous and powerful, and were certain to rally round him. Sedgwick was away at Norwich, and in his absence a joint paper would be read and discussed at a great disadvantage. For a time Murchison was in favour of an adjournment; but, after discussion with common friends, he concluded that it would on the whole be best to place what he called "our whole view" before the public at once, in order to secure priority for it. So strongly did he feel this, that he persuaded himself that Sedgwick must be of the same opinion, and wrote as follows:

"I presume you are rapidly preparing the concern. Now what I propose is this. That the present communication be rigorously restricted to a clear general view of the classification and succession which we intend to adhere to '*per fas et nefas.*' Such a condensed sketch, I mean, as would be fit for publication in the *Transactions*, and yet such a one as might this very spring be followed by details (*pièces justificatives*) the production of which would give you an opportunity of coming out *vivâ voce* in defence of the grand positions provided they were attacked. Lonsdale and Lyell both agree in the wisdom of this arrangement. Let, therefore, the forthcoming bolt be ringing, clear, and sharp, but not encumbered with the hundredweight of culm and sandstone which I could ornament it with[2]."

Sedgwick was much occupied with clerical and domestic engagements, and unwilling to be disturbed: "My servants are ill of the influenza" he wrote, 23 January 1837, "and for the last ten days I have myself been out of sorts, though never quite laid up, and should I have no new access of the malady, I will do my best during the hours of leisure I have this week to write out something for our meeting on Wednesday the 1st. What we say on that occasion can only be the prelude of larger details....I have for this week refused every engagement, and hope to send my work off about next Friday or Saturday. In the meantime I will think of some materials for the sections south of Dartmoor, and send them when

---

[1] From H. T. de la Beche, 11 December, 1834.

[2] From R. I. Murchison, without date, but evidently written early in January, 1837.

ready." Two days later he proceeded: "My house is a
hospital; but, thank God, I am almost well, and yesterday
made a start. If the influenza don't shed again its baneful
influence on me I shall have matter enough for our next
meeting. Your book you will find in the parcel. Your maps
shall follow soon, but I cannot yet part with them, as I
mean to write with the map spread before me, and without
consulting my MSS., which in a synopsis would only bother
me." But, before the day was over, the draughts in the
huge empty cathedral had given him a chill, and he ended
with the ominous words : "I will send your book and this
note by the mail to-night. I am wretchedly out of sorts,
but after tea I hope to write a few pages. I really can't
make out a section to-day." On the same day Murchison
reported progress, but sent some queries which shewed
Sedgwick that his colleague was not yet quite sure of his
ground. On the day he received this missive (26 January)
he sent three sections roughly coloured, with a few notes,
concluding with what he intended to be an intimation, as he
said afterwards, that he had "struck work." "Under all
these circumstances we should be mad to bring on our paper
before we have made our minds up. Throw the blame on me
as much as you like, or blame the influenza. My stomach is
sadly out of order, and for the last two days I have been
tormented by a palpitation of the heart, which, though not
acutely painful, is very distressing....Though I think we
ought not, in our present misty state, to bring on our paper,
yet we might put in, along with the sections, a kind of
resumé, just enough to explain our views of the classification.
This might be done in five or six pages, and best of all *vivâ
voce*. I read prayers myself in the Cathedral this morning.
All the minor Canons, and most of the singers, are laid
up in bed." In the interval between this letter and the
meeting Sedgwick's ailments increased. A last appeal from
Murchison[1] reached him on Saturday, 28 January, and "then",

[1] This letter has not been preserved.

1837.
Æt. 52.

1837.
Æt. 52.

he says, "I was unable to work." Finally, on the day before the paper ought to have been read, he was well enough to send a long letter containing an outline of the speech he wished Murchison to deliver. It is evident that he was conscious that he had used him hardly, for he begins by a confession: "Had I not been in a miserable condition from the effects of this malady which is raging in the heart of Norwich, I should be ashamed of myself for failing you in the hour of need. But the influenza destroys all feeling for others." Murchison, meanwhile, had convinced himself that a paper to be read would accompany the sections to be exhibited. These latter he got ready, and then "waited patiently like a lamb for the sacrifice." The misadventure of the next day, vexatious as it must have been to Sedgwick's unfortunate colleague, wears a comic rather than a tragic aspect. Murchison shall relate it in his own words. All things considered he wrote very good-humouredly:

3 BRYANSTON PLACE, 2 *February*, 1837.

My dear Sedgwick,

The part of Hamlet being omitted, the play was not performed, and all the scenic arrangements which I had laboured at were thrown away, though the room looked splendid. The morning's arrivals certainly surprised me !! Ten o'clock brought me your double letter; eleven o'clock by the same mail the maps, and a little note to Lyell, but in vain I looked through the parcel for the document to be read. I read and re-read your letter, and still I could not understand it. One thing I clearly perceived, and with great regret, that you were seriously out of sorts, and had been suffering; so, after waiting till two, I journeyed down to the Society, still thinking that a third package with the paper might be sent to Somerset House—not so however. These things going on; the whole room decorated for the fight; Buckland arrived; Fitton present; and a large meeting expected; what was to be done? Fitton and Lonsdale, considering what had been said and done *covertly* on the other side, and looking to the fact of the non-arrival of the despatch, counselled me to give up the thing, which I resolved to do, to the very great annoyance of the President[1], and of all the others who came to hear....

I am mortified that the memoir did not come; of course I blame myself somewhat for having thrown in doubts on some points,

[1] Mr Lyell.

because I see that ill as you have been and without the power on   1837.
my part of talking the case over, we mutually misapprehended each   Æt. 52.
other.   But enough of what is past.   The thing now to consider is
when to have this paper out.   I should certainly not wish to have it
done *till you were present,* because we must have a fair stand-up
fight...

Ever yours,

Rod. I. Murchison.

Sedgwick's answer gives no hint of contrition.   Evidently
he had been too ill either to remember what he had written,
or to care very much what happened.   "Your letter this
morning" he writes, "astonished me as much as the want of
my paper surprised you.   I was very much out of sorts at the
time I received your letter (last Wednesday week), and
instantly struck work.   I thought I had said as much in a P.S.
to the letter with the sections.   Soon afterwards I was incapable of writing to any good purpose; and for the last
three days I have been almost confined to my bed.   This
morning I am beginning to take up, and have taken a short
walk in the sunshine.   But I have still a pain in my brow,
and a swimming in my head, and am still slightly feverish....
On the whole it is well our Devonshire affair stands where it
is[1]."

Sedgwick had constantly deprecated Murchison's notion of
organised opposition to their views respecting Devonshire, and
had urged him, on more than one occasion, to avoid any personal
controversy with De la Beche.   In the interest of peace he
now laid the whole matter before Lyell in a letter which
contains besides a good deal of valuable information respecting himself.

Close, Norwich, *February 4th,* 1837.

My dear Lyell,

A long letter I received yesterday morning from
Murchison gave me great uneasiness.   It seems you had a
warm altercation in the Council of Wednesday when our
conduct and motives were taxed in no measured terms.   I

---

[1] To R. I. Murchison, 3 February, 1837.

fear these repeated sparring matches will end in ill blood, and
ruin the harmony of our Society.   Fortunately I have been
away during the late exhibitions of them, and therefore cannot be responsible for their consequences.

Unfortunately De la Beche has published a map of Devon,
bad in its details, (at least bad for an Ordnance survey,) and
destitute of any principles of classification.   Murchison re-
monstrated against some parts of it so loudly, that during my
first Residence here (December 1834) De la Beche sent me
a caricature of him, in which he was represented as arguing
against the existence of his (De la Beche's) nose.   The map
was, however, published I believe, in spite of Murchison's
most urgent remonstrances.   Again, I know that Murchison
intended over and over again to go to Devonshire and work
out the Culm case for himself; and, had the time served, we
purposed to have undertaken the task together after our Irish
tour in the autumn of 1835.   Independently of the Culm
question (which was brought before the British Association
at Dublin in 1835 by Mr Williams, who treated the Geological Society with very little ceremony), I always intended
to spend a summer in Devonshire and Cornwall before I
attempted to publish my general views of the relations of our
older British formations.   My first paper was on the structure
of those countries.   I read a second paper on the same
subject early in 1830 to the Cambridge Philosophical Society,
which was never published; and finally I read a supplement
to the second paper in November last, which, if I live, will be
published this coming spring.   Are all these labours to stop,
because an official person has had the misfortune to publish
a bad map?   The map ought to be withdrawn without loss
of time.   It cannot, by any tinkering, be brought into order.
It must start on new principles.   Many of the details no
doubt may stand, but even the details are very far indeed
from what they ought to be.   But is it not a piece of great
stupidity to call our memoir on the physical structure of
Devonshire an attack on De la Beche?   It might just as well

be called an attack on Greenough, Buckland, and Conybeare.
De la Beche complains that Murchison does not wish to
cooperate with the Trigonometrical Survey. How is it
possible to cooperate? In our opinion Devonshire is radically
wrong, as it is now published. Are we to shut our mouths,
and let the error continue to be propagated?

But you may ask why I did not send up my paper for
Wednesday. In truth I was not well when I started with it;
and a letter I received from Murchison on Wednesday week
disgusted me with my task, so that I struck work. Two
posts after he wished and urged me to go on and finish my
work : but I was then so much out of sorts that I was unable
to hold up my head, and all the early part of this week I
have been almost confined to my bedroom. On the whole I
am glad our paper did not come forward. We have a good
case, and want to steal a march on no one. In my opinion
De la Beche will be disgusted when he hears the paper has
been put off, because, when we produce it, in a few weeks,
fortified by sections, fossils, and details, no one can then
throw out any sneer of indecent hurry, of attacking a man
behind his back, etc. etc., and thus turn the attention of the
meeting from a geological to a personal question.

Murchison tells me you mean to allude to the Devonian
case in your speech. We could, either of us, easily give you
a condensed synopsis of our general idea of the structure of
the county so as to bring it into comparison with the other
parts of England. But on this matter I can give you no
advice; you must act for yourself. Thank God I am now
very much better; two days since I could not have written
so long a letter. My best regards to Mrs L.

Yours ever,

A. SEDGWICK.

P. S. Let me hear from you. My friend R. I. M. was at
a white heat on Wednesday, if I mistake not. A very
flippant and ill-mannered letter from De la Beche seems to

have nettled him not a little. I was far more amused than vexed with it, when I saw it; it was so very characteristic of its author. What was the provocation I don't know.

The difficulties with which the production of this unfortunate paper were beset by no means ended with Sedgwick's illness at Norwich. On his return to Cambridge in February he was at first hopeful of being ready by the end of March. Then came "paralysis of one half of the optic nerve of the left eye[1]," which, though it passed away in a few hours, left the eye weak, and as he said "the loss of one side of my only eye would be no joke." This was succeeded by a fresh attack of influenza with the usual result. Indeed he vowed that he had never been so ill since he was an undergraduate. His spirits, however, survived the general wreck of his intellectual powers, to judge from the letter in which he described his condition to Ingle.

GOOD FRIDAY EVENING, 24 *March*, 1837.

My dear Ingle,

Did ever mortal see such a Good Friday? The weather is cold as Christmas, and the Great Court of Trin. Coll. is literally mid-leg deep with snow. That old hoary brute winter has come, like a Proctor, when little looked for and less welcome, with his two bull-dogs Pain and Pestilence; and many civil gentlemen, besides myself, have been sorely mouthed by them. In plain Yorkshire, I have had a sore relapse of the influenza, and for ten days have been confined to my sittingroom and bedroom ; in both of which, by Haviland's[2] order, there is, and has been for the last ten days, a great roasting fire. Under such treatment I ought before this to be well done on both sides, and to want nothing but a good kitchen basting to make me fit to serve up. But alas! instead of a roaster I am still a very starveling, and fit for nothing in the world except to stop a cracked pane of glass.

---

[1] To R. I. Murchison, 6 March, 1837.
[2] John Haviland, M.D., Regius Professor of Physic.

Thank God I am a little better than I was. For a whole
week, even old Shakspeare and Walter Scott acted only as
sounding-lines for the depth of my intellectual vacuity. But
now I can read them for an hour together, and fancy, at least,
that I am refreshed and built up by them. Should Haviland
be able to draw me from my hole by Wednesday I shall then
have to corroborate my faculties among thousands of scholarship examination papers; and if they won't cure me I must
be out of the reach of physic. So much for my case; having
reached the utmost limits of the ablative, it is high time for
me to turn back and get into a better case. When I was in
the dumps before, you told me I described my symptoms well.
Having then a good diagnosis, pray prescribe for me forthwith. A long letter from you will do all my inwards good;
therefore in mercy send me one, and soon. This request is
the main object of my handling the pen, while my head is
ready to split, and my lungs wheezing like the cracked
bellows in the organ of St John's Chapel. A cup of tea has
produced and is producing a partial thaw among the hydraulic
pipes of my thoracic regions; and the boheasian vapours
have even reached some of the lower organs of my pericranium.

What think you? Charles Musgrave was here this morning, looking all glorious to behold in a new shovel-hat, and
his little sweet-looking boy was with him. He has been
brought here by the promotion of his brother[1], the Bursar,
Professor, and Baron. Before long he may become a Baron in
more senses than one, for even now he is a Dean—Dean of
Bristol to wit, with all the appurtenances thereunto belonging.
For his sake I rejoice, but for my own, I sorely lament. He
is a friend of thirty years standing, for whom I have always
felt great kindness; and I only say the truth when I add that
an unkind word, or an unkind thought, never passed between
us since we were first acquainted. Such a friend is not to be
replaced. Still I rejoice, and so will you. But when are you

---

[1] Thomas Musgrave. See above, p. 419, *note.*

to mount a shovel? How well would it set off those venerable locks, and that grave visage of thine. By the way I
meant to have asked a question or two *at you*, as they say
over the Tweed, touching that false varlet who gave you such
a lying account of my hustings speech at Dent; but I have no
room for so big a subject. So I will conclude by telling
you—a work of supererogation as you know it already—that
I am your sincere and affectionate friend

A. SEDGWICK.

The influenza cured, and the Scholarship Examination
despatched, Sedgwick was beginning to feel in working order,
when a fresh mishap occurred. "On Friday" he wrote "I
took a short ride; and though I returned from it much
fatigued, I was certainly refreshed by the exercise. Yesterday
I again started on my horse; and unfortunately (whether
from my own fault or not I hardly know) he fell with me. My
face is a good deal cut, and much disfigured, and my knee
received a severe contusion. Two University men were near at
the time, and one of them very kindly galloped to Cambridge,
and sent a fly to bring me home. My bruises were severe,
and I was much shaken by the shock of the fall. Last night
I was very miserable, and did not close my eyes in sound
sleep. But to-day, thank God, I am much better; all fever
has left me...You never saw a more extraordinary *phiz* than
mine is at this moment. I have a great black patch running
horizontally across my face under my eyes, and my nose
is as red as flame, and my chin and cheeks scarified[1]."

Towards the end of May, after a week at the seaside,
he was able to announce: "I am almost myself again;" the
finishing touches were put to the long-expected paper, *On the
Physical Structure of Devonshire, and on the Subdivisions and
Geological Relations of its older stratified Deposits*, and it was
read to the Geological Society, 31 May and 14 June. It was
followed by a discussion which was evidently lively, but the

---

[1] To R. I. Murchison, 16 April, 1837.

few words that Sedgwick wrote about it do not convey the impression that it was at all hostile. On the following day he told Canon Wodehouse: "We had a grand battle at the Geological Society last night, in which I bore the brunt on our side; but, though well banged, I was not beaten[1]."

A few words must be bestowed on the pains with which Sedgwick criticised Mr Babbage's *Ninth Bridgewater Treatise* in the spring of this year. The proofs had been shown to Mr Lyell and Dr Fitton, who both praised the work generally, but advised the omission of certain passages. To this the author would not consent, and when other friends were suggested as referees, he objected to all except Sedgwick[2]. The proofs were accordingly sent, and Sedgwick—who appears to have had a genuine regard for the eccentric inventor of the calculating machine—went through them carefully, and returned them to Lyell with the following criticism: "I have gone over the slips, except the last page, with some care; and I think what I have thrown out may be of some use. Don't show the paper to Babbage if you think he will be offended at my freedom. But he ought not, I am sure; for I do to his sheets precisely what I do to Whewell's, or those of any other friend, whenever they fall in my way. If I can be a means of preventing Babbage from publishing any of the expunged passages I shall have helped you in doing him a service. The whole is too ambitious in its style of writing, and the condemned passages I think in dreadful taste, and also quite out of place. In the whole there is too much attempt at swell and amplification. But in that respect the author must of course have his own way. Only his proof-men must try to make him reef a few of his *studding sails, spankers,* and *sky scrapers.*"

Lyell's next letter shows that Babbage had the good sense to take Sedgwick's advice:

[1] To Canon Wodehouse, 15 June, 1837.
[2] From Charles Lyell, 5 April, 1837.

31—2

My dear Sedgwick,   21 *April*, 1837.

When I saw the outside of your letter, I said at once, that I ought before to have thanked you for having so immediately, and when out of sorts, complied with my wishes. Had I not fully expected to see you last meeting I really should have written to say that Babbage had prized the two capital pages of *critique* as they deserved, and I hardly know anything else which would have induced him to leave out the most offensive passages, on which you, Fitton, and myself had vented our chief displeasure. The coincidence outweighed the flattery of a certain popular preacher, I forget his name, and some others (John Murray ! included), who thought those very flights the finest things in the whole. Samuel Rogers, at whose house we were last night, told us he had kept back the said *Bridgewater* two months, and observed that, as usual, the author was most attached to the most far-fetched and extravagant parts in the whole. I told Babbage the *critique* was by you. He took it all in excellent part, and had you been much more severe, as Fitton was, he would not have been out of humour, though it would have influenced him less. B. told me that when he had left out much of what you had cut out all he could get from Fitton was, that "he then believed the book would not *disgrace* him ;" which B. thought a marvellous relaxation of his former sentence. It has been a great want of tact in Fitton that he has been so unmerciful, and has scarcely done justice to the good parts which preceded what you saw. If your letter had come two months earlier, before Fitton's, every sentence struck out by you would have been omitted, but I dread still to see the thing in print, as he has grown obstinate by too much sweeping contradiction.

I suppose you read my *Anniversary Address*, and I hope you approved of what I said of the Devon affair....I wish much you were more and oftener in town. It is rare even in one's own pursuits to meet with congenial souls, and Darwin is a glorious addition to my society of geologists, and is working hard and making way, both in his book and in our discussions. I really never saw that bore —— so successfully silenced, or such a bucket of cold water so dexterously poured down his back, as when Darwin answered some impertinent and irrelevant questions about S. America. We escaped fifteen minutes of a vulgar harangue in consequence. Whewell does famously in the chair. He will tell you of Owen's paper on Darwin's *Toxodon*....

We were very sorry to hear of your fall, and have every day since had news of you from Whewell, Murchison, and others. Pray write again if disposed, and believe me, Yours most truly,

CHA. LYELL.

Another, and a very different, matter, which occurred at about the same time, gave Sedgwick no little anxiety. When he first went to Norwich the see was occupied by Bishop

Bathurst. He was then ninety years of age, and could have taken but little part in the affairs of the city or diocese. Moreover, for some time before his death he had resided almost continuously in London. In fact, though much respected for his personal character and amiable disposition, Bishop Bathurst had been throughout life a Bishop of the old school, a man of letters and a politician rather than a churchman, and a devoted whist-player. A good story is still current respecting Sedgwick's first dinner with his diocesan. The whist-table was set out as usual in the drawing-room, and Sedgwick was asked to take a hand. He regretted his inability to do so, protesting his complete ignorance of the game. The Bishop said nothing, but afterwards lamented his melancholy position in the following pathetic words: " I have consistently supported the Whigs all my life—I believe I am called the only liberal Bishop—and now in my old age they have sent me a canon who does not know spades from clubs!" The Bishop died in April 1837, and some of Sedgwick's friends were anxious that he should be his successor. One gentleman let him know that he had pressed upon the government "the benefits which would accrue to this diocese, the Church at large, and the Ministry, by appointing *you.*" Sedgwick was much annoyed. "I found by a letter yesterday" he said, "that a friend of mine had made a move as he supposed in my behalf. But, unknown to himself, he was trying to do me as great an injury as he could inflict upon me[1]." Before a week was over he was relieved by the news that his friend had failed, and that the Rev. Edward Stanley had accepted the Bishopric. He was not at first quite satisfied with the appointment; but, before many months were over, the Bishop and all his family became his most intimate friends, and the palace was quite as much his home in Norwich as his own residence.

By the middle of May Sedgwick had left Cambridge, and for the next five months led an unusually wandering life.

[1] To R. I. Murchison, 9 April, 1837.

1837.
Æt. 52.

1837.
Æt. 52.

When he was back again at the end of October, he wrote: "Since I last saw you I have had no resting-place for my feet. From Norwich to London, from London to Westmoreland, from Westmoreland to Cornwall, from Cornwall back again to Westmoreland, from Westmoreland to Yorkshire, from Yorkshire to Cumberland, from Cumberland to Liverpool (where I halted one week among the flesh-pots and sections of the British Association), from Liverpool to the Warwickshire and Leicestershire coal-fields, out of which I finally emerged, and once again am enjoying the light of the sun in the atmosphere of Cambridge. Is not this enough to make a man's head turn round[1]?" These journeyings to and fro are further described in the following letters:

To Canon Wodehouse.

TRIN. COLL., *July* 8, 1837.

My dear Wodehouse,

I only reached Cambridge (on my way back from Dent) about the middle of the day yesterday; and I should endeavour to leave it this afternoon were it not for the King's funeral, which prevents all work from being done. Now a man starting on a tour in Cornwall has need of certain sartorial and sutorial helps, which put me in some perplexity, and the end of it will be that I shall not be off the stocks before Monday. Should you ask why I am in such a hurry, I should reply that I have much work to do, and little time to do it in. I have to examine a corner of Cornwall, and to be in Yorkshire in time for the contest for the West Riding. In short I have to do things which require a 40-horse steam-power to be done well. This is a power much beyond my muster; but I must do my best, and many a time and oft shall I have to wipe my brow if I do all I hope for during the next three weeks.

I left you in a very husky condition, and I continued so till the weather fairly changed, when my sweet voice came

---

[1] To S. Woodward, 26 October, 1837.

back again. After attending the Geological and Royal
Societies, I scampered down to Manchester by the break-neck
day-coach in eighteen hours, and the day following found my
way to Dent. The country on the way looked most charmingly, and the crops among my native mountains were
almost as forward as I had left them near London. This
never happens except when a very severe spring destroys the
difference of climates, and makes them all start together.
You would have laughed at my solemnity had you seen me
for three days looking over papers, casting up accounts,
and making dividends among a set of legatees who were
anxiously waiting my arrival, and meanwhile solacing themselves by deep and long potations in the beershops. After
emancipating myself from this bondage, I was detained a
great part of another week in order that I might have the
happiness of laying the foundation-stone of a little chapel in
the upper part of the valley of Dent. The day was glorious,
the face of nature beautiful, and all parties in good humour
and charity. About seven hundred mountaineers, including
nearly two hundred Sunday-school children and about one
hundred strangers, some of whom came from the distance of
twenty miles, made a curious mixed procession in the wild
glen where the little chapel is now rising from the ground.
It is built upon the solid rock which forms the bed of a
mountain stream that washes the churchyard side, and over
which the waters descend in a long succession of rapids and
falls; and it will be surrounded by birch, mountain-ash, and
other wild trees of the country. I trust God will bless the
undertaking which begins so smilingly. We began by making
the rocks echo back the old hundredth Psalm; my brother
read one or two short prayers from our liturgy; Mr Wilson
of Casterton made a short address; I handled the trowel, and
laid the stone, and then addressed my countrymen, after
which we again uncoiled ourselves into a long string to the
tune of *God save the King;* and the strangers, school-children,
and some others went down to Dent and had cold meat and

1837.
Æt. 52.

coffee at the old parsonage. My sister made thirty-six gallons of coffee in a brewing-vessel. Among the unexpected strangers was that strange, wild, but very clever person Hartley Coleridge. I must honestly say that I was a little afraid of him, for he not only possesses the poetic powers of his father, but he is an incomparable mimic. I believe, however, the impression produced on him by the whole scene was such as to save us from all risk of mockery. With all his faults, and strange wild habits, he is a kind-hearted man, and I believe by no means devoid of religious feeling, however imperfectly it may in some instances have influenced his life.

On Monday I hope to be in town, and in two days more to be set down in Cornwall. My best address for the next fortnight will be Launceston, and if you or any of your young people would only take up the pen, it would be charity. About the end of this month I shall probably be facing about to the north again. The British Association meets this year at Liverpool on the 11th of September. When the hurly-burly is over I hope to spend a week or two in Cumberland.

Yours ever

A. SEDGWICK.

To Mrs Lyell.

TRINITY COLLEGE,
October 16th, 1837.

My dear Mrs Lyell,

I returned to my den this day week; having been absent (with the exception of one day in passing from Yorkshire towards Cornwall) ever since the middle of May... In June I ran down to Yorkshire and paid away £8000 one morning among some countrymen of mine for whom I have been made trustee. I also laid the foundation-stone of a little chapel in a wild part of my native valley, and for the first time in my life turned field-preacher, as I addressed about eight hundred wild people for more than an hour, having a large rock of mountain-limestone for my pulpit, and the vault

of heaven for my sounding board. Then I turned my face to
the south, halted in London just long enough to take in water
and get up my steam, whence by another move I was transported to the eastern flank of Dartmoor. I spent a delightful
week or two in battering its sides and cracking its crown, and
then I made an attack on Rough Tor and Brown Willy, and
might, for aught I can tell, have reached Land's End but for
the abominable election. But I had promised to return, and
head my radical countrymen against a combination of the
rural squires. So I packed up bags and hammers, and
(halting only one day with Conybeare) went back almost with
the speed of the wind to my native valley. A few hours after
myself arrived a cousin of mine at my brother's house, bent on
the same purpose. He heard of the election while in the
northern extremity of the Highlands, and moved southward
with the same speed as I had done northwards. Does not
this prove a little good whig leaven to be lodged in the blood
of the Sedgwicks?

Having *done* the Squires to their hearts content, I went to
Cumberland—a country of charms to every one who has a
germ of feeling, and a thousand times more charming to me
from being associated with the recollections of early life. I
dare say you have heard of the incursion old ocean made last
summer into Mr Curwen's collieries. He became indignant at
the thought of their lighting fires under his lower extremities ;
so he took a most effectual way of putting them out for ever.
By the way an old Irishman, ycleped Dan Brennan, acted a
most gallant part during the rush of waters, and saved the lives
of four fellow-labourers. I told the story in the geological
section at Liverpool in so moving a way that I brought a
shower, not of tears, but of half-crowns, shillings, and sixpences
amounting to £37, which I sent as a solace to the old hero. I
should never have succeeded so well, but for my previous
lesson in field-preaching. But I will not torment you any
more with the dismal atmosphere of a coal-pit. Let me then
transport you to Liverpool, among mountains of venison and

1837.
Æt. 52.

oceans of turtle. Were ever philosophers so fed before?
Twenty hundred-weight of turtle were sent to fructify in the
hungry stomachs of the sons of science ! Well may they body
forth, before another returning festival, the forms of things
unknown! but I will not anticipate the monsters of philosophy
which such a seed-time portends. The crop no doubt will be
of vast dimensions.

After a very laborious week, a large party adjourned to
Sir Philip Egerton's. We had one glorious day in one of the
Northwich salt-mines. Conceive a chamber of twenty-six
acres, with a flat roof supported by rows of rude pillars of salt
arranged in perfect symmetry; conceive this monstrous and
almost interminable perspective traced by 2500 candles ;
conceive all this represented to the sense of sight by a kind
of darkness visible, converted, ever and anon, into actual light
by the coruscations of fireworks; lastly conceive my attempting
to get upon stilts to describe such wonders, and then falling flat
on my face and breaking the nose of my imagination. When
you have done all this you will know so little about the
matter that it will be better for us both to shift the subject.

I have only time to say that I started with Greenough[1] and
two Cambridge[2] friends for the Warwickshire coal-field. G. B.
G.'s paces and mine did not suit ; so we parted with mutual
good-will, after going one day in the same harness. Tell
Mr Lyell that I have also been working in the Leicestershire
coal-pits and in Charnwood Forest. The poor miners are
really to be pitied. At one place they are soused in old
Ocean's watering-pots ; at another they are broiled by Pluto's
kitchen-fires. I descended one pit about 1100 feet deep, and
in two hours was baked to the very marrow of my bones.
N'importe ! here I am, with vigour enough to torment you
with a very long rambling letter. I have just room for my

---

[1] George Bellas Greenough, one of the founders and first President of the
Geological Society.

[2] Sedgwick tells Murchison, 9 February, 1838, that one of these friends was
Mr J. B. Jukes, of St John's College, B.A. 1836. He was one of Sedgwick's
geological pupils.

kind remembrances to all your family and to assure you, with a long face, and a penitent heart[1], that I am most truly yours

1837.
Æt. 52.

A. SEDGWICK.

To R. I. Murchison, Esq.

TRIN. COLL., *Oct.* 29, 1837.

"Pray what news? where are you in your book? I think I told you that Greenough and I separated amicably after the first day. His paces, and mode of working, did not suit me. I made out all the tricks of the Nuneaton field. The coal-field passes into the New Red series; has beds of limestone (fresh-water I suppose) near the separation; and the lower part of the New Red has calcareous portions that are burnt for lime.

The Leicestershire coal-field astonished me, but it is very obscure. By the way, we found the Warwickshire coal-field brought out by a synclinal dip several miles to the S. W. of the line marked on the maps. Greenough, by his precipitation, overran this phenomenon. And, what delighted us, we found a single patch of the old slate rock tangling out by a riverside to the west of this western flap; and just where it showed itself it set the coal strata so much on edge that in one place they had been worked in a vertical position by gallery under gallery, like a lead vein. Charnwood Forest I knew before; but was delighted with a second visit.

They got up a dinner for me at Leicester[2], and I tried to pay them by a kind of evening lecture on the structure of the neighbourhood, endeavouring to prove that the money they are spending near the town in sinking for coal is so much thrown away. Some of them did not thank me for this damper; but honesty is the best policy in geology as in every other thing."

The chapel of which Sedgwick joyfully records the foun-

---

[1] Mrs Lyell's last letter was dated 21 April, and had lain unanswered in a drawer during Sedgwick's summer excursion.

[2] The dinner was given by the Philosophical Society.

1837.
Æt. 52.
dation, and in which he ever afterwards took the liveliest interest, is situated in the upper and contracted part of the dale of Dent called Kirthwaite.   The ancient name was Cogill or Coegill, but this, by long usage, has become Cowgill, though the correct pronunciation still survives.   The circumstances which led to the building of the chapel, have been narrated by Sedgwick in the *Memorial* from which we have already made long quotations.   After describing how it had come to pass that the inhabitants of Dent had sunk into "a state of comparative poverty," he proceeds: "the hamlet of Kirthwaite partook of this change, and of the unhappy moral consequences which gradually followed.   In the first quarter of this century many of the poorer inhabitants of the hamlet, especially those in the remoter parts of it, were without instruction, of reckless life, and without the common comfort and guidance of social worship in the house of God.   To meet these evils Mrs John Sedgwick, the wife of the incumbent of Dent, personally devoted the best efforts of her life.   Year after year she worked on in good hope ; and her pious work had its blessing.   For she gradually drew together an united body of Christians, who were ready to sink out of memory all points of dissent or difference, and with true hearts to join in common worship, and in prayer for the erection of a chapel to be lawfully consecrated to the services of the Church of England.

"A site for a chapel and a chapel-yard was the first object of inquiry ; and Mr Bannister of Cowgill gave generous help in the hour of need.   For he offered to convey to trustees the materials of an old chapel[1], with such addition from his family freehold as would form a beautiful and convenient site and burial-ground for a new chapel, which might become for ever a

---

[1] Sedgwick says in a note (*Memorial*, p. 34) that the older chapel had been built, so far as he had been able to ascertain by tradition, "by a member of the family of Cowgill, who had while in Scotland adopted the doctrine and discipline of the Presbyterian Church.   For some years, while he lived, the chapel was zealously attended, and the yard in which it stood was used as a burial-ground for the congregation."   After his death the congregation melted away, and the chapel became a ruin.

chapel-of-ease to the old church of Dent. This offer was met
with heartfelt gratulations on the part of the inhabitants of
Dent. In conformity with such feelings, and in good hope, a
circular letter was published in July, 1836, calling upon all
who had a pious interest in the spiritual and temporal well-
being of the valley of Dent, to subscribe for the erection of a
chapel, to be called Cowgill Chapel. The public generously re-
sponded to the call, and on June 30th, 1837, the foundation
stone was laid amidst demonstrations of joy, which everyone
in Dent, whatever might be his name or form of worship,
seemed with a full heart to share[1]."

At the Liverpool meeting of the British association Sedgwick was President of the Geological Section. He protested
that he was almost worked to death, but his various letters
show that he did not find the work disagreeable. His only
paper was a *Notice of an Incursion of the Sea into the collieries
at Workington.* It was at the close of this paper that he
told the tale of Brennan's intrepidity. His narrative is said
to have been extraordinarily dramatic—a happy combination
of humour and pathos—often expressed in the language of
the man himself[2]. A few days afterwards he sent the money
subscribed to Mr Williamson Peile, Lord Lonsdale's agent,
with the following letter[3];

LIVERPOOL, *September 18th*, 1837.
My dear Peile,

We have had a glorious week at Liverpool. On
Wednesday last I described before the Geological Section the
geological position of the great main band in the sub-marine
portion of the Workington colliery, and the circumstances
that led to its submersion. Before I sat down I also stated

---

[1] *A Memorial by the Trustees of Cowgill Chapel*, pp. 4—6. Sedgwick sent
one hundred guineas to the building-fund.

[2] *Report* of the British Association, 1837, p. 75. *The Athenæum*, p. 697.

[3] This letter, of which the original cannot now be traced, was fortunately printed
in *The Cumberland Pacquet*, for 26 September, 1837. Our copy is due to the
kindness of J. H. Kendall, Civil Engineer, of Whitehaven.

the facts you and I heard from the mouth of Daniel Brennan. The account produced such a sensation in the room that a subscription was instantly set on foot, and hats were handed round for the brave fellow, who, under Providence, has been the means of saving the lives of four of his fellow-creatures. The fact of his turning back to save the old man in the stable (after he himself and the three others whom he had conducted safely through the air-course were out of danger) excited a loud burst of admiration. I now enclose a bill for the sum of £37. 2. 0, which I request you to pay over to Daniel Brennan, being (with the exception of £2 or £3 since received) the amount of the subscription raised on the spot, and handed over to me in the Chair.

I am sorry to write in such a hurry, but the glad tidings I send you, with my own hearty good wishes, are better things by far than a more formal letter. I should have written to the man himself had I known how to address him. Pray give him good advice; tell him to do again what I hope he has done already, to thank God for the deliverance, and for this further good that has come to him. Tell him also that those who subscribed hope he will not spend on folly and sin this overflow of Christian charity on his behalf.

Make what use of this letter you please. The substance of it should appear in the papers, as the fact it states is a lesson of kindness, and may do good.

Believe me, my dear Peile,

Very truly yours,

A. SEDGWICK.

Mr Peile lost no time in assembling his workmen, and informed Brennan of his good fortune in their presence, "to his great joy and indescribable astonishment[1]." The money was not wasted, as such gifts too often are; it was prudently given to Brennan as he wanted it, so that his debts were paid, and his position substantially bettered. Six months afterwards

---

[1] To Professor Agassiz, 5 March, 1838.

Sedgwick could write: " I am happy to hear that the gift has
improved both his fortune and his morals. How much good
may be done by kindness! and alas! how seldom men even
think of it.[1] "

In a summer so fully occupied there was but little time
left for geology. In Devon Sedgwick was joined by Mr
Austen, but they did not complete any part of their survey,
both being unexpectedly called away. In Cumberland we do
not know that he did more than examine the drowned shafts at
Workington. The explorations in Leicestershire are described
with some detail in a letter to his brother. From Coventry,
he writes, "I walked to a small village near Bedworth, and next
morning commenced work along with three companions. In
the course of five days we worked our way, through a very
interesting country, as far as Tamworth. It is highly beautiful
to look upon ; and we did not shut our eyes to its loveliness ;
but the vast beds of coal underneath the surface (some of
them 12 feet thick) occupied the greater share of our attention.
From Tamworth we removed our head-quarters to Ashby de
la Zouche in Leicestershire ; and again began to dive underground, descending one pit more than 350 yards deep. The
temperature at the bottom made the perspiration run off in
streams. To reward our toil we found them working a magnificent coal-bed full 12 feet thick. After experiencing much
kindness at Ashby, we again removed our quarters to the
skirts of Charnwood Forest, and gradually worked our way
to Leicester. Finally we reached Cambridge on Monday
the 9th October[2].

During the summer of 1837 the condition of Sedgwick's
house at Norwich had given him much trouble. The drains
had made themselves so unpleasant that he declared: "People
cross the street to avoid me, and hold their noses when I pass
them." In consequence certain indispensable alterations

---

[1] Ibid. Mr Peile wrote, 17 February, 1838: "The poor fellow is extremely
grateful, and frequently asks after the *Purfessor*. He is sober and industrious."

[2] To Rev. John Sedgwick, 16 October, 1837.

were carried out under the direction of Canon Wodehouse, and many letters passed between him and Sedgwick respecting them. The first paragraph of the next letter refers to this subject; the rest, on a more delicate matter, explains itself. Sedgwick's reply reads like a fragment of an autobiography, and explains much that without it appears contradictory and almost unintelligible in the detailed history of his life.

From Canon Wodehouse.

NORWICH, *October 10th*, 1837.

My dear Sedgwick,

      I had better tell you at first, for fear of forgetting it at last, that your house is well nigh finished in every respect: that it looks respectable, if not beautiful, without, and promises much comfort within.

      I hope you will open this letter in a quiet, sit-still, leisurely, contemplative mood, because I want you not only to read patiently, but to digest it properly, not like a Heron with an Eel (as the fable goes) but "*more ruminantium.*" I have not forgotten a conversation we had in this house somewhere about midnight, on the subject of some sermons. You gave me a sketch, or rather a bold clear outline, of a short course, which was to be printed, and preached at Norwich Cathedral. Now I have strongly set my mind upon your realizing this excellent scheme, and in your approaching Residence, and the only object of this letter is earnestly to beg you not to disappoint me. You have now time before you to do this in a manner worthy of yourself. Your lectures exact no toil as to preparation, you have only therefore to throw aside other matters for a few weeks, and devote some spare hours to this, and the thing is done. I have many reasons to urge in behalf of this proposition. I persuade myself that you are not only a lover of truth, but will give me credit for being the same, and that, when I assure you that I write very much from a jealous regard for your reputation here, you will not find fault with me for speaking openly. You have done much for us here as a Geologist. I am now most anxious you should give us some good, well-considered Theology from the Prebendal Chair. "Sedgwick won't take the trouble to compose sermons," say some. "He can't", say others. "He gives us nothing but a few scraps of paper written with a pencil in church, and he'll teach all the young ones to do the same." "If he really has not time to prepare properly a few sermons, why was he made a Prebendary?" "How much good might he do if he would turn his talents this way, and give us every Residence a good course of good sermons?"

      Now I should not detail these waspish speeches to you, but that, professing to have within me a few sparks of friendship, never, I trust, to be extinguished, I am considerably nettled by them, and want you to furnish me with the *best* answer to such of them as may be in any degree true. It is not however merely on such grounds that I write. "*How much good might he do &c.*" is a motive of a very different kind; one which you, I am assured, will not refuse seriously to consider and respond to....

I will not add any apologies or explanations to what I have written, because I am assured that to you any such expletives are unnecessary. I rather expect to be called to London for a few days before very long, and shall try to look in upon you going or returning. All here are well, and Alice has not forgotten you. With every kind wish from the whole party, always most sincerely yours,

CHARLES N. WODEHOUSE.

TRIN. COLL. *October 12th,* 1837.

My dear Wodehouse,

I have to thank you for the letter which reached me yesterday morning, though some of its sentences made me wince a little. As to the waspish speeches you mention I care not for them; but the other remark *How much good might he do* &c. &c. &c. I agree with you, I ought to respond to as far as I am able. And were I to give myself up to theological studies exclusively no doubt I might do more than I have done. But it is at present impossible for me to do so. And were the alternative given me this day of resigning my professorship or my stall, because I could not do the duties of both properly, I should instantly give up the latter, and not the former. For I have a great accumulation of original matter piled up during observations directed to one main point, and not to turn this to use and profit would I think be a shame and a sin. In a few years I may perhaps be emancipated from the rubbish which surrounds me. A museum is rising from the ground[1]; were my collection once arranged, and my geological books and papers written, I should wash my hands, and try to pass my office to a younger and lustier person. And when is this to be done? Perhaps in three or four years, if it please God to spare my life and faculties. At this very moment I have no less than five papers on the stocks:

---

[1] The present Geological Museum, part of the new Library begun 1837.

two for the Cambridge Philosophical Society, and three for
the Geological; and of these, three are joint works in which
the reputation of others is concerned. These arrears arise
partly from my own procrastinating temper, and partly from
last year's long-continued indisposition, which prevented me
from putting pen to paper during the spring months. Under
these circumstances I have assuredly no time to prepare a
course of sermons for the press during the coming winter.
But pray come here, and talk over this and a hundred other
things. Illnatured things are seldom said without some
semblance of foundation. Have I then neglected my clerical
duties at Norwich? In answering this question I ought to be
the last person appealed to. But in my first year's Residence
I had during the greater part of the time a double duty in the
Cathedral, and during three or four successive Sundays I did
duty thrice,—which was fair hack work. During my second
Residence I worked hard at the Museum lectures, and preached
on every occasion on which I was apparently called on. The
Dean was away, and I was called on at least six or seven times.
Last winter I was direly out of sorts for six or seven weeks,
and utterly incapable of much intellectual labour. This is all
the apologetic matter I dare muster in my defence. My endeavour always was to avoid everything like rhetoric or fine
writing, and above all to fly from that which always tempts
me—metaphysical disquisition, of which I was passionately
fond when a young man. In short I tried to be off the stilts,
and to speak truth plainly and practically. Whether I hit the
mark I aimed at I hardly know; but I used to think that the
people listened well to what I said. As to tickling people's
ears with fine sermons, I never could condescend to do it were
it in my power. Some of my friends expect more from me
than I can do. Because on some occasions I speak fluently,
it by no means follows that I write readily. Experience tells
me the very contrary. I write with pain to myself in every
sense of the word, for a very few hours' writing will generally
bring on a fit of indigestion and swimming in the head, which

puts an end to all rational continuity of thought. I was, ever since my childhood, utterly incapable of doing more than one thing at a time to any purpose. To pass from one thing to another frequently makes me utterly worthless. This peculiarity of my mind (and I believe it is what I might almost call an organic peculiarity) has sometimes led me into dire misconduct, and once or twice in my life brought me nearly into a condition of monomania. I think I have more than once hinted to you about the melancholy depression of my mind before I was appointed to the Woodwardian Chair. My summers' rambles, more than any other thing, brought me round; and I very much doubt whether even now I could long hold my head up without them. This consideration has naturally had some influence with me in determining my choice in one or two rather trying occasions. But, after all, many of my friends expect more from me than I have the power of doing were I to try. I know that much of my life has been dreamy and worthless; but I am constitutionally incapable of much sedentary exertion. My friends also think me a thousand times better than I am. For my conscience tells me, in language my soul cannot misinterpret, that in the hourly conduct of my thoughts, and in the daily actions of my life I have not only much to repent of, but that which ought to sink me to the earth, and fill me with humiliation and shame. I am now, at least, writing seriously. And, if it torments you, you must blame yourself for writing so frankly to me; and I am sure I am, and ought to be, deeply thankful to you for what you have done. How different the tone of your letter from the doses of nauseous flattery I received at some of the public meetings this summer! Their quantity was an antidote to their quality, as no human stomach would hold them, and they acted as an emetic. I bear that about with me which ought to make me humble, whatever persons may say flattering to my face, or waspish behind my back. This letter is not fit to be seen. Therefore after reading it (and I don't call on you, as you do on me, to ruminate on it) pray burn it. First,

however, give my kind remembrances to your family, beginning with Lady Jane, and ending with Boppity. My dear
Wodehouse,

affectionately yours

A. SEDGWICK.

During the Michaelmas term Sedgwick tried the experiment "for the first time in my Professorial life," of giving six
lectures in each week—but the strain was too great, and brought
on palpitation of the heart, and sleepless nights. In addition
to this labour he was reprinting his *Syllabus*, and bringing it
up to date. "I want to be correct in my synopsis of the
Silurian System," he writes to Murchison, "have you one of
your printed outlines or synopses? If you have such a paper,
don't fail to send it me by *return of Post;* for remember I am
in the Press." The following letter shows that Murchison
was asking his advice on the elaborate description of his own
peculiar domain which he was now beginning to print. In the
previous year Sedgwick had criticised the *Introduction* with
some severity, but his opinion on the body of the work was
evidently far more favourable.

My dear Murchison,

I send you the three sheets, which I looked over last
night as carefully as I could. It is a good joke for you to tell
me to just give half an hour to your three sheets. Three sheets
like yours are three honest hours' work, and so I found them
last night. I have altered as little as possible. One or two
sentences I did not understand. After all, your revises seem
to want no rasping. My suggestions, small as they are, are
perhaps changes for the worse, and I will venture to say that
not one man in a thousand would mark any difference. After
all your book is only a book for geologists. Natives of the
country will read and pick out parts of it, as Jack Horner ate
Christmas pies; but, as a whole, it is far too good and deep
for any but a true geologist. And what geologist will care,

one grain of trap, how a sentence is written provided he un-
derstands it. Yours in a hurry,

1838.
Æt. 53.

A. SEDGWICK.

Many thanks for your Silurian papers[1].

Sedgwick's occupations at the beginning of 1838 are
described in the following letters:

To the Rev. John Sedgwick.

NORWICH, *January 6th,* 1838.

My dear John,

It is too late for this day's post, but still I will try
to put myself out of your debt before to-morrow....

On the 1st of December I began my Residence, and a day
or two afterwards I was elected President of the Norwich
Museum for the next two years. At the time of my appointment I undertook to give a few public lectures. My second
was given on Thursday to an audience of about 400, and
I hope to give two more. My friends here received me with
their usual kindness, and I was delighted to come again
among them; and really my new house is quite charming. I
have one very good bed-room fit for a married couple; and
three spare beds for bachelors. One Cambridge friend is with
me now; and before long I am expecting three more. I have
a capital housekeeper who provides and cooks for me—the
same person whom Lady Jane Wodehouse recommended to
me—a housemaid, and a young lass to help her. My own
servant comes and acts as my butler and waiter and *factotum;*

---

[1] This letter is undated, but docketed by Murchison, "November, 1837."
He further notes: "Revision of the three sheets (the only ones) of my work
which he undertook. R. I. M." This statement, as Mr Geikie observes in a
pencilled sentence appended to the above note, "is not quite accurate. Sedgwick
revised the Introduction to the *Silurian System,* and expressed great dissatisfaction
with the original, making many alterations and suggestions, most or all of which
Murchison seems to have adopted. Probably the three sheets above referred to
were the only three in the body of the work." To this may be added the fact that
in the spring of the following year (7 March, 1838) Murchison writes to Sedgwick:
"I send you the concluding chapter of my Part I., and a portion of the opening
chapter of Part II., which *must* be seen by you before they see the light." It is
of course possible that Sedgwick returned these proofs without looking at them.

1838.
Æt. 53. and I have an assistant and occasional waiter to rub down my
horse. Such is my establishment. Before long I hope some
of your family will come to see me. When Isabella and
Emma are out, they must come their first winter and keep
house for me during a two months' Residence. It will be
a good start for them: and I shall be able to show
them excellent society. Indeed I have more of it than I like.

The Bishop's family is an admirable addition to us. The
Bishop I knew before I came hither. His wife is a charming
and sensible woman. His eldest son was frozen up in the
North Sea with Captain Back, and is just appointed com-
mander of a gun-brig going out to the South Seas. The
second son took a very distinguished degree at Oxford, and
the youngest son has just joined the Engineers at Woolwich—
all of them have been here. The two daughters are also very
clever and agreeable young ladies. One of them is not yet
come out. The whole family dined with me one day. But
the most noisy party I have had consisted of about twenty
children, whom I made very happy for a few hours. Don't
suppose that I have been quite idle in my profession. · I have
been preaching almost every Friday, and often sitting up very
late at night, and rising very early in the morning, to write
my sermons. I preached last Sunday week in the Cathedral
for the benefit of the County Hospital. I rise very early,
and read prayers to my servants frequently by candle-light.
I then breakfast about eight....A happy new year to your
wife and children.

Affectionately yours,

A. SEDGWICK.

To Professor Agassiz.

TRINITY COLLEGE, CAMBRIDGE,
March 5th, 1838.

My dear friend,

After an absence of three months I only returned
to Cambridge last week. My winter has been spent in
Norwich; and many of my letters were detained in Cam-

bridge for a long time, and yours among the rest. This will
explain my long delay in replying to your very kind and
delightful letter. In one respect, however, it gave me pain, as
it deprived me of the hope of seeing you in England next
summer. But I rejoice that Mr Dinkel is coming, as he is a
very kind-hearted man, as well as a very accomplished artist,
and will I believe do your work better than any other person.
I will give him every facility in my power in making drawings
from the specimens in the Cambridge Museum. Some I will
with great pleasure convey to town for him ; others he will
have to copy at this place, and I will procure lodgings for his
use. In short I will assist him, and you, in every way I can,
and more I cannot promise. The specimens of Agostino
Scilla shall also be put at his disposal.

I only reached Cambridge on Friday, when I found your
eighth and ninth *Livraisons* on my table, and I cannot tell you
how much I am delighted with them. The engravings speak
to the senses like the originals; and they have a great
additional interest to an English geologist, in being chiefly
derived from the formations with which he is familiar in his
own country. Pray put me down as a subscriber to your
work on Echinoderms, the successive parts of which may
be sent me through Baillière along with the *Livraisons* of
your great work on fossil fish.

Till I see your memoir on the erratic blocks of the Alps I
don't know how I can offer any opinion, as I don't at present
know exactly what is your hypothesis. Where has it been
printed ?...On the subject of the erratic blocks of Switzerland
it strikes me that no one can possibly account for them
without the aid of the carrying power of ice. Without
knowing what it is, I am, therefore, favourably disposed
towards at least a part of your hypothesis. A great deal of
evidence, both positive and negative, has been advanced
in favour of the iceberg theory. For example, Mr Darwin
has shown that throughout South America erratic blocks are
found within the limits of latitude where glaciers are, or may

have been, down to the level of the sea; and that they are
wanting in the tropical latitudes, where ice could never have
existed near the sea level.   In England (where everything is
on so small a scale, yet where we have such a fine succession
of phenomena, illustrating almost every important point in
the geological history of the earth), we have a most interesting
series of erratic blocks.   I don't think the iceberg theory can
be applied to them, because they go in almost all directions,
and not towards any prevailing point of the compass, and
because they follow the exact line of waterworn *detritus* and
comminuted gravel.   Such blocks I attribute to currents
produced during periods of elevation and unusual violence.
There are many instances of rocks grooved deeply, and
partially rubbed down, by the currents of what we formerly
called *diluvium*, a word which is passing in some measure out
of use in consequence of the hypothetical abuse of the term
by one school of geologists.   There are very fine examples of
this kind near Edinburgh.   Stones transported in this way
are always rounded by attrition, and in every question about
the origin of erratic blocks we ought to regard their condition
(viz. whether rounded or not), as well as their geographical
relation to the parent rock.

Since I had the pleasure of meeting you in Ireland I have
not published any memoirs.   In the spring of 1836 I was out
of health and spirits.   In the summer of 1836 I had a
laborious and successful tour in Devonshire, Cornwall, and
South Wales, partly with my friend Murchison, and we tried
to reduce the stubborn old formations to some order, and
I think they were a little more obedient when we left them.
Our labours have caused much discussion in the Geological
Society.   The winter and spring of 1837 I was again sadly
out of health, and literally confined to my chamber for more
than a quarter of a year.   The sad consequence was that all
my intellectual labours were suspended.   I turned out, however, again last summer, finished my observations on the S.W.
counties of England, and ended the vacation in a survey

of the coal-field of Cumberland. Thank God I am now well, 1838.
and before my summer's tour hope to pass four memoirs Æt. 53.
through the press. All this will, however, depend on my
health, which generally gives way under sedentary labour.
We had a glorious meeting of the British Association at
Liverpool, and did ourselves the *great pleasure* of voting
a hundred pounds in aid of your work. I wish other
societies would follow so good an example.

So now, my dear Professor, I have sent you a very long
letter, which I hope you will be able to read. Most heartily
do I wish you health and happiness, and a rich harvest of
honour.

Believe me, your sincere and affectionate friend,

A. SEDGWICK.

We see from this letter that Sedgwick accepts the theory
of transport by ice to account for the position of erratic
blocks. At the same time, like many geologists at the
present day, he doubted whether the ice in question was
floating-ice, or land-ice (except in certain particular cases);
and further, whether stranding icebergs, or the passage of
glacier-ice, ought to be invoked as the cause of the striation
of rocks. Agassiz had stated his own views on the subject of
erratic blocks in the following passage :

NEUCHATEL, 30 *November,* 1837.

" Il est une autre question sur laquelle je désirerais ardemment
connaître votre opinion. C'est la question des blocs erratiques
répandus sur les flancs du Jura. Depuis plusieurs années je me
suis appliqué à observer avec le plus d'exactitude possible ce
phénomène si remarquable, et si diversement expliqué par les
géologues. Aucune des explications qui en ont été données jusqu'à
ce jour ne me paraît résoudre le problème, les unes étant en
contradiction avec une foule de phénomènes non moins généraux
et importants que les blocs eux-mêmes, les autres les passant sous
silence. Au nombre de ces phénomènes il en est un surtout dont on
n'a tenu aucun compte dans toutes les théories, et dont vous avez
constaté la présence dans le nord de l'Angleterre ; je veux parler des
surfaces polies sur lesquelles reposent en partie les blocs erratiques.
J'ai taché de rattacher ce phénomène ainsi que plusieurs autres, au

1838.
Æt. 53. transport des blocs, en supposant qu'ils sont le résultat de masses de
glace qui auraient comblé à une certaine époque la grande vallée de
la Suisse, et sur lesquelles les blocs auraient été charriés, depuis le
sommet des Alpes jusque sur les flancs de nos montagnes.   Je fis de
cette question le sujet de mon discours d'ouverture au congrès des
naturalistes Suisses à Neuchatel.   Mon opinion fut vivement combattue
par MM. de Buch et Elie de Beaumont, qui assistaient à notre réunion,
et je ne doute pas qu'elle ne sera encore plus amèrement critiquée
lorsqu'elle sera livrée au public.   J'attendrai votre réponse avec
impatience, heureux si je puis voir mon opinion appuyée par le
géologue illustre en la science duquel j'ai le plus de confiance."

In April we find Sedgwick on a visit to Lord Braybrooke,
at Audley End, to witness the opening of one of the four
large barrows situated on the estate of Viscount Maynard,
near Bartlow in Essex.   Popular belief had ascribed the
erection of these conspicuous artificial mounds to the Danes,
in commemoration of a battle; but when three smaller
mounds, distant about eighty feet from the former, were
opened in 1832, the discovery of Roman antiquities, with a
coin of the Emperor Hadrian, shewed that they at least
belonged to the Roman period.   In 1835 the largest and
loftiest barrow was explored under the superintendence of
Mr John Gage, Director of the Society of Antiquaries.   As the
sepulchral chamber was probably in the centre of the mound,
which was 144 feet in diameter, and formed of chalk and soil
in alternate layers, of the firmest consistency, it was pierced
by a gallery.   After ten days labour, the central chamber was
reached, and a large party of visitors from the neighbourhood
and friends from Cambridge, among whom were Sedgwick,
Lodge, and Whewell, assembled to see the contents removed.
They proved to be Roman, like those of the smaller mounds,
and were presumably of the same period.   On this occasion
Whewell wrote *An Eclogue*, in which Sedgwick is supposed
to maintain that the tumuli were due to natural causes.

April 21st, 1835.

Mr. Gage.   My antiquarian bosom burns to explore
These relics of the art of men of yore.
Professor Sedgwick.   Stay, my good sir; control your zeal, or lose it.
This is no work of art; 'tis a *deposit.*

1838.
Æt. 53.

Gage. Geologist, avaunt! and hide your head:
Ne'er was deposit thus deposited.
  Sedgwick. I hold, despite your antiquarian pride,
That Bartlow's tallest hill is stratified.
  Gage. Your theory of strata, sir, is rickety:
'Tis a Romano-Dano-Celt antiquity.
  Sedgwick. Sir, your antiquity's a joke to me:
'Twas left here by "the last catastrophe."
  Gage. I tell you, sir, that Queen Boadicea
Killed fifty thousand men, and put them here.
  Sedgwick. Sir, throw your queens and battles to the dogs:
'Twas when the Deluge made the Gogmagogs.
  Lady Braybrooke. O gentle swains! be for a moment mute,
For here is that will settle your dispute.
The spade proceeds, the earth is outward thrown,
And now at last we find a bit of bone.
  Gage. Ha! give it me. It is, upon my word,
A British heel chopped by a Roman sword.
  Sedgwick. No; with your idle tales no longer weary 'em:
'Tis a new fossil beast—the *Bartlotherium.*
  Dr. X. Now, gentlemen, since bones are my affair,
I, as anatomist, the truth declare:
The bone is a heel-bone—observe it thus—
The beast, the *Asinus domesticus.*
No theorist is safe from trifling ills:
So to the Lord and Lady of these hills
Pay, as becomes you, thanks and reverence due,
And then proceed to theorize anew.

In 1838 the explorations were resumed, and a second
barrow, situated next to the largest on the south, was investigated in the presence of the same party, with the addition of
Professor Henslow. Mr Romilly has preserved an account of
what then took place:

" 19 *April.* Sat a long while with Lady Cotton, hearing Lodge's
account of the opening of the second of the Bartlow Barrows, which
took place last Tuesday. He, Sedgwick, Whewell, and Henslow went
from Audley End with a large party. They found a box (some four
feet long, three broad, and three high) in which were sundry Roman
glass bottles, pateras, bronze jugs, and a lamp; also some small
bones (supposed to be chicken bones). The urn containing the
dead man's bones was not yet found. Sedgwick exhibited to the
mob a pot, which he declared had belonged to Julius Caesar.
Whewell wrote a copy of humorous verses on the occasion, viz. a
complaint of the dead man for being disturbed."

Whewell's verses have no special connexion with Sedgwick's life; still, as they are amusing in themselves, and

besides were written to commemorate an exploration in
which Sedgwick took an active part, we may be excused for
printing them.

April 17th, 1838.

WHERE Bartlow's barrows of wondrous size
Stand side by side to puzzle the wise,
In a certain year, on a certain day,
A voice was heard in the morning grey:
'Twas a grumbling, growling, muttering din,
Like a man who talks a box within;
And it seemed to come, to the standers by,
From the center of one of the tumuli.
The language, as well as the ear could take it,
Was Latin,—but such as a Briton would make it.
And this is a close translation, penn'd
For Carolus Neville of Audley End:

"Brother Icenius Crispus Caius!
Close together our friends did lay us,
Seventeen hundred years ago,
And our two cousins, all in a row:
Tell me Caius, how do you lie?
Do you find any change as the years go by?
Are you still in your quarters narrow,
Snug in the mould of the tall green barrow,
With the tears of your friends around you lying
In tiny jars, to console you for dying?
I've an awkward feel that the outward air
Is making its way to my bones so bare;
It seems as if the sharp north-west
Were somehow getting within my *chest:*
And, if the cold very much increases,
I shall sneeze my barrow all to pieces.
Are you cold too? I feel, by Bacchus!
An epidemic disease attack us:
And I really fear, as learned men say,
'A touch of a tumular influenza.'"

And another voice, from another hill,
Replied in a hoarser grumble still:

"What! O Jupiter! cousin Verus,
Haven't you heard what pass'd so near us?
Poor Icenius! don't you know
They carried him off three years ago?
Certain robbers, call'd antiquaries,
Came and disturbed his quiet Lares;
Bored his barrow, and stole, alas!
His urns and bottles, his bronze and glass:

His worship's chair, that he used to sit in
At the quarter-sessions for Eastern Britain;
His handsome funeral *præfericulum;*
His wife's new-fashion'd enamel *ridiculum*[1] *;*
Bagg'd the whole!—it did not matter a
Pin whether vase, or lamp, or patera.
Even his bones, though stript of their clothing,
They took away, and left him nothing.
All are gone,—and the world may see 'em
Making a show in the *Maynard Museum.*

1838.
Æt. 53.

"And now I fear these folks intend
To rob you too, my respected friend ;
And, following up their barbarous custom,
They've dug a hole to your very *bustum ;*
And that's the reason, or I'm mistaken,
You feel so *bored,* and so sadly shaken.

"It is really hard that one's very great age
Can't save one from prying *Fellows* like *Gage*[2] *;*
When one comes to ones *teens of centuries,* clearly
One should not be treated so cavalierly.

"But since it is so, and the move's begun,
I trust we shall meet when all is done.
So, when near Caius you're set on the shelf,
Tell him I hope to be there myself;
And say the thing which I doubt the least on
Is our coming together again at *Easton*[3]*,*"

Before starting on his summer excursion Sedgwick read a very elaborate paper to the Geological Society (23 May) entitled *A Synopsis of the English Series of Stratified Rocks inferior to the Old Red Sandstone, with an attempt to determine the successive natural groups and formations.* In this communication he linked together most of his previous work, in the Lakes, in Wales, in Devonshire, and in Cornwall, and ended with a tabular arrangement of the classes and subdivisions into which the rocks might be conveniently sorted.

---

[1] Resembling the modern *reticule.*

[2] John Gage, Esq. F.S.A.

[3] Easton Lodge, the seat of Viscount Maynard, the proprietor of the Bartlow Hills. [These poems are printed in *Sunday Thoughts and other Verses,* a collection of poems on various subjects printed for private circulation in 1847, according to Mr Todhunter (*William Whewell,* p. 167). The successive explorations of the Bartlow tumuli are described in *Archæologia,* xxv, xxvi, xxviii.]

1838.   The occupations of the summer are described in the next
Æt. 53.   letters.   The first was written from Mr Conybeare's house.

To Miss Emma Sedgwick.

AXMINSTER, *July 6th*, 1838.
My dear Emma,

        I sent a short note to your father a day or two
since; but had no time to tell him what I had been about
for the last month.   Nor indeed have I time now; but I will
do my best before the dinner-bell rings to give you a kind of
outline of my movements.

        On the 14th of last month I left Cambridge, and the day
following attended a great public dinner given to Sir John
Herschel on his return from the Cape.   He has been spending
five years there, making discoveries in the southern hemisphere,
and has returned laden with astronomical treasures.   I was
called on to make a speech, a rather nervous business among
so many distinguished strangers[1].   The day following (16th)
I went with Professor Henslow to St Albans to visit his
father's family.   It was a long-promised visit, made partly for
the purpose of seeing my friends, and partly for the purpose
of examining the interesting antiquities of the neighbourhood.
On the outskirts of the town is a pretty little church with the
vault of the Verulam family, and the statue of Lord Bacon.
Tradition says that it is an admirable likeness, and assuredly
it is an exquisite work of art; indeed it was this monument
(of which I had seen a drawing) which first induced me to
think of visiting St Albans.   I also, during this pilgrimage,
visited, as you may well suppose, the ruins of Lord Bacon's
house, among which I rambled for an hour " chewing the cud
of sweet and bitter fancy," as Shakspeare says in quaint
phrase, pregnant however with meaning.

        After this charming visit I returned to town in time to
attend the Queen's levee on Wednesday the 19th.   It was a

---

[1] Sedgwick proposed the health of the chairman, H. R. H. the Duke of Sussex.
There is a full account of the dinner in *The Athenæum*, 16 July, 1838.

splendid show, as all the foreign ambassadors, who had come to grace the Coronation, were present. When it was over, a party of the Royal Society were introduced to the Queen in her private closet. We had an excellent opportunity of seeing her, and hearing her speak, as she received the Duke of Sussex without any of the formality of a Court, and seemed only to remember that he was her uncle. After presenting the Statutes of the Royal Society, and obtaining her signature to the book, he offered to bend his knee and kiss her hand (which is the regular form on such occasions) ; but she immediately stopped him, put her arm round his neck, and kissed his cheek. He afterwards presented us all in turn, and we had a most gracious reception. So, my dear Emma, when I next give you a kiss I dare say you will find my lips all the sweeter for having touched our youthful Queen's hand.

The day following (the 21st) I went to Oxford to stand godfather to Dr Buckland's youngest son. He was christened *Adam Sedgwick;* so you see my name is to be perpetuated, though as yet I have no child of my own. My visit to my friends at Christ Church was delightful; and though I have now seen Oxford so very often, yet I think I can say that I see it each succeeding time with renewed and increasing pleasure. It is one of the most beautiful cities in the world, and so full of historical interest. The whole party one day made a trip of about six or seven miles down the Thames, and then had an *al fresco* dinner in the park of the Archbishop of York.

On Monday the 26th I returned to London by the line of the Birmingham railroad; and such streams of population were floating into London along the lines of all the great roads, to see the festivities of the Coronation, that even the steam-engines could hardly drag them along the tram-roads. The train by which I reached town conveyed numerous carriages and horses besides 900 passengers, and was dragged along the rails by three engines which puffed and groaned most piteously under their extraordinary load. But what can I say of the Coronation week? The papers are all full of it, and the accounts you have been able to read are infinitely more in detail than anything I can tell you. All London seemed to be mad, and half of England seemed to be packed in the streets of London. I was in the Abbey during the grand and solemn ceremonial; and in what place think you? In the Queen's private box. I know Colonel and Lady Isabella Wemyss very intimately, and one of them is Queen's Equerry, and the other a Lady of the Bedchamber. So the night before the Coronation they sent me a Queen's private ticket. I wish you and your sister could have been with me. It was the best place in the whole Abbey, just on one side of and overlooking the two thrones. I certainly never saw before so grand, dazzling, and solemn a ceremonial, and I trust in Providence that I shall never have an opportunity of seeing it repeated. The little Queen performed her part with grace, dignity, and good feeling; and all hearts seemed to swell with delight. In the evening there was in the parks the finest display of fireworks I ever beheld. You may judge of the scale when I tell you that 850 large rockets were shot vertically into the air at one single explosion. When they began to burst it seemed as if a regiment was giving a platoon fire a quarter of a mile above our heads; and the instant after, the whole sky was lighted up with red, green, and blue lights; which gradually descended in the form of a gorgeous canopy towards the surface of the earth. There were seven other discharges of 300 or 400 rockets at a time, besides many other gorgeous displays of different kinds.

The Sunday after the Coronation I went with the Sub-Dean to the Chapel Royal. He introduced me through the vestry, so I got in without any press. Hundreds were waiting at the door without being able to find a place. The chapel is small and plain. The Bishop of London preached a good plain sermon; and the young Queen, the Queen Dowager, and the Duchess of Kent sat in front of the Royal pew. They were all dressed as plain as plain could be, stopped to receive the Sacrament together, and seemed anxious to throw away all the pomp and pride of greatness as much as they could. I was now anxious to get away from London. But how was this to be effected ? Every coach and means of conveyance were engaged for several days to come. Last Wednesday evening, however, I succeeded in securing a place in one of the night coaches, and in course of the following day found my way to this place, where I am halting a day or two before I fairly set to work upon the rocks. Mr Conybeare's family, whom you have often heard me speak of, now live at this place. The country is delightful. Now have I not sent you a long gossiping letter, my dear Emma ? I seldom have time to write such. Ever my best love to my brother and sister and your sister. Believe me, with prayers for your health and happiness, your most affectionate uncle

A. SEDGWICK.

ALNWICK CASTLE, *September 4th,* 1838.

My dear Isabella,

I am beginning a letter which I fear I shall not have time to finish, as I expect to be called off by a party going to start for Chillingham Castle; but I will do my best during the few minutes they allow me.

Since the letter I sent to Emma I have been in almost continual movement. First I went round the greater part of the south coast of Devonshire. It is a most smiling and lovely country, and full of interest connected with my own favourite pursuits. I then cut through the eastern extremity of Cornwall and passed along the north coast of Devon. The cliffs along that part of the coast are perhaps the finest in all England, reaching in some places the height of eight or nine hundred feet, and they are here and there ornamented with foliage among all the broken ledges, which gives great richness and picturesque effect to the rugged elevations. From North Devon I tracked my way through some exquisite country in Somersetshire, and from thence went up the Bristol Channel to Clifton, where I met my friend Professor Whewell. We spent the Sunday there with a Cambridge friend, and then went through a most delicious country to Gloucester, Cheltenham, and Tewkesbury. The last place was very famous during the wars of York and Lancaster ; and you may perhaps remember that the battle which finally decided the fortunes of Queen Margaret and her son was fought close to the town. Among some exquisite monuments of the Middle Age, is a small marble slab to the memory of the unfortunate young Prince who was murdered soon after the battle ; and a shrine of exquisite tabernacle-work to the memory of Clarence, who was buried there after his murder by Richard the Third. Every house, and every field, seems to savour of the poetry of Shakspeare.

Tuesday Evening. Let me see. I left off at Tewkesbury and Shakspeare, but let them pass, and in imagination travel with me first to Birmingham, and thence by the railroad to Liverpool, where I arrived almost in the dark, and while the rain was descending in torrents. But for these unfortunate circumstances I should have gone across to Woodside[1] on the chance of seeing you, though I had not heard of your return from Dent. Next day at 7 a.m. I started by coach for Preston, and thence by the canal-boat to Burton in Kendal, from which I posted to Dent. They were all excellently well, and very happy to see me. After remaining in Dent about a week, I went over Stainmoor to Newcastle, where I arrived last Saturday fortnight, to attend the meeting of the British Association. Our reception was truly noble, and our numbers have increased enormously, so that I fear before long no place will be able to hold us. The Duke of Northumberland was President, and went through all the formal business with great dignity and good temper, keeping a rather ungovernable body in excellent order, without seeming to press hard on any one. I had no office this year ; at which I greatly rejoiced, as it left me free to amuse myself. I found a nice quiet lodging

---

[1] Sedgwick's nieces were at school at Woodside.

in a good central situation, and my landlady procured cold
fowls, tongue, and two or three bottles of white wine for me;
so I became a snug housekeeper, and had several parties
to eat luncheon with me during the intervals of the week's
turmoil. On the Friday of the Association week I went
to the mouth of the Tyne with a geological class of several
hundreds, and nearly all the population of Tynemouth turned
out to join us. You would have been amused at the picturesque group, clustering among the rugged precipices of a
noble sea-cliff, or congregating on the sand below, while I
addressed them at the utmost stretch of my voice six different
times, from some projecting ledge that served as a natural
pulpit. Every one was in high spirits, the day was glorious,
and we all returned up the river in steam-boats, so as to join
in the work of the evening meetings. I must conclude—the
bell has rung five minutes since.

Ever, my dear Isabella,

most affectionately yours

A. SEDGWICK.

Sedgwick would have us believe that he played a very
subordinate part at Newcastle; but Sir John Herschel, in a
letter to his wife, gives a vivid impression of his eloquence:

NEWCASTLE, *August* 7, 1838.

"All the show here is over. It has been by far the most brilliant
meeting of the Association, and in all the public proceedings perfect
good taste has reigned. Sedgwick wound up on Saturday with a
burst of eloquence (something in the way of a sermon) of astonishing
beauty and grandeur.

"But this, I am told, was nothing compared to an out-of-door
speech, address, or lecture, which he read on the sea-beach at
Tynemouth to some 3000 or 4000 colliers and rabble (mixed with
a sprinkling of their employers), which has produced a sensation
such as is not likely to die away for years. I am told by ear and eye
witnesses that it is impossible to conceive the sublimity of the scene,
as he stood on the point of a rock a little raised, to which he rushed
as if by a sudden impulse, and led them on from the scene around
them to the wonders of the coal-country below them, thence to the
economy of a coal-field, then to their relations to the coal-owners and

33—2

capitalists, then to the great principles of morality and happiness, and last to their relation to God, and their own future prospects....

"And, by the bye, though one should not tell one's own good things, here is one *so* good, that you must have it! Sedgwick, in his talk on Saturday, said that the ladies present were so numerous and so beautiful that it seemed to him as if every sunbeam that had entered the windows in the roof (it is all windows) had deposited there an angel. Babbage, who was sitting by me, began counting the panes, but, his calculation failing, he asked me for an estimate of the numbers. 'I can't guess' was my answer, 'but, if what Sedgwick says be true, you will admit that for every little pane there is a great pleasure!'"

The next three letters carry us to the end of 1838:

To Rev. John Sedgwick.

ST BEES, *September 14th*, 1838.

I remained at Newcastle over Monday the 26th to attend the anniversary dinner of the Natural History Society, of which I am a member, and the day following went to Belsay Castle to meet a son of Sir Charles Monk, who was an old pupil of mine; after remaining two nights I moved on to Sir John Trevelyan's, and thence to Alnwick Castle, where I remained till Thursday the 6th of this month. Nothing could be more kind than the reception I had from the Duke and Duchess, who live in their magnificent old feudal fortress in almost regal state. It stands on nearly six acres, in the centre of which rise the clustering towers of the great keep, in which are the family apartments. They form two circular suites, arranged one above the other, round the central court; and you approach them through three great gothic portals, each communicating with a separate court; and you emerge into the streets by a vast gateway, imposing from its mass, and in ancient days most formidable for its defences.

While I was at Alnwick Whewell joined the party, and we made a day's expedition to Lord Tankerville's Castle, mainly for the purpose of seeing his celebrated wild bulls—the legitimate descendants of the old inhabitants of the Caledonian forests. Like all wild animals, they are all exactly alike; they are white as snow, with black hoofs, and black eye-lashes, and horns tipped with black. We approached very near them by the shelter of a wood which was to leeward; had it been to windward they would have smelt us, and been off long before we came in sight. On leaving Alnwick I returned by coach to Newcastle, and from thence found my way to the house of my old pupil Mr Wharton, who lives close to Durham. The weather was so dreadful that for two days I could not stir from his door. Sunday was fine, and on Monday and Tuesday we effected our purpose of visiting the new harbours of Hartlepool and Seaham, and of examining some coal-works in the southern part of the county of Durham. On Wednesday I left Wharton's, and yesterday worked my way to Whitehaven, passing from Newcastle to Carlisle by the railroad in two hours and fifty minutes. We came the first four miles in six minutes exactly by my watch. The country was lovely, especially along the Tyne; but the objects seemed literally to fly past us. Ainger is not looking well, though better than he was last year.

To R. I. Murchison, Esq.

DENT, *October 11th,* 1838.

I spent several days delightfully with my friend Wharton, in rambling along the coast of Durham, and examining the enormous excavations which have been made through the magnesian limestone since my former visits to the county. Nothing can be more beautiful than the sections made by the railroads, laying bare the Lower New Red and the fish-beds, which in some places are mineralogically identical with the copper-slate of Germany. The want of conformity between the coal-measures and the overlying series is also most strikingly exhibited in several of the new sections.

From Durham I went to Whitehaven, where I spent ten days in collecting materials for the Whitehaven paper. They accumulate so much upon me that I am afraid of being overlaid by them....

From Whitehaven I went by Lowther Castle (where I halted two days) to Kendal, and there I collected some magnificent mountain-limestone fossils.

To Rev. W. Ainger.

TRIN. COLL. *October 27th,* 1838.

My dear Ainger,

I only reached this place on Wednesday evening. The day following I met George[1] in the street. This morning I have been calling on him, and giving him good advice—to read hard and walk hard. There is no fear about the former; and I hope, for health's sake, he will follow the latter. If there were any meaning in the proverb about cutting the coat to the cloth (a proverb by the way quite *à propos* to a tailor's box) I ought to send you a very long epistle, but, to tell the truth, I don't use scribbling-paper because of my fulness of matter, but because I have no other just now in my room. George tells me he is going to send off a box to St Bees with furniture for your outer man. So I mean to instruct him to stuff this sheet into the pocket of your new pair of breeches; and scribbling-paper will do quite well enough for such a roosting-place.

We parted, if I remember right, at Shap Wells[2]; and soon after I started with a batch of the Kendal magistrates. The following morning was spent in packing fossils. They ought to be precious things, as the carriage of the box comes to more than two pounds sterling. I then went on to Dent and found them all quite well. Jane talks of nothing but Cowgill Chapel. I think she has a right to be proud of it. It is a very pretty, correct, building, and admirably fitted for its purpose. The consecration is to take place on Tuesday next, a day unfortunately too late to allow of my being present. I

---

[1] George Henry Ainger, of St John's College, B.A. 1842, afterwards Fellow.

[2] Dr Ainger had accompanied Sedgwick to Lowther Castle, on leaving which they went together to Shap Wells "to take another look at the granite hills." To Rev. John Sedgwick, 26 September, 1838.

spent a day or two with Mr Farrer of Clapham, who has been
making great discoveries under Ingleborough. He has blown
away a great deal of rock with gunpowder, and so formed
communications between a succession of very beautiful caverns
richly adorned with stalactites, some of which reach the ground,
and form beautiful white pillars. We endeavoured (at the
distance of about three-quarters of a mile from the entrance of
the cave), to make some new advances. But to effect this we
were forced to use our abdominal muscles as sledges, and our
mouths as candlesticks. On, however, we went (serpent-wise,
though not perhaps wise as serpents), and wriggled our way
about two hundred yards, when the roof became more lofty,
and the water more deep. We were provided with a cork
jacket, which one of the party mounted, anxious, like Hotspur,
"to pluck up drowned honour by the locks;" and so equipt
(and over and above provided with a long cord fixed to said
jacket) ventured on a voyage where man had never before
floated. Meanwhile we sat on a low ledge of rock, each of us
in the very position of Law's man when he was stitching, on
his board, the back-seam of your new breeches; and we solaced
ourselves with a cigar, a luxury our floating friend could not
enjoy while he held a large candle in his mouth to light his
way over the waters. After running out 100 yards of rope
the chamber closed, and the water seemed to escape through
many narrow sink-holes. So the voyager came back, and we
all returned as we could, with our clothes almost peeled off
our bodies, and our knees and elbows the colour of damaged
indigo. I halted a day or two on my way back with our old
school-fellow, Welsh. He gave a sort of *al fresco* party one
day in Ingleton fells; which I enjoyed much, for its own sake,
and from the remembrance of past days that it stirred up.
After halting a few days more at Dent I was at length forced
to run away by my Cambridge engagements; but I halted a
day at Liverpool, partly to see my nieces, and partly to see
the footsteps of a preadamite beast that used to walk over
the sands of Cheshire between high and low water mark. The

rock with the impressions is as old as St Bees Head. By the
way, next year the lasses will have left school, and I will try
to give them a run through the Lakes, and bring them as far
as St Bees. Will you take us in? If so, Done! And the
word 'Done' is *à propos;* for my sheet is done, and I have
done wonders, considering that this is the 28th letter I have
scrawled since my return. Thank God I have now broken
the neck of my arrears in the way of letter-writing. It always
takes me a good part of a week to work myself right after my
return, as no one knows my address while I am away for the
summer.

Yours ever,

A. SEDGWICK.

Sedgwick's enforced absence from the consecration of
Cowgill Chapel gave him much annoyance. "Is it not a
pity," he wrote to his brother, "after being so often dragged
to the North by odious money transactions, that I cannot
now attend when it would gladden my heart to be with you?
But it would not do for me to be away from College at this
time of year[1]." It was some consolation to him to learn that
the ceremony was favoured by fine weather; that the congregation was large and sympathetic; and that his own absence
had been generally regretted. He testified his unabated
interest in the good work by a further donation, by which
the whole amount of his subscriptions was raised to two
hundred guineas.

Murchison's great work *The Silurian System*, on which
he had been so long engaged, was published in the first days
of 1839. It was dedicated in the most cordial terms to
Sedgwick, and one of the first copies was sent to him at
Norwich. The gift was promptly followed by a request that
he would write a "little article, if only half a column," in *The
Times*, to which he returned the following decisive answer:

---

[1] To Rev. John Sedgwick, 24 October, 1838.

8 *a.m. January* 10, 1839.

My dear Murchison,

I am very deeply in your debt. First, your magnificent present, for which I know not how to thank you. Then two letters to answer, and no time to answer them at any length. Accept, however, my best thanks, and warmest congratulations on the new birth, full-grown and strong, as it ought to be, after so long a period of gestation. And now I hope you are doing as well as can be expected!

In regard to your second letter and its request, I cannot comply with it, for two or three reasons. First, I have not time to write to newspaper editors any letters fit to be read. Secondly, I have not time (while here) even to skim the book, and yet I ought to do that if I dare to sit in the reviewer's chair. Thirdly, I will not condescend to write in a paper which I think calumnious and dishonest, though very clever. Fourthly, you have the assistance of Mantell, who having a little followed the trade of puffing can blow a better blast than myself. Fifthly, because I think your work of such solid merit as not to need any artifice of publication; and in my own case I had much rather the means you point out were not used at all. Reasons! why, I have sent you reasons enough to sink a three-decker!

I am very busy as usual, and in addition to a great many gastronomic labours, I have day after day been sitting on committees of the Norwich Museum. After morning service I have one to attend to-day, so I now am writing by candle-light. Pray how are you all going on in London? Have you heard anything about Jukes, who is willing to go to Newfoundland as surveyor? I think he will do the work very well, and that we are fortunate in having so good an offer. Pray give him a shove if you can. I have written about him to Lonsdale and Darwin. Yours always,

A. SEDGWICK.

But, though Sedgwick declined to take up his pen in

1839.
Æt. 54.
commendation of *The Silurian System*, his admiration of it was cordial and sincere. Even at the close of his life, after long estrangement from the author, he could say that this year "formed an epoch in the history of European Geology;" and that whatever assistance might have been rendered by others, "the chief honour will ever be given to the author of the System, who brought the materials together and arranged them in that manner in which they are seen in his splendid work. Under his hands the older Palæozoic Geology had assumed a new and a nobler type[1]."

During the winter and early spring Sedgwick and Murchison were busy with a second joint paper on Devonshire and Cornwall. On this occasion, though Sedgwick's working-days were clouded, as usual, by business, lectures, and illness, the work proceeded more smoothly than might have been expected, and just before the date appointed for the reading of the paper (24 April) the weather cleared up, if we may use such a metaphor, and he could write: "Yesterday I had my field-lecture, and a scamper (often at full gallop) of thirty-five miles did me a great deal of good, so that my head is clearer, and my gouty symptoms are, for the nonce, almost gone[2]." The paper, when read, "excited some pugnacity, and we had a debate upon it that lasted nearly to midnight[3];" but the result was evidently satisfactory, for Sedgwick good-humouredly passes on in the next sentence to lighter topics. The passage is interesting as shewing the affection he had already begun to feel for the Stanley family:

"On Friday morning I called at Brook Street on the Bishop. All well. K. T. [Miss Catherine Stanley] all glorious to behold, with a gorgeous plume, just going to be presented at the Drawing-room. So she is now fairly out— no longer a mere chrysalis, a creature without sensibility, having neither organs of nutrition, nor wings, nor legs. But,

---

[1] *Preface* to Salter's *Catalogue*, etc. p. xxii.
[2] To R. I. Murchison, 20 April, 1839.
[3] To Canon Wodehouse, 30 April, 1839.

on the contrary, she has glorious antennæ waving over her
head, organs of sense to discourse sweet music, arms for offence
and defence, and wings glittering like sunbeams on a May
morning.   I hope in this transformation she has not acquired
a sting, as some winged creatures have.   Be this as it may, I
shall hereafter treat her with great deference, now that she
is transformed, by the magical touch of the royal hand, into
a young lady."

We have in the next place to speak of Sedgwick's joint
labours with Murchison on the Continent, which occupied four
months of the summer and autumn of 1839.   Their object in
undertaking this tour can fortunately be stated in a passage
from their subsequent paper, read to the Geological Society
in May, 1840—a passage which, from internal evidence, we
feel inclined to ascribe to Sedgwick:

"When we entered, during the summer of 1836, upon an examination of the structure of North Devon, our sole aim was to determine
the position and relations of the culmiferous strata, about which
there had been much controversy.   Before our task could be completed, we were led to an examination of the lower groups of strata
both in North and South Devon, and to follow them in their prolongation into Cornwall; and we at length arrived at the conclusion
that nearly all the older stratified rocks of both counties belonged to
one epoch, and must be included under one common designation.
We proposed the name of *Devonian System* for this great series of
deposits, and we placed it in a position intermediate between the
Lower Carboniferous and the Upper Silurian groups; and therefore on
the exact parallel of the Old Red Sandstone, of which it was assumed
to be the equivalent under a new mineral type.   We need not
inform the Society that this classification was strenuously opposed at
the time it was first brought forward, and that it has been combated
in some public journals since the appearance of our abstracts; nor do
we now deny that it was encumbered with great difficulties.   To
believe that a limestone (like that of South Devon) overlaid by many
thousand feet of slate rocks almost devoid of fossils, was of the date
of the Old Red Sandstone, and that the greater part of the slates of
Cornwall (heretofore regarded, from their crystalline structure, as
primary formations of great antiquity), also belonged to the same
epoch, required no common confidence in the weight of evidence
offered by groups of fossils.   With good hopes, but not without
considerable anxiety, we therefore resolved to examine some of the
continental localities which seemed likely to throw light on our
proposed classification; conscious that it could never meet with

general acceptance unless confirmed by some analogous development
of the upper transition rocks of France and Germany[1]."

It was at first intended to begin with Brittany, "which
from its geographical position, might be expected to offer
some analogies of structure with Devonshire and Cornwall."
In consequence we find Sedgwick corresponding with M. Elie
de Beaumont and other French geologists, with the result
that Brittany was presently abandoned, on learning from
them "that the connexion between the carboniferous system
and the inferior strata was obscurely exhibited in that region,
and that the evidence offered by the fossils of the lower groups
was meagre and unsatisfactory[2]."

Murchison, meanwhile, was suggesting more distant fields
of exploration ; and urged Sedgwick "to shake off Norwichian
trammels," and start with him for Norway about the middle of
May. To this tempting proposal Sedgwick replied: "I
should delight in a tour in Norway and the country round the
Baltic. My only fear is, that my engagements in our Chapter
(and remember I am junior) may make the tour impossible at
the early time you fix on. The tour you mention, and
another tour among the old rocks of France and the corresponding rocks of the South of Ireland, I am very anxious to
effect ; and then, as far as hard-working geology is concerned,
I will shut up shop[3]." Finally, they tell us, "We resolved to
begin with the transition rocks of the Rhenish provinces,
knowing that on both banks of the Rhine we should be
conducted through a true carboniferous series, based on
mountain limestone, into still lower groups of strata."

Murchison, eager to begin his holiday as soon as possible,
started alone at about the time originally suggested. His
colleague departed, Sedgwick set to work to prepare their
joint papers on Devonshire for the press. This labour
brought on "gout, mental prostration, and entire loss of

[1] *Transactions* of the Geological Society, Second Series, vi. 222.
[2] Ibid. p. 223.
[3] To R. I. Murchison, 4 February, 1839.

intellect," a condition which Chapter business did not improve. At last, however, in the middle of June, he too
effected his escape, and joined Murchison at Bonn. They set
to work at once, ailments vanished, and before a fortnight was
over Murchison could report that "Sedgwick is as well as
ever I knew him—eats, drinks, and digests like a Hercules,
and is in great force[1]." If he wrote letters during his long
exploration of Rhineland and its neighbourhood, they have
not been preserved; and for the route followed we must
content ourselves with a brief summary sent to Ainger after
his return.

"We threaded our way through all the country on the
right bank of the Rhine between Westphalia and the Taunus
inclusive, then we crossed the plains of Hessia to the Hartz,
where we disported ourselves as in duty bound among
goblins and witches, with whom you know all that region is
peopled. Afterwards we crossed through the dominions of
the Prince of Sondershausen (did you ever hear of him
before?), and so to Gotha and Weimar. From the latter
place we plunged into the vast forests of Thuringia, and so
through Coburg (the breeding-ground of kings) and on by
the Franken Wald to Hof and Bayreuth, the old capital of the
Margraves of Anspach. From the last-named place we
doubled back to the Rhine by the great road of Bamberg and
Wartzburg, halting a day or two at Frankfort. What I have
written, though you trace my route on a map, will give you a
very inadequate notion of the extent of our tour, as all our
way was made in traverses and zigzags after the manner of
geologists. After we reached the Rhine again, we performed
similar evolutions on its left bank, working all the country
between the Hundsruck and the southern portion of Belgium;
and, while my companion was away for a month in England[2],
I extended my traverses to the very limit of the forest of the

---

[1] Geikie's *Life of Murchison*, i. 276.
[2] Murchison went to England to attend the Meeting of the British Association
at Birmingham.

1839.
Æt. 54.

Ardennes, and threaded my way, partly on foot, through the gorges of the Meuse between Mezières and Namur. When Murchison rejoined me we had a second run through West-phalia, and finally crossed from Dusseldorf to Aix-la-Chapelle, and so to Liège. We then went on a visit of three days to Baron d'Halloy, a geological friend in Belgium. From his house we found our way to Brussels, where we sold our carriage. A railroad gave us a ready run to Antwerp, and a steamer conveyed us to Tower Stairs about the end of October[1]."

These laborious investigations did not at first lead to any clear result. In fact Sedgwick felt inclined to doubt whether the rocks he had been studying on the Continent could be classed as Devonian at all. " To-day," he wrote, soon after his return to Cambridge, " I began my lectures; and as I had last year stated my views of the Devonian case to my pupils, and told them that I was going to join you in Brittany (for so I thought), to put our new views to the test, I felt obliged to explain to them the results of our summer's work, of which I gave them a brief abstract. This was quite necessary, and it can do no harm, as there is no report of what I say in my lecture-room. I told them plainly that I gave up the Devonian case, and that I considered the whole of the old rocks of Devon and Cornwall (excepting the Culms) as inferior to the Derby limestone; but I added that the matter was still *sub judice*, and that our fossils had not been examined. My class was a good one; two Heads of colleges, many Masters of Arts, and about sixty undergraduates; and many who attend me are not yet come up for the term[2]."

While Sedgwick was lecturing at Cambridge, Murchison was examining with Sowerby and Lonsdale the fossils collected on their tour, and corresponding with M. de Verneuil, who had accompanied them for a portion of their journey. Fortified by these authorities, he made up his mind

---

[1] To Rev. W. Ainger, 18 February, 1840.
[2] To R. I. Murchison, 28 October, 1839.

more rapidly than Sedgwick, and before the end of January, 1839.
1840, we find him pleading for an interview with his collabo- Æt. 54.
rator, "when I might explain in words what it would require
volumes to write[1]." ⸱ Soon after this the interview took place,
and in February he was able to announce "the identity of the
uppermost greywacké of the Continent and the Devonian as
defined by Sedgwick and myself. I have arrived at this
conclusion for many months, and only waited the coming to
town of my colleague to open the campaign. Now that he has
been here, and that we are all agreed, the course is clear, and
we shall soon give a grand memoir to shew, etc.[2]" The
production of this memoir was, however, retarded, as usual,
by Sedgwick's ailments, the catalogue of which we know by
heart. This year he had more than common difficulty in
shaking them off, and during the early months of the year
was quite unable to work. He could still, however, joke over
his sufferings. Writing to Mrs Lyell he says:

"Should you think it worth while to ask me what I am
doing, I can describe all my most active employments by the
word *grumbling*. All winter I have been a cross, crusty,
crabbed, careworn, caitiff. In outer looks I am sour, and
ill-favoured ; and, as I used to be rather vain of my person, I
have been ashamed of my ill-looks, and have not called on
any one, otherwise I might before this have tapped at your
door. My daily bread is made of calomel and colchicum ; and,
besides Medea's drugs, they have given me every day a stew in
her caldron. It would melt your heart, could you see me sitting
for half an hour each day in brimstone vapour at a temperature of 110°! And to crown all my misfortunes I have lost my
favourite dog[3]. The very beasts you see run away from me!"

Towards the end of April he was ordered to Cheltenham.
The waters agreed with him, and he was soon able to report
some slight improvement.

[1] From R. I. Murchison, 23 January, 1840.
[2] Geikie's *Life of Murchison*, i. 286.
[3] Enclosed is a hand-bill offering £1. 0s. 0d. for the restoration of *Shindy*, a
Spitzhund brought from Germany.

1839.
Æt. 54.

"I am better than I was," he writes, "but far from good working condition. All I can say is, I will do my best. Alas! during this winter and spring, my best has been bad indeed. I have lately been taking a shower-bath against some very uncomfortable symptoms in the nerves of my left side; but on this subject *mum*[1]."

Meanwhile Murchison was preparing to start for Russia, and could only be present on the evening when their paper was begun (13 May). It fell to Sedgwick's share to complete it. This he did, coming up from Cheltenham on purpose (27 May). The interest shewn when the new classification was first propounded had now died away, and all that he had to tell Murchison was, "Our paper went off well. Nobody made any fight." A summary of the conclusions arrived at will fitly close this chapter:

"An examination of our fossil evidence has now led to a satisfactory conclusion, viz. that there is a group of rocks, characterized by an appropriate group of fossils, in a position, geologically as well as zoologically, intermediate between the Carboniferous and Silurian systems. The supposed 'Ludlow rock' of Belgium passes into the mountain limestone, and even in its lower portions does not, as far as we know, contain a single Silurian fossil: and our Eifel mollusca (about which we were most anxious, and of which we had procured a fine series) were pronounced by Mr Sowerby, the moment he saw them, to form a group analogous to that of South Devon.... At the same time we submitted the Eifel corals to Mr Lonsdale, who with equal confidence pronounced them Devonian[2]."

In the next chapter we shall review the geological work accomplished by Sedgwick between the year 1828 and the year which we have now reached.

[1] To R. I. Murchison, 1 May, 1840.
[2] *Transactions* of the Geological Society, Second Series, vi. 226.

CHAPTER XI.

WHEN we last interrupted the general narrative of Sedgwick's life to discuss the value of his geological work[1], it happened that the papers reviewed had appeared during the ten years which succeeded his election to the Woodwardian Chair in 1818. In the present chapter it will be convenient to consider in the first place those covering the next decade, from 1828 to 1838.

This second decade is certainly not the least important in the history of Sedgwick's work. In the first he had travelled far, and often; had examined on the ground the work of many of the pioneers of geology; had discussed the interpretation of many difficult structures with the leading physicists of the age; had made excursions with Conybeare and Murchison, of whom the first taught him much, the second learnt much from him; and had met many good local observers who one and all record the keenness of his observation, the shrewdness of his generalisation, and the inspiring charm of his enthusiasm.

His first syllabus, drawn up early in his first decade, shows how clearly he realised even then the problems to be solved. His second syllabus, published at the beginning of his second decade, has on it the impress of his own observations made in the interval.

[1] See above, pp. 284—297 ; 325—328.

In the same way we find in the first decade but one great original paper, that on the *Lower New Red and Magnesian Limestone Series*, with several notes of less importance, such as those on Devon and Cornwall, the Isle of Wight[1], the Trap dykes of the north of England, Alluvial and Diluvial deposits, and the strata of the Yorkshire coast. The larger papers of this period are those which give an account of his tours with Murchison in Scotland and its Western Islands, and in the Alps. These, as we have seen, were important contributions in their way, and added much to what was already known of the structure of the districts examined. When, however, these papers are carefully analysed it will be found that the chief value of those on Scotland consists in the development and extension of the work of Macculloch ; of that on the Alps that it brings to the notice of English geologists for the first time the work of Boué, criticised and epitomized by two such skilful handlers of the hammer and the pen as Sedgwick and Murchison. But, after all, the principal result of all this exploration is to be sought in its effect upon Sedgwick's own subsequent work.

At the commencement of the second decade we find him President of the Geological Society, and his two addresses tell us what new matter he thought of greatest importance and interest to bring before the Fellows at their anniversary meetings in 1830 and 1831. But he was now taking in hand a far larger task, and within the next ten years he had laid before the world sketches of the structure of the Cumbrian and Cambrian mountains which have formed the foundation of all subsequent work, not only in those districts, but wherever Cambrian rocks occur throughout the world.

---

[1] His observations in the Isle of Wight were of considerable value, though he did not himself fully realise their bearing till some years afterwards. We know from the collection of fossils made by him in 1820, and still preserved in the Woodwardian Museum, that he had discovered the marine beds at the top of the Hampstead Cliff which are characterised by *Voluta geminata* Sow. [*V. Rathieri*, Heb.], *Potamites plicatus* Sow. [*Cerithium plicatum* Sow.], *Corbula rugosa* Lam. [*Corbula pisum* Sow.].

Although it might appear from a glance down the list of papers written by Sedgwick in his second decade that his work was very desultory, closer examination reveals that there was method in it, and that he was grappling with some of the greatest problems of stratigraphical and dynamical geology. The older rocks, among which nestle most of the lakes of the north-west of England and which form the central part of the Lake District, are wrapped round by newer deposits, and stand out like a huge boss protruding through a lawn. Sedgwick set himself to work to make out, first, what were the outside newer rocks, and how they behaved with regard to the ancient central mass round which and over which they had been thrown down as shingle and sand upon the shore of an encroaching sea. All the while he was watching the fragment of a still more ancient world which is revealed in the central part of the district, after having been buried up for ages, and comparing the character and structure of the older rocks with those observed in the newer beds ; and in fact we find his papers on the North of England of this date beginning with an *Introduction to the General Structure of the Cumbrian Mountains*. In this there is a short sketch of the ancient central mass, but the principal part of the paper is taken up with a description of the great dislocations by which the surrounding Carboniferous rocks are now separated from it. Then, to make sure that the character and succession of the newer rocks was clearly established, in order to understand how they behaved when they crept up to and round the central mass, he made traverses across them in various directions, and published the results in a paper read before the Geological Society in March, 1831. In this he gave sections across the highest mountains of the Yorkshire moorlands, carrying them over Penyghent to the Craven district, and over the Pennine Range into the Eden Valley near Kirkby Stephen.

In the Eden Valley, however, there was evidence of another great interruption in the continuity of deposit. As the

Carboniferous rocks had been thrown down upon the ancient
land-surface seen in the central Lake District, so here in the
valleys that surround it there is clear evidence that the
Carboniferous Series in its turn had been uplifted from the
depths, and moulded by rain and rivers into an irregular
land-surface, which had again in its turn gone down to
receive the debris of the wasted land, the shingle-beach made
up of pebbles from the coast, and the sand and mud with
drifted plants which we see along the river-cliffs and hillsides
of the beautiful Eden Valley. This called for a separate paper,
which was read before the Geological Society on Feb. 1st, 1832.
Sedgwick was here on the equivalents of the Lower New Red
Sandstone which he had so well worked out on the other side
of the backbone of England and described six years before.
In the same year, 1832, he read before the Society a paper
on the *Geological Relations of the Stratified and Unstratified
groups of Rocks composing the Cumbrian Mountains.* This was
a further step in the prosecution of the work of which he had
given what he called the introduction in January 1831. In
this there is more about the pre-Carboniferous rocks. He
describes the structure of the district more in detail, and,
with full acknowledgment of the excellent work done by
the local guide and fossil-collector Jonathan Otley, gives
a general classification of the rocks composing the central
mass. The opening words of the introduction explain the
place of this paper in his great plan for working out the
district. In it he

"first shews that the limits of the region to be described are
defined by a zone of Carboniferous limestone based here and there
upon masses of old red conglomerate. This zone is described as
entirely unconformable to the central system, and for the phænomena
presented at the junction of the two great classes of rocks, he refers
to previous memoirs read before the Society."

In 1832, the 4th year of his second decade, he had given
a very good sketch of the sequence of rocks in and around the
Lake district.

Seeing the advantages of this mode of treatment, Sedgwick

now worked out more of the details and subdivisions of the newer series as they approached close up to the Lake district, and described the rocks which generally lie at the base of the Carboniferous Series in that country, immediately above the old shingle which is found collected in the deeper hollows all round the ancient pre-Carboniferous mountain-land. In the same paper, which was read before the Geological Society 10 November, 1835, he gave a description of some of the coal-bearing strata on the N.W. coast of Cumberland, having been assisted in the examination of that district by Mr Williamson Peile. This, however, required further working out, so they returned to it, and in June of the following year Sedgwick read another paper giving the results of their work. As soon as he had touched any district he was of course likely to feel an interest in all scientific questions connected with it, whether they bore directly upon the original object of his investigation or not; and so we find him in 1837 reading a note before the British Association on the incursion of the sea into the colleries of Workington.

In the midst of all this work in the North of England, and well prepared by it to grapple with the difficulties of a complicated district, Sedgwick plunged into the heart of North Wales in 1831. We have seen that in 1821 he had sketched out in his first syllabus how he would set to work on the classification of the Transition Rocks, or Greywacké. He had done much among them during the ten years that had elapsed since that was printed, so it was as no novice that he undertook the task of disentangling the folded and crumpled masses that lie in the wild mountain region of North Wales, where he set to work in such a masterly manner that, " before the working-season was over (in 1831)" he had " completed an approximate Geological Map from actual survey of the whole of Caernarvon-shire[1]." In the following year, 1832, he returned to the district, and undertook what proved to be, in his own words, " the severest summer's task of my Geological life, namely, the

[1] *Preface* to Salter's Catalogue, 4to. Camb. 1873, p. xvi.

interpretation and partial delineation of the order and principal
flexures of all the older deposits of the counties of Merioneth,
Montgomery, and Denbigh[1]." Nothwithstanding the difficulties of the task, he was prepared with "a brief synopsis,
illustrated by sections," of what he "had effected in Caernarvonshire," by the time the British Association met at Oxford
in June, 1832. The meagre account of this synopsis in the
Report of the Association does not enable us to understand
his mode of procedure, or what were the results arrived at,
but fortunately we are not dependent upon that alone for
our information. He constantly wrote long letters when out
on such excursions, put down his first impressions, and discussed all his difficulties. "I never had a geological secret in
my life" he once exclaimed in his later years. He very often
made a rough sketch of what he saw, sometimes to amuse a
child, sometimes to explain a point to a scientific friend, and,
though he much decried his own sections, they always give a
very good notion of the really important features in the
structure of the country. So we learn from his letters to
Murchison, and from Darwin's all too short recollections of
the trip, how Sedgwick tackled a new bit of country in those
days. From peak to peak, in traverse after traverse, he
always sought some more easily recognised strata as base
lines on which to build up the succession, and with which to
correlate adjoining sections. The amount of work he did was
marvellous.

Even the unscientific reader will see that if Sedgwick is
right in the sketch of what he afterwards called the Cambrian
System which he made in 1832, and sent to the British
Association and to Murchison in the same year, the question
of priority of discovery of the succession of the rocks of North
Wales is very much simplified. So, with the view of fixing
the date upon the minds of our readers, we have already
printed the letter which Sedgwick wrote to Murchison from
North Wales, soon after he had begun work there, together

[1] *Preface* to Salter's Catalogue, 4to. Camb. 1873, p. xvii.

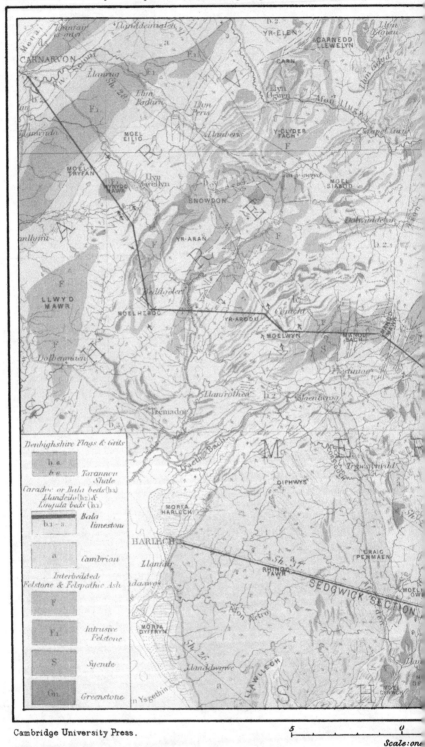

Denbighshire Flags & Grits

b.s.

b.e.   Tarannon Slate

Caradoc or Bala beds (b.2)
Llandeilo (b.2) &
Lingula beds (b.1)

b.1 - 2   Bala limestone

a   Cambrian

Interbedded
Felstone & Felspathic Ash

F

F.1   Intrusive Felstone

S   Syenite

Gn.   Greenstone

Cambridge University Press.

Scale: one

MAP OF PART OF NORTH WALES SHEWING LINES OF SECTI

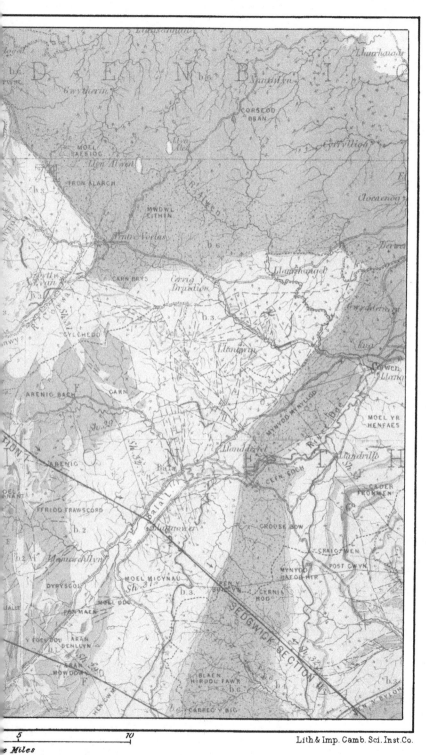

BY SEDGWICK IN 1832.

with a reproduction in facsimile of the pen-and-ink section across the district which accompanies it[1], and in this chapter we give a part of the index-map of the Geological Survey, which happens to be on about the same scale of distances as Sedgwick's section. From this it will be seen that the Survey thirty years afterwards gave no more subdivisions of the stratified rocks of this area than Sedgwick had made in 1832, and that their principal sections were drawn approximately along the very same lines as Sedgwick had then suggested to Murchison.

Though Sedgwick did not dwell much upon Cosmogonies and Theories of the Earth, except so far as they had to be explained in his educational work, yet such enquiries as those upon which he was engaged forced him to give attention to some of the important questions of dynamical geology. He could hardly have passed years in endeavouring to unravel the gnarled and twisted Cambrian rocks, and in trying to make out the dip where there were all sorts of divisional planes, and the bedding turned out to be the least obvious, without enquiring into the causes of those superinduced structures which obscured the original simplicity of the sedimentary rocks. Hence we find him towards the end of his second decade putting together his notes *On the Structure of large Mineral Masses*, and especially on the chemical changes produced in the aggregation of stratified rocks during different periods after their deposition. The explanations given of these obscure phenomena were considered of such importance for the right understanding of Sedgwick's paper on the origin and structure of the older stratified rocks, that the Council of the Geological Society wisely decided to publish this memoir before its turn, because they considered it to be introductory to other papers by Sedgwick, some of which were already printed. In this paper Sedgwick was evidently drawing upon his own original observations, for we find many references to districts on which we know he had

[1] See above, pp. 391—394.

been at work, as for instance to the concretions in the
Magnesian Limestone of Durham, and the nodular felsites
of North Wales, while in illustration of the cleavage and
other divisional planes his sections are all taken from North
Wales. New and better theories as to the cause of cleavage
may have been put forward, but the facts connected with its
mode of occurrence were once for all clearly established by
Sedgwick, whatever he may have understood by the polar
forces which, according to him, played so large a part in the
production of the phenomena. The columnar structure of
basalt he rightly referred to shrinkage, and showed that some
of the curious forms produced among the Granite Tors
should be referred, not to concretionary action, but to the
peeling off of the exposed surfaces, urging in illustration the
fact that "ancient pillars of granite have been known to exfoliate in cylindrical crusts, parallel to the axes of the pillars ;
and even pillars of oolitic limestone, which unquestionably
have no spheroidal structure, sometimes exfoliate (e. g. in
the second court of Trinity College, Cambridge) in crusts
parallel to the axes of the several pillars." There is no
doubt that this paper of Sedgwick's is a grand classic work,
treating of questions which must be grappled with by any one
who would work profitably among the older and altered rocks.

About the close of this period (1828—38), Sedgwick, in
conjunction with Murchison, commenced those investigations
in England and on the continent which resulted in the establishment of the Devonian System. The course of Sedgwick's
biography has already indicated the steps by which they
were gradually led to lines of enquiry beyond their first
intentions, and to inferences very different from their first
impressions ; but, as even those conversant with the history of
geology have probably forgotten the details of the subject,
and the controversy which arose over it, some further explanation is necessary.

Looking back it does not seem difficult to explain how such
men as Sedgwick and Murchison on the one hand, and De la

Beche on the other, should have so long differed as to the interpretation of the geology of a district like Devon and Cornwall.
Neither side had as yet all the data, and the facts which
had been forced on the notice of one, were not the same
as those that had come under the observation of the other.
It is a controversy very characteristic of a certain stage
of geological enquiry. In the crumplings that have been
going on in the crust of the earth, large areas that were once
uneven land-surfaces have been submerged, and, more than
that, one side of a country has often gone down sooner, or
more rapidly, than another; so that, although the debris of
the land may have been continuously deposited all the while
it of course gathered more quickly and more thickly in the
hollows, and formed vast accumulations in one place, while,
not far off, it may not have begun to be laid down at all.
So the basement-bed in one district is much newer than the
basement-bed of what seems to belong to the same age of
submergence and deposition in an adjoining area. Sand and
shingle gathered in one place, silt in another; here the water
was turbid, and creatures that loved the mud enjoyed their
life; while round the headland there was clear and perhaps
deep water, and creatures abounded that loved the deep blue
sea, or the crisp white breakers. If in after-ages we were to
come upon isolated patches of deposits accumulated under
such various conditions it is obvious that there might be
great difficulty in placing them in chronological order. This
is the difficulty which lies at the bottom of the Devonian
Question.

The rocks of the northern part of the Devon promontory
lie in a trough-like fold running east and west, so that the
older rocks turn up along the Bristol Channel, and again on
the south along a line running roughly west from Exeter to
the sea. The newer rocks, therefore, lie in the middle of this
trough. Neither of these two series was found to be exactly
like the rocks which occurred on the other side of the channel
about the horizon where these might be expected. The older

and the newer were supposed at first to be of the age of some part of the Cambrian or Silurian of Wales deposited under somewhat different conditions, and like them were spoken of as greywacké; yet they were different from the greywacké of Wales. Moreover the upper series contained plants and coal, but was not like the rocks of the South Wales coal-field, nor was the lower series like what was there found below the coal. It therefore became necessary to consider whether any of these Devon beds were the geological equivalents of deposits thrown down under different conditions in the adjoining area of Wales, and, when there was a likeness, whether some of them might after all have only an accidental resemblance to the strata with which they had been hitherto identified.

Sedgwick and Murchison first removed the upper series from the greywacké, and identified it with the coal-measures. At this stage De la Beche objected, as he could not see such a strong line of separation between the upper and lower series of Devon as this implied, either palaeontological or stratigraphical. Sedgwick and Murchison urged the similarity of the fossils of the Culm beds to those of the Carboniferous Rocks, and showed how the coal-measures of South Wales changed as they were traced west, and the coal of the South Wales Field passed into the Culm of Pembrokeshire; while De la Beche pointed out that plants similar to those of the culmiferous beds are found also in the subjacent rocks into which the Culm Measures passed down in Devonshire. Whewell in his presidential address to the Geological Society (1838) states the case very fairly, and expresses the opinion that, whatever views with regard to the correlation of the rocks of Devonshire might eventually prevail, they had then sufficient clear and positive knowledge to separate the culmiferous from the subjacent strata in their lists and on their geological maps.

But there was something in De la Beche's contention that the lower set of beds in Devonshire also contained some plant remains, and that the newer beds were for various reasons more closely connected with them than the true coal measures were known to be with the greywacké in any area which had been worked out. The next step in the enquiry therefore was to make out whether even the lower set of beds of Devonshire were greywacké; and when they had been re-examined, and their fossils had been submitted to Lindley, Sowerby, and Lonsdale, and the Devon rocks and fossils had been compared with those which occurred on the continent in a corresponding position, it was found that the lower series of Devon rocks must also be taken out of the greywacké, and considered as something older than the carboniferous beds, but much newer than any beds of greywacké age. When this view had been clearly formulated, and Murchison had given up his identification of part of the Devonian with the Caradoc sandstone, no further serious opposition was offered. The work had grown gradually. De la Beche, Austen, Weaver, and Williams had helped with facts or useful criticism; and, as far as the points then in dispute are concerned, the question was finally settled, and the now received classification of the Devon rocks remains as Sedgwick and Murchison left it: Culm Measures above, and Devonian below: the base of the Devonian being there unknown.

END OF VOLUME I.

Printed in the United States
By Bookmasters